MACHINE DESIGN DATA BOOK

DR. SADHU SINGH

Professor & Head, Mechanical Engineering Department (Former)

&

Dean, Faculty of Engineering & Technology (Former)

Govind Ballabh Path University of Agriculture & Technology,

Pantnagar – 263145 (Uttarakhand)

and

(Director Colleges), Punjab Technical University, Jalandhar (Former)

KHANNA BOOK PUBLISHING CO. (P) LTD.

Publisher of Engineering and Computer Books

4C/4344, Ansari Road, Darya Ganj, New Delhi-110002

Phone : 011-23244447-48 **Mobile:** +91-9910909320

E-mail : contact@khannabooks.com

Website : www.khannabooks.com

Dear Readers,

To prevent the piracy, this book is secured with HIGH SECURITY HOLOGRAM on the front title cover. In case you don't find the hologram on the front cover title, please write us to at contact@khannabooks.com or whatsapp us at +91-99109 09320 and avail special gift voucher for yourself.

Specimen of Hologram on front Cover title:

Moreover, there is a SPECIAL DISCOUNT COUPON for you with EVERY HOLOGRAM.

How to avail this SPECIAL DISCOUNT:

Step 1: Scratch the hologram
Step 2: Under the scratch area, your "coupon code" is available
Step 3: Logon to www.khannabooks.com
Step 4: Use your "coupon code" in the shopping cart and get your copy at a special discount
Step 5: Enjoy your reading!

Price: ₹ 360.00

ISBN: 978-93-82609-51-3

Machine Design Data Book
by Dr. Sadhu Singh

First Edition: 2014
Reprint: 2022, **2023**

To view complete list of books,
Please scan the QR Code:

Published by:
Khanna Book Publishing Co. (P) Ltd.
CIN: U22110DL1998PTC095547

Visit us at: **www.khannabooks.com**
Write us at: **contact@khannabooks.com**

Printed in India by: Vaibhav Products, Delhi.

Preface

While designing machine elements lot of data is required in the form of factor of safety to be used, allowable stresses for the materials selected, etc. Using design data at random shall lead to a fictitious design. Therefore, there is a need to have authentic design data while designing any machine component. Keeping this in view, the present book has been prepared for the ready reference of designers as per the recommendations of Bureau of Indian Standards (BIS).

The book contains 25 chapters on Basic Design, Engineering Materials, Design for Static Loading, Design for Variable Loading, Riveted Joints, Welded Joints, Screwed Joints, Power Screws, Transmission Shafts, Keys and Couplings, Cotter Joints, Knuckle Joint, Pipe Joints, Mechanical Springs, Flat and V – Belts, Chains and Ropes, Spur Gears, Helical Gears, Bevel Gears, Worm Gears, Sliding Bearings, Rolling Contact Bearings, Flyweel, Clutches and Brakes and Internal Combustion Engine Parts.

The book shall be useful to the students and teachers of all Indian Universities and Institutions in the branches of Mechanical Engineering, Production Engineering, Aeronautical Engineering, Agricultural Engineering, Chemical Engineering and other allied branches.

Suggestions for the improvement of the book shall be welcome and duly acknowledged.

The support provided by the publishers is highly appreciated. My thanks are due to Mr. Puneet Khanna and Mr. Mukul Seth in particular. My family members and grand – children have been a great moral support during the preparation of the manuscript of the book.

—Sadhu Singh

Contents

Basic Design Data

Table 1.1: Physical constants of some common materials

Material	*Modulus of elasticity* E *GPa*	*Modulus of rigidity* G *GPa*	*Poisson's ratio* ν	*Density* ρ kg/m^3
Aluminium	69 – 72	24 – 27	0.33 – 0.35	2700
Brass	90 – 102	34 – 35	0.30 – 0.40	8450
Carbon steel	200 – 210	80 – 90	0.25 – 0.31	7850
Cast iron (grey)	100 – 110	35 – 53	0.23 – 0.31	7200
Copper	110 – 129	34 – 36	0.25 – 0.28	8920
Phosphor bronze	111	41.4	0.35	8160
Wrought iron	190 – 220	77 – 83	0.27 – 0.29	7850

Table 1.2: Properties of plane sections

Form of section	*Area and centroid, A and (x, y)*	*Area moment of inertia* I	*Distance of axes from farthest fibres, c*	*Section modulus* z	*Radius of gyration squared* K^2
Triangle	$A = \frac{1}{2}bh$ $\bar{x} = \frac{2}{3}b$ $\bar{y} = \frac{1}{3}h$	$I_1 = \frac{bh^3}{36}$ $I_2 = \frac{b^3h}{36}$ $I_x = \frac{bh^3}{12}$ $I_y = \frac{b^3h}{12}$	$C_1 = \frac{2}{3}h$ $C_2 = \frac{2}{3}b$	$Z_1 = \frac{bh^2}{24}$ $Z_2 = \frac{b^2h}{24}$	$K_1^2 = \frac{h^2}{18}$ $K_1^2 = \frac{b^2}{18}$

Square	$A = a^2$ $\bar{x} = \bar{y} = \dfrac{a}{2}$	$I_1 = I_2 = \dfrac{a^4}{12}$ $I_x = I_y = \dfrac{a^4}{3}$	$C_1 = C_2 = \dfrac{a}{2}$	$Z_1 = Z_2 = \dfrac{a^3}{6}$	$K_1^2 = K_2^2$ $= \dfrac{a^2}{12}$
Rectangle	$A = bh$ $\bar{x} = \dfrac{b}{2}$ $\bar{y} = \dfrac{h}{2}$	$I_1 = \dfrac{bh^3}{12}$ $I_2 = \dfrac{hb^3}{12}$	$C_1 = \dfrac{h}{2}$ $C_2 = \dfrac{b}{2}$	$Z_1 = \dfrac{bh^2}{6}$ $Z_2 = \dfrac{hb^2}{6}$	$K_1^2 = \dfrac{h^2}{12}$ $K_2^2 = \dfrac{b^2}{12}$
Circle	$A = \pi r^2$ $\bar{x} = \bar{y} = r$	$I_1 = I_2 = \dfrac{\pi r^4}{4}$	$C_1 = C_2 = r$	$Z_1 = Z_2 = \dfrac{\pi r^3}{4}$	$K_1^2 = K_2^2$ $= \dfrac{r^2}{4}$
Hollow Circle	$A = \pi(r_0^2 - r_i^2)$ $\bar{x} = \bar{y} = r_0$	$I_1 = I_2$ $= \dfrac{\pi}{4}\left(r_0^4 - r_i^4\right)$	$C_1 = C_2 = r_0$	$Z_1 = Z_2$ $\dfrac{\pi\left(r_0^4 - r_i^4\right)}{4r_0}$	$K_1^2 = K_2^2$ $= \dfrac{r_0^2 + r_i^2}{4}$
Semi-Circle	$A = \dfrac{\pi h^2}{2}$ $\bar{x} = r$ $\bar{y} = \dfrac{4r}{3\pi}$	$I_1 = 0.10976r^4$ $I_2 = \dfrac{\pi r^4}{8}$	$C_1 = 0.5756r$ $C_2 = r$	$Z_1 = 0.1907r^3$ $Z_2 = 0.3927r^3$	$K_1^2 = 0.0699r^2$ $K_2^2 = \dfrac{r^2}{4}$
Ellipse	$A = \pi ab$ $\bar{x} = a$ $\bar{y} = b$	$I_1 = \dfrac{\pi}{4}ab^3$ $I_2 = \dfrac{\pi}{4}a^3b$	$C_1 = b$ $C_1 = a$	$Z_1 = \dfrac{\pi}{4}ab^2$ $Z_2 = \dfrac{\pi}{4}a^2b$	$K_1^2 = \dfrac{b^2}{4}$ $K_2^2 = \dfrac{a^2}{4}$

Table 1.3: Preferred Numbers

R 5	*R 10*	*R 20*	*R 40*
1.00	1.00	1.00	1.00
			1.06
		1.12	1.12
			1.18

	1.25	1.25	1.25
			1.32
		1.40	1.40
			1.50
1.60	1.60	1.60	1.60
			1.70
		1.80	1.80
			1.90
	2.00	2.00	2.00
			2.12
		2.24	2.24
			2.36
2.50	2.50	2.50	2.50
			2.65
		2.80	2.80
			3.00
	3.15	3.15	3.15
			3.35
		3.55	3.55
			3.75
4.00	4.00	4.00	4.00
			4.25
		4.50	4.50
			4.75
	5.00	5.00	5.00
			5.30
		5.60	5.60
			6.00
6.30	6.30	6.30	6.30
			6.70
		7.10	7.10
			7.50
	8.00	8.00	8.00
			8.50
		9.00	9.00
			9.50
10.00	10.00	10.00	10.00

1. Number of Steps

Let Range ratio, R_n = Maximum value of a quantity/Minimum value of the quantity

Common ratio = ϕ

Number of steps, $z = 1 + \text{In } R_n/\text{In } \phi$ ...(1.1)

Table 1.4: Conversion formulae for Units

From	*To*	*Multiply by*
Length:		
cm		1×10^{-2}
mm	m	1×10^{-3}
foot		0.3048
inch		0.0254
yard		0.9144
mile		1.609344×10^{3}
Area:		
acre	m^2	4046.86
hectare		10^4
Volume:		
litre		10^{-3}
gallon(UK)	m^3	0.0045461
barrel		0.158987
Mass:		
gram		1×10^{-3}
Ibm		0.4536
quintal	kg	1×10^{2}
tonne		1×10^{3}
slug		14.59
Moment of inertia:	$kg.m^2$	9.80665
$kgf.m.s^2$		
Power:		
BTU/s		1.05435×10^{3}
Cal/s		4.184
ft.lbf/s		1.35582
kgf.m / s	W(J/s)	9.80665
erg / s		1×10^{-7}
HP (metric)		735.499
HP. (FPS)		745.7

Pressure:		
bar	Pa (N/m^2)	1×10^5
cm of mercury (0°C)		1.33322×10^3
cm of water (4°C)		$9.80638 \times 10^{+5}$
dyne/cm^2		1×10^{-4}
Ibf/in^2		6895
kgf /cm^2		9.81×10^4
Velocity:		
ft/s	m/s	0.3048
km/h		2.77778×10^{-1}
mile/h		4.4704×10^{-1}
knot		5.14444×10^{-1}
Work:		
erg	J	10^{-7}
kgf.m		9.80665
kWh		3.6×10^5
Force:		
Ibf	N	4.44822
dyne		10^{-5}
kgf		8.80665
poundal		0.138
tons		9806.65
Energy:		
BTU	Joule (J)	1.05435×10^3
Calorie		4.184
ft.lbf		1.3558179
kWh		3.60×10^6
kgf.m		9.80665
erg		10^{-7}
Watt/cm^2		1.602×10^{-19}
BTU/Ib	kJ/kg	2.326
Density:		
gram/cm^3	kg/m^3	1×10^3
Ibm/in^3		2.76799×10^4
Ibm/ft^3		16.018463
Acceleration:		
ft/s^2	m/s^2	0.3048
amis^2		0.01

1 HP (FPS) = 1.014 HP (metric) = 0.7457 kW = 33000 ft.lbf / min

1 HP (metric) = 75 kgf.m/s = 0.73548 kW

CHAPTER

2

Engineering Materials

Tabled 2.1: Typical Properties of Grey Iron Castings

Properties	*Grade*						
	FG150	*FG200*	*FG220*	*FG260*	*FG300*	*FG350*	*FG400*
Tensile strength, N/mm^2	150	200	220	260	300	350	400
Compressive strength, N/mm^2	600	720	768	864	960	1080	1200
Shear strength, N/mm^2	173	230	2253	299	345	403	460
Torsional strength, $N.mm^2$	173	260	253	299	345	403	460
Modulus of Elasticity :							
Compression, Tension, GN/m^2	100	114	120	128	135	140	145
Modulus of Rigidity, GN/m^2	40	46	48	51	54	56	58
Poisson's Ratio	0.26	0.26	0.26	0.26	0.26	0.26	0.26
Fatigue limit, N/mm^2 (Wohler, unnotched, 8.4 mm dia)	68	90	99	117	135	149	152
Density, kg/m^3	7050	7100	7150	7250	7250	7300	7300

IS: 210—1977

Table 2.2: Properties of Spheroidal Graphite or Nodular Graphite Iron

Grade	*Tensile strength N/mm^2*	*Proof stress, 0.2 o/o N/mm^2*	*Elongation o/o*	*Impact Strength Charpy Nm*	*Brinell Hardness*	*Predominant Structure*
SG 800/2	800	480	2	–	245-350	Pearlite or tempered
SO 700/2	700	440	2	–	225-300	Pearlite
SG 600/3	600	370	3	–	190-270	Pearlite and ferrite
SG 500/7	500	320	7	–	170-240	Ferrite and pearlite
SG 400/12	400	250	12	6	max-200	Ferrite
SG 370/17	370	230	17	15	max-180	Ferrite

IS: 1865-1974

Table 2.3 Tensile Properties of White Heart Malleable Iron Castings

Grade Designation	*Sectional Thickness mm*	*Diameter of Test Bar mm*	*Tensile strength Min N/mm²*	*0.5 per cent proof stress, Min N/mm²*	*Elongation (L_0 = 3D) Min o/o*
WM 410	8 and under	9	350	200	10
	over 8 upto and including 13	12	390	230	6
	over 13	15	410	250	4
WM 340	8 and under	9	270	–	7
	over 8 upto and including 13	12	310	–	4
	over 13	15	340	–	3

*As per IS: 4843-1968 **IS : 2107-1977**

Table 2.4: Tensile Properties of Blackheart Malleable Iron Castings

Grade Designation*	*Brinell hardness HB Max*	*Dia of test bar mm*	*Tensile strength Min N/mm²*	*0.5 percent proof stress, Min N/mm²*	*Elongation (L_0 = 3D) o/o*
BM340	150	15	340	200	12
BM310	150	15	310	190	10
BM290	160	15	290	–	6

* As per IS: 4843-1968 **IS: 210-1977**

Table 2.5: Tensile Properties of Pearlitic Malleable Iron Castings

Grade Designation*	*Brinell hardness HB Max*	*Dia of Test Bar mm*	*Tensile strength Min N/mm²*	*0.5 perc ent proof stress, Min N/mm²*	*Elongation (L_0 = 3D) Min o/o*
PM 690	241-285	15	690	540	2
PM 590	200-248	15	590	420	3
PM 540	180-230	15	540	350	4
PM 490	170-229	15	490	310	5
PM 440	149-201	15	440	270	7

* As per IS: 4843-1968 **IS: 210-1977**

Table 2.6 Mechanical Properties of Steels Specified by Tensile and/or Yield Properties

Designation	*Tensile Strength N/mm²*	*Yield Stress Min N/mm²*	*Elongation Percent, Min Gauge Length = 5.65 $\sqrt{A_0}$*
Fe 290	290	170	27
Fe E 220	290	220	27
Fe 310	310	180	26

Fe E 230	310	230	26
Fe 330	330	200	26
Fe E 250	330	250	26
Fe 360	360	220	25
Fe E 270	360	270	25
Fe 410	410	250	23
Fe E 310	410	310	23
Fe 490	490	290	21
FeE 370	490	370	21
Fe 540	540	320	20
Fe E 400	540	400	20
Fe 620	620	380	15
Fe E 460	620	460	15
Fe 690	690	410	12
Fe E 520	690	520	12
Fe 770	770	460	10
Fe E 580	770	580	10
Fe 870	870	520	8
Fe E 650	870	650	8

IS : 1570 (Part-I)—1978

Table 2.7: Typical Uses of Steels Specified by Tensile Properties

Steel Designation	*Typical Uses*
Fe290	Structural steel sheets for plain drawn parts, tubes for oil well casing, steam, water and air passages, cycle, motor cycle and automobile tubes, rivet bars and wire.
Fe 310 Fe 330	Steels for locomotive carriage and car structures, screw stock.
Fe 360	Structural steel for chemical pressure vessels.
Fe 380	Structural steel for ships, chemical vessels, air receivers and fasteners.
Fe 410	Structural steel for bridges and building construction, railway rolling stock, screw spikes, and oil well casing.
Fe 430	Structural steel for railway rolling stock, pressure vessels, fasteners and valve fittings for compressed gas cylinders.
Fe 460	Structural steel for railway rolling stock, pressure parts of marine and land boilers, and rivets for air receivers.
Fe 490	Structural steel for mines, forgings for marine engines, and machine parts.
Fe 510	Steel for railway wheels, disc wheel centres for railway and electric tramway cars, and seamless tubes.
Fe 540	High tensile steel for locomotive, carriage wagon and tramway axles, bolts, and seamless and welded tubes.
Fe 570	High tensile steel for bridges and general construction, and bars and wire for concrete reinforcement.
Fe 620	High tensile steel for tramway axles and seamless tubes.

Table 2.8: Typical Uses of Carbon Steels

Steel Designation	*Typical Uses*
4 C 2	Dead soft steel generally used in electrical industry.
5 C 3, 7 C 3, 10 C 4	Used for cold forming and deep drawing. Rimming quality used for automobile body and fender stock, hoods, lamps, oil pans. They are also used for heading wire and rivets and low carbon wire products. Killed quality is used for forgings and heat treating applications.
10 C 4, 14 C 6	Case hardening steels are used for making cam shafts, cams, light duty gears, worms, gudgeon pins, selector forks, spindles, pawls, ratchets, chain wheels, tappets, etc.
15 C 4	Used for lightly stressed parts. It is suitable where cold web, such as bending and riveting may be necessary.
15 C 8, 20 C 8, 25 C 4 and 25 C 8	General purpose steels for low stressed components.
30 C 8	Used for cold formed shift and brake levers. Hardened and tempered steel is used to make parts such as socket, tie rods, yolk adjustable control lever cable, shaft fork and rear hub, sprocket, lever, cams, rocker arms and bushes, tubes. Also used for tubes for aircraft, automobile and bicycle.
35 C 4	Steel for low stressed parts, automobile tubes and fasteners.
35 C 8	Steel for making low stressed parts in machine structures, cycle and motor cycle frame tubes, fish plates for rails and fasteners.
40 C 8	Steel for crank shafts, shafts, spindle, automobile axle beams, push rods, connecting rods, studs, bolts, lightly stressed gears, etc.
45 C 8	Steel for spindle of machine tools, bigger gears, bolts and shafts.
50 C 8	Steel for making keys, shafts, cylinders, machine components required for moderate power resistance. Hardened steel is used for large pitch worms and gears.
50 C 12	Rail steel. Also used for making spike bolts, gear shafts, rocking levers and cylinder liners.
55 C 6 and 55 C	Steels used for making gears, cylinders, cams, keys, crankshafts, sprockets and machine parts requiring moderate wear resistance for which toughness is not of primary importance.
60 C 6	Steel for making spindles for machine tools, hardened screws and nuts. Couplings, crankshafts, axles and pinions.
65 C 6	High tensile structure steel for making locomotive carriage and wagon tyres. Used also to make engine value springs, small washers and thin stamped parts.
70 C 6	Steel for making baffle springs, shock absorbers, rail tyres, unhardened gears and worms.
75 C 6	Used for making light flat springs formed from annealed stock. When properly heat treated it is used for making plough shears, rake teeth scrappers and cultivator's shovels.
80 C 6 and 85 C 6	Used for making flat and coil springs for automobiles and railway vehicles. After suitable heat treatment these steels are also used for making scraper blades, discs and spring tooth harrows.
98 C 6 and 113 C 6	In oil hardened and tempered condition it is used for coil or spiral springs.

IS: 1871—1965

Table 2.9: Typical Uses of Alloy Steel

Steel Designation	*Typical Uses*
10 C 8 S 11	Small parts to be cyanided or carborised.
14 C 14 S 14	Parts where good macfainability and finish are important.
25 C 12 S 14	Bolts, studs and other heat treated parts of small section.
40 C 10 S 18	Heat treated bolts, engine shafts, connecting rods, miscellaneous gun carriage and small parts not under high stresses and wear.
3 C 10 S 25	Components under light stresses and not under shock. Suitable for components for production on automatic lathes.
40 C 15 S 12	Heat treated axles, shafts, small crankshafts and other vehicle parts.
55 Si 7	Oil hardening spring steel for road and railway vehicles.
20 Mn 6,27 Mn 6	Welded structures, crankshafts, steering levers, shafting spindles, etc.,
37 Mn 6	Axles, shafts, crankshafts, connecting rods, etc.
35 Mn 6 Mo 3	General engineering components such as crankshafts, bolts, wheel studs, axteshafts, levers and connecting rods.
5 Mn 6 Mo 4	General engineering components such as crankshafts, connecting rods, axle shafts, levers, bolts and wheel studs.
17 Mn 5 Cr 4	Small gear wheels and shafts, cardan joints and steering regulators.
40 Cr 4	Gears, connecting rods, stub axles, steering arms, wear resistant plates for earthmoving and concrete handling equipment.
50 Cr 4	Spring steel.
105 Cr 4	Roller bearing steels.
105 Cr 5, 50Cr 4 V 2	Laminated, coil and volute springs.
40 Cr 4 Mo 3	Axle shafts, crankshafts, connecting rods, gears, high tensile bolts and studs, propeller shaft joints, etc.
15 Cr 4 Mo 6	Boiler superheater tubes, plates, pipefittings, forging and bolting material for petroleum and chemical industries.
15 Cr 13 Mo 6, 25 Cr 20 Mo 6	Crankshafts, cylinder liners for aero and automobile engines, gears and machine parts with high surface hardness and wear resistance.
40 Cr 13 Mo 10 V 2	Components requiring high tensile properties, heavy stress and severe wear.
40 Cr 7 Al 10 Mo 2	Components requiring maximum surface hardness of a nitraded case and high core strength.
40 Ni 14	Parts requiring high toughness, parts working at low temperatures, heavy forgings, turbine blades, severely stressed screws, bolts and nuts.
16 Ni 4 Cr 3	Gear box and transmission components, cams, camshafts, etc.
35 Ni 5 Cr 2	Crankshafts, connecting rods, gear shafts, chain parts, clutches, flexible shafts for planetary gears, camshafts of aircrafts and heavy vehicles.
35 Ni 6 Cr 5	Highly stressed gears, components requiring high tensile strength.
15 Ni 5 Cr 4 Mol, 1 15 Ni 7 Cr 4 Mo 2	Heavy duty gears, heavy vehicles and automobile transmission components.

40 Ni 6 Cr 4 Mo 2	General machine parts such as bolts, nuts, gear, axles, connecting rods, shafts etc.
40 Ni 6 Cr 4 Mo 3	High strength machine parts, collets, spindles, screws, high tensile bolts and studs, gears, pinions, axle shafts, tappets crankshafts, connecting rods, boring bars, arbours etc.
31 Ni 10 Cr 3 Mo 6	Highly stressed bolls and nuts, shafts, crankshafts, connecting rods, gears, axles mandrel bars, components for aircraft power units, airframes and under carriages, for low temperature service.
40 Ni 10 Cr 3 Mo 6	Highly stressed gears, connecting rods, crankshafts, components for all types of aircraft power units, axles, airframes and under carriages, die blocks, mandrel bars, etc., for low temperature service.

IS: 1871—1965

Table 2.10: Mechanical Properties (Core) of Case Hardening Steels in the Refined and Quenched Condition

Steel Designation	*Tensile Strength N/mm²*	*Minimum Elongation, Percentage (gauge length) = $5.65\sqrt{So}$)*	*Izod Impact Value, Min (if specified) N.m*	*Limiting ruling section, mm*	*Brinell Hardness, Max, HB.*
10 C 4	490.3	17	54	15	130
10 C 6	490.3	17	54	over 15 upto 30	143
10 C 8 S 11	490.3	17	54	30	143
14 C 14 S 14	588.4	17	40.2	30	154
11 Min 6	588.4	17	54	30	154
15 Cr 3	588.4	13	47	301170	
17 Mn 5 Cr 4	784.5	10	34.3	30	207
20 Mn Cr 5	980.7	8	37.3	30	217
16 Ni 3 Cr 2	686.5	15	40.2	90	184
16 Ni 4 Cr 3	833.6	12	40.2	30	217
	784.5	–	–	60	217
	735	–	–	90	
13 Ni 13 Cr 3	833.6	12	47	60	229
	784.5	–	–	100	229
15 Ni 16 Cr 5	1223.9	9	34.3	30	241
	1176.8	–	–	60	
	1127.8	–	–		
20 Ni 8 Mo 3	833.6	12	60.8	30	
	686.5	–	–	60	207
20 Bi Cr Mo 2	882.6	11	40.2	30	

	784.5	–	–	60	213
	735.5	–	–	90	
15 Ni 5 Cr 4 Mo 2	980.7	9	40.2	30	
	931.6	–	–	90	217
	931.6	–	–	90	
20 Ni 8 Cr 6 Mo 2	1323.9	9	34.3	30	
	1176.8	–	–	60	224
	1127.8	–	–	90	

IS : 4432 - 1967

Table 2.11: Mechanical Properties (Core) of Case Hardening Steels in the Refined and Quenched Condition

Designation	*Condition*	*Tensile Strength* N/mm^2	*Elongation, Percentage (gauge length)* $= 5.65\sqrt{So}$	*Izod Impact Value, Min N.m*	*Brinell Hardness HB*	*Limiting Ruling Section, mm*
20 Mn 6	H + T	588.4-753.5	18	47	217 Max	63
15 Cr 3	R + Q	588.4 Min	13	47	–	–
17 Mn 5 Cr 4	R + Q	748 Min	10	34.3	–	–
20 Mn 5 Cr 5	R + Q	980.7 Min	10	37.3	217 Max	60
21 Cr 4 Mo 3	A	588.4–735.5	–	–	217 Max	100
07 Cr 4 Mo 6	N + T	441.3	–	–	–	–
10 Cr 9 Mo 10	N + T	490.3-627.6	20	53.9	143-187	–
13 Ni 13 Cr 3	R + Q (Core properties)	833.6 Min	12	47	–	–
15 Ni 16 Cr 5	R + Q	1323.9	9	43.3	–	–
15 Ni 5 Cr 4 Mo 1	R + Q	980.7 Min	9	40.2	–	–
15 Ni 7 Cr 4 Mo 2	R + Q	1078.7 Min	9	34.3	–	–
16 Ni 8 Cr 6 Mo 2	R + Q	1323.9 Min	9	34.3	–	–
36 Si 7	H + T	784.5–931.6	–	–	217 Max	100
37 Mn 6	H + T	686.5-833.6	18	47	248 Max	100
35 Mn 6 Mo 3	H + T	784.5-931.6	16	54	217 Max	100
40 Cr 4	H + T	686.5-833.6	18	54	210-248	100

40 Cr 4 Mo 3	H + T	784.5-931.6	16	54	229-177	100
35 Ni 5 Cr 2	H + T	784.5-931.6	16	54	229-277	100
40 Ni 6 Cr 4 Mo 3	H + T	980.7-1127.8	13	47	285-341	100
40 Ni 10 Cr 3 Mo 6	H + T	1520 Min	8	40.2	444 Min	100
25 Cr 20 Mo 6	H + T	1078,.7-1225.8	12	40.2	311-365	100
55 Si 7	H + T	1274.9-1471	–	–	235 Max	–
50 Cr 4 V 2	A	784.5 Max	–	–	235 Max	63

IS : 4367-1957

Code: H = Hardned, T = Tempered,
R = Refined, Q = Quenched
A = Annealed, N = Normalized.

Table 2.12: Mechanical Properties (Core) of Case Hardening Steels in the Refined and Quenched Condition

Steel Designation	*Tensile Strength N/mm²*	*0.2 Per cent Proof Stress, N/mm²*	*Elongation Min, Lo = $5.65\sqrt{A}$*	*Izod Impact Value J*	*Limiting Ruling Section mm*
30 C 8	600-750	400	18	55	30
35 C 8	600-750	400	18	55	63
40 C 8	600-750	380	18	41	63
45 C 8	700-850	480	17	35	30
	600-750	380	17	41	100
50 C 8	700-850	480	15	35	30
	700-850	460	15	–	63
55 C 8	800-950	540	13	–	30
	700-850	460	15	–	63
40 C 10 S 18	800-950	540	13	–	30
	600-750	380	18	41	60
40 C 15 S 12	700-850	480	17	49	30
	700-850	500	18	35	60
	600-750	440	18	48	30
	700-850	500	16	48	15
27 C 15	600-750	440	18	48	63
	700-850	500	16	48	30
37 C 15	600-750	440	18	48	100
	700-850	540	18	48	63
	800-950	600	16	48	30
	900-1050	700	15	41	15

Table 2.12: *continued*

Steel Designation	*Tensile Strength* N/mm^2	*0.2 Per cent Proof Stress,* N/mm^2	*Elongation Min,* $Lo = 5.65\sqrt{A}$	*Izod Impact Value J*	*Limiting Ruling Section mm*
35 Mn 6 Mo 3	700-850	540	18	55	150
	800-950	600	16	55	100
	900-1050	700	15	55	63
	1000-1150	800	13	48	30
35 Mn 6 Mo 4	800-950	600	16	55	150
	900-1050	700	16	55	100
	1000-1150	800	15	48	63
55 Cr 3	900-1050	660	12	35	63
	1000-1150	740	10	17	63
40 Cr 4	700-850	540	18	55	100
	800-950	600	16	55	63
	900-1050	700	15	55	30
40 Cr 4 Mo 3	700-850	540	18	55	150
	800-950	600	16	55	100
	900-1050	700	15	55	63
	1000-1150	800	13	48	30
15 Cr 13 Mo 6 and 25 Cr 15 Mo 6	700-850	540	18	55	150
	800-950	600	16	55	150
	900-1050	700	15	55	150
	1000-1150	800	13	48	150
	1100-1250	880	12	41	100
	1550 Min	1300	8	14	63
40 Crl 13 Mo 10 V 2	1350 Min	1120	8	21	63
	1350 Min	1300	8	14	30
40 Cr 7	700-850	540	18	55	150
AI 10 Mo 2	800-950	600	16	55	100
	900-1050	700	15	48	63
40 Ni 14	800-950	600	16	55	150
	900-1050	700	15	55	100

Table 2.12: *continued*

Steel Designation	*Tensile Strength* N/mm^2	*0.2 Per cent Proof Stress,* N/mm^2	*Elongation Min,* $Lo = 5.65\sqrt{A}$	*Izod Impact Value J*	*Limiting Ruling Section mm*
35 Ni 5 Cr 2	700-850	540	18	55	150
	800-950	600	16	55	100
	900-1050	700	15	55	63
30 Ni 13 Cr 5	1550 Min	1300	8	14	63 (air hardened) 150 (oil hardened)
40 Ni	800-950	600	16	55	150
6 Cr 4 Mo 2	900-1050	700	15	55	100
Cr 4 Mo3	1000-1150	800	13	48	63
	1100-1250	880	11	41	30
40 Ni	800-950	600	16	55	150
	900-1050	700	15	55	150
	1000-1150	800	18	48	100
	1100-1250	880	11	41	63
	1200-1350	1000	10	30	30
	1550 Min	1300	6	11	30
30 Ni 10	900-1050	700	15	55	150
Cr 3 Mo 6	1000-1150	800	12	48	150
	1100-1250	880	11	41	100
	1200-1350	1000	10	35	63
	1550 Min	1300	8	14	63
40 Ni 10	1000-1150	800	12	48	150
Cr 3 Mo 6	1100-1250	880	11	41	150
	1200-1350	1000	10	35	150
	1550 Min	1300	8	14	100

IS : 5517 - 1978

CHAPTER 3

Design for Static Loading

1. Stresses in Machine Members

P = tensile or compressive force, N

F = shear force, N

A = area of cross-section, m^2

M = bending moment, N.m

z = section modulus, m^3

T = torque, N.m

z_p = polar section modulus, m^3

(*a*) Simple Stresses

Tensile/compressive stress,

$$\sigma_t \text{ or } \sigma_c = P/A \text{ N/mm}^2 \text{ or MPa} \quad ...(3.1)$$

Shear stress, τ = F/A MPa ...(3.2)

Bending stress, σ_b = M/z MPa ...(3.3)

For a circular shaft of diameter d,

$$\sigma_b = 32\ M/(\pi d^3) \quad ...(3.4)$$

Torsional shear stress, $\tau = T/z_p$ MPa ...(3.5)

For a circular shaft of diameter d, $\tau = 16\ T/(\pi\ d^3)$ ...(3.6)

(*b*) Eccentric Axial Loading (Fig. 3.1)

$$\sigma = P/A \pm Pe/z \quad ...(3.7)$$

(*c*) Principal Stresses

Normal stresses, $\sigma_{1,\,2} = (1/2)\,(\sigma_x + \sigma_y) \pm (1/2)\,[(\sigma_x - \sigma_y)^2 + 4\tau^2_{xy}]^{1/2}$ ...(3.8)

Maximum shear stress,

$$\tau_{max} = + (1/2)\,[(\sigma_x - \sigma_y)^2 + 4\tau^2_{xy}]^{1/2} \quad ...(3.9)$$

For a shaft subjected to combined bending and torsion,

$$\sigma_{1.2} = [16/(\pi d^3) [M \pm (M^2 + T^2)^{1/2}] \quad ...(3.10)$$

$$\tau_{max} = [16/(\pi d^3)][\pm (M^2 + T^2)^{1/2}] \quad ...(3.11)$$

2. Simple and Principal Strains

If σ_1, σ_2 and σ_3 are the principal stresses at a point in a body, then the principal strains are:

$$\varepsilon_1 = (1/E)\,[\sigma_1 - \nu(\sigma_2 + \sigma_3)] \quad ...(3.12)$$

$$\varepsilon_2 = (1/E)\,[\sigma_2 - \nu(\sigma_3 + \sigma_1)] \quad ...(3.13)$$

$$\varepsilon_3 = (1/E)\,[\sigma_3 - \nu(\sigma_1 + \sigma_2)] \quad ...(3.14)$$

3. Relationship between Elastic Constants

G = shear modulus

K = bulk modulus

$$G = E/[2(1 + \nu)] \quad ...(3.15)$$

$$K = E/[3(1 - 2\nu)] \quad ...(3.16)$$

Whore ν = Poisson's ratio

4. Factor of Safety

Factor of safety, n = Ultimate tensile strength of material/Allowable stress, for brittle materials

= Yield strength of material/Allowable stress, for ductile materials

For steady loads, the factor of safety "n" may be calculated from the following formula:

$$n = a \times b \times c \quad ...(3.17)$$

Where Factor, 'a' = elastic ratio

= Ultimate strength/Yield strength, for static loading

= Ultimate strength/Endurance limit, for dynamic loading

Factor, 'b' = shock or service factor

= Stress produced by a shock or suddenly applied load / Stress produced by a gradual load

$= [1 + (1 + 2h/\delta_{st})^{1/2}]$ for linear impact

$= 1 + \{1 + 2h/(r\,\theta_{st})\}^{1/2}]$ for angular impact

Where h = height fallen by the load

δ_{st}, θ_{in} = static linear or angular deflection respectively

For sudden loading, $h = 0$ and $b = 2$

b = 1.25 – 1.5 for light shocks

= 2 – 3 for heavy shocks

= 5 for extreme cases

Factor, 'c' = real margin of safety to account for unreliable material and to account for unforeseen causes of failure.

Average values of 'c' are:

$$c = 1.5 - 2 \text{ for ductile materials of uniform structure}$$
$$= 2.5 - 3.0 \text{ for brittle materials}$$
$$= 3 - 4 \text{ for timber}$$

$$\text{Design stress, } \sigma_d = S_{ut}/n \quad \text{...(3.18)}$$

$$\text{Margin of safety} = n - 1 \quad \text{...(3.19)}$$

$$\text{Factor of utilization} = 1/n \quad \text{...(3.20)}$$

Where S_{ut} = ultimate tensile strength

Refer to Table 3.3 for suggested values of factor of safety for various applications.

5. Theories of Failure

(i) Maximum Normal Stress Theory (Fig. 3.2)

If $\sigma_1 > \sigma_2 > \sigma_3$ are the three principal stresses at a point, then according to this theory, we have

$$\sigma_1 = \sigma_{ut} \quad \text{or} \quad \sigma_1 = \sigma_{uc} \quad \text{...(3.21)}$$

Whichever is applicable.

(ii) Mohr's Theory (Fig. 3.3)

$$\sigma_1 = \sigma_{ut} + \sigma_2/\sigma_{uc} = 1 \quad \text{...(3.22)}$$

(iii) Maximum Shear Stress Theory (Fig. 3.4)

If $\sigma_1 > \sigma_2 > \sigma_3$ and $\sigma_3 = 0$, then according to this theory, we have

$$(\sigma_1 - \sigma_2) = \sigma_y \quad \text{...(3.23)}$$

For the torsion case,

$$\sigma_2 = -\sigma_1 = \tau_y$$
$$\tau_y = 0.5\,\sigma_y \quad \text{...(3.24)}$$

(iv) Maximum Distortion Energy Theory (Fig. 3.5)

$$(\sigma_1 - \sigma_2)^2 + (\sigma_2 - \sigma_3)^2 + (\sigma_3 - \sigma_1)^2 = 2\sigma^2 y \quad \text{...(3.25)}$$

If $\sigma_3 = 0$, then

$$\sigma_1^{\,2} + \sigma_2^{\,2} - \sigma_1\,\sigma_2 = \sigma^2{}_y \quad \text{...(3.26)}$$

For the torsion case,

$$\sigma_2 = -\sigma_1 = \tau_y$$
$$\tau_y = 0.577\,\sigma_y \quad \text{...(3.27)}$$

(v) Maximum Strain Theory

$$\sigma_1 - \nu(\sigma_2 + \sigma_3) = \sigma_y \quad \text{...(3.28)}$$

(vi) Maximum Strain Energy Theory

$$\sigma_1^{\,2} + \sigma_2^{\,2} + \sigma_3^{\,2} - 2\nu(\sigma_1\sigma_2 + \sigma_2\sigma_3 + \sigma_3\sigma_1) = \sigma_y^{\,2} \quad \text{...(3.29)}$$

6. Buckling of columns

(i) Euler's buckling load,

$$P_e = \pi^2 EI/l^2{}_e \quad \text{...(3.30)}$$

Where l_e = equivalent length of column (Table 3.4)

The column shall buckle about that axis about which slenderness ratio, l_e/K, is minimum,

(*ii*) Johnson's parabolic formula

$$P_{cr} = \sigma_c A[1 - \sigma_c (l_e/K)^2/(4\pi^2 E)] \quad ...(3.31)$$

(*iii*) Rankine – Gordon formula

$$P_{cr} = \sigma_c A/[1 + a(l_e/K)^2] \quad ...(3.32)$$

Where 'a' = Rankine's constant (Table 3.5)

7. Hertz's Contact Stresses

Refer to Table 3.6.

Table 3.1: Formulae for shear force, bending moment, deflection and slope

Structure	*Shear force*	*Bending moment*	*Deflection*	*Slope*
W, x, A, B, C, a, x, b, l	$F_A = \frac{Wb}{l}$	$M_C = \frac{Wab}{l}$	$Y = \frac{1}{EI}\left[\frac{Wbx^3}{6l} + \frac{W(x-a)^3}{6} + \frac{Wab}{6}\left(\frac{l+b}{l}\right)x\right]$	$\theta_A = \frac{Wab}{6\,EI}\left(\frac{l+b}{l}\right)$
	$F_B = \frac{Wa}{l}$	$M_x = \frac{Wbx}{l} - W(x-a)$	$Y_C = \frac{Wa^2b^2}{3EI}$ For Y_{max}, $x = \sqrt{\frac{a(l+b)}{3}}$	$\theta_B = \frac{Wab}{6\,EI}\left(\frac{l+a}{l}\right)$
W, x, A, B, l/2, l/2, x, x, l	$F_A = \frac{W}{2}$	$M_C = \frac{Wl}{4}$	$Y = \frac{1}{EI}\left[\frac{-Wx^3}{12} + \frac{W}{6}\left(x - \frac{l}{2}\right)^3 + \frac{Wl^2x}{16}\right]$	$\theta_A = \theta_B\ \frac{Wl^2}{16\,EI}$
	$F_B = -\frac{W}{2}$		$Y_C = \frac{Wl^3}{48\,EI}$	
w/unit length, x, A, B, x, x, l	$F_A = \frac{wl}{2}$	$M_{max} = \frac{wl^2}{8}$	$Y = \frac{1}{EI}\left[-\frac{wlx^3}{12} + \frac{wx^4}{24} + \frac{wl^3x}{24}\right]$	$\theta_A = -\theta_B\ \frac{wl^3}{24\,EI}$
	$F_B = -\frac{Wl}{2}$	at $x = l/2$	$Y_{max} = \frac{5}{384}\frac{wl^4}{EI}$	
W, W, C, D, A, B, a, a, x, x, l	$F_A = W$	$M_C = M_D = Wa$	$Y_C = \frac{Wa^2}{6\,EI}(3l - 4a)$	$\theta_A = -\theta_B$
	$F_A = W$		$Y_{max} = \frac{Wa}{24\,EI}(3l^2 - 4a^2)$ at $x = \frac{l}{2}$	$= \frac{Wa}{2\,EI}(l - a)$

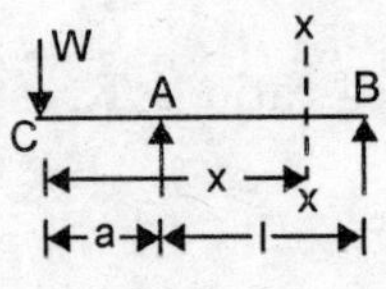	$F_A = \frac{W(a+l)}{l}$ $F_B = -\frac{Wa}{l}$	$M_A = Wa$ $M_B = 0$	$Y_C = \frac{Wa^2}{3\,EI}(l+a)$	$\theta_C = -\frac{Wa}{6\,EI}(2l+3a)$
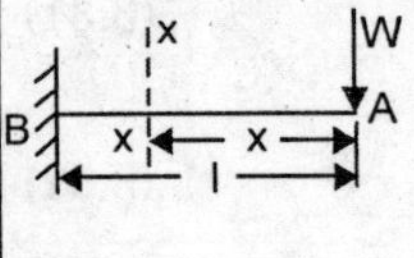	$F_A = F_B = W$	$M_B = -Wl$	$Y = \frac{W}{6\,EI}(x^3 - 3l^2x + 2l^3)$ $Y_A = \frac{Wl^3}{3\,EI}$	$\theta_A = -\frac{Wl^2}{2\,EI}$
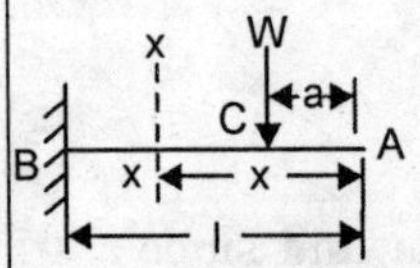	$F_B = F_C = -W$	$M_B = -W(l-a)$	$Y = \frac{W}{6\,EI}[(x-a)^3 - 3(l-a)^2x + (l-a)^2(2l+a)$ $Y_C = \frac{W}{3\,EI}(l-a)^3$	$\theta_C = \frac{W(l-a)^2}{2\,EI}$
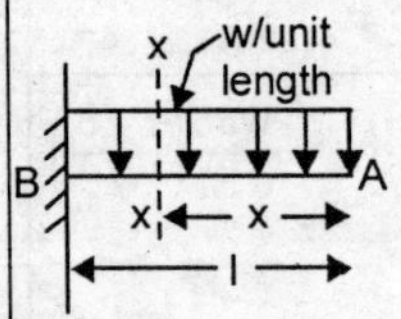	$F_B = -wl$	$M_B = \frac{wl^2}{2}$	$Y = \frac{w}{24\,EI}(x^4 - 4l^3x + 3l^4)$ $Y_A = \frac{Wl^4}{8\,EI}$	$\theta_A = -\frac{wl^3}{6\,EI}$
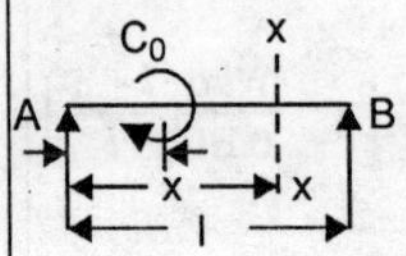	$F_A = \frac{C_0}{l}$ $F_B = -\frac{C_0}{l}$	$-C_0$	$Y_C = \frac{C_0}{EI}\left[-\frac{2}{3}\frac{a^3}{l} - \frac{al}{3} + a^2\right]$	$\theta_A = \frac{1}{EI}\left[\frac{C_0l}{6} - \frac{C_0(l-a)^2}{2l}\right]$ $\theta_B = \frac{C_0}{EI}\left[\frac{l^2-a^2}{2l} - \frac{l}{3}\right]$
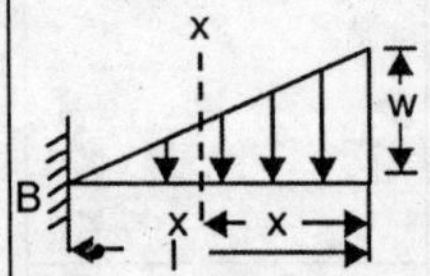	$F_B = -\frac{wl}{2}$	$M_B = -\frac{wl^2}{2}$	$Y_A = \frac{11}{120}\frac{Wl^4}{EI}$	$\theta_A = -\frac{wl^3}{8\,EI}$

Table 3.2: Formulae for torsional shear stress

Cross - section	*Polar area moment of inertia, J*	*Maximum shear stress,* τ_{max}
Solid circular rection	$\frac{\pi r^4}{2}$	$\frac{2T}{\pi r^3}$
Hollow circular rection	$\frac{\pi}{2}(r_0^4 - r_i^4)$	$\frac{2\,T}{\pi r_0^3 (1-k^4)}, k = \frac{r_i}{r_0}$
Solid elliptical rection	$\frac{\pi\, a^3 b^3}{a^2 + b^2}$	$\frac{2\,T}{\pi\, ab^2}$ at end of minor axis
Thin tube	$2\pi r^3 t$	$\frac{T}{2\pi r^2 t}$
Solid rectangular reaction	$ab^3\left[\frac{16}{3} - 3.36\frac{b}{a}\left(1 - \frac{b^4}{12a^4}\right)\right]$	$\frac{T(3a + 1.8b)}{8a^2b^2}$ at mid-point of longer side

Table 3.3: Suggested values for factor of safety

Working conditions	*Factor of safety*
1. Exceptionally reliable materials subjected to well defined loads under controlled conditions	1.25 – 1.5
2. Well known materials subjected to known loads under reasonably constant environmental Conditions.	1.5 – 2
3. Average materials subjected to known loads under ordinary environmental conditions.	2 – 2.5
4. Less tried or brittle materials under average conditions of environment and load.	2.5 – 3

5. Untried materials under average conditions of environment and load.	3 – 4
6. Better known materials under uncertain environment and load.	3 – 4
7. Repeated loads.	1 – 6
8. Impact loads.	3 – 6
9. Brittle materials.	Twice the value in 1 to 6

Table 3.4: Equivalent length of a column

Equivalent length ratio	*End condition*			
	Both ends hinged	*Both ends fixed*	*One end fixed other hinged*	*One end fixed other free*
l_e/l	1	1/2	$1/\sqrt{2}$	2

Table 3.5: Values of Rankine's constant, '*a*'

End conditions	*Steel*	*Cast iron*	*Wrought iron*	*Timber*
Both ends hinged	1/7500	1/1250	1/900	1/750

Table 3.6: Hertz's contact stresses

Condition	*Dimensions of contact area*	*Maximum stress*
Sphere on a flat plate (P, D, a°)	$a = 0.721\left[PD\left\{\frac{1-\nu_1^2}{E_1}+\frac{1-\nu_2^2}{E_2}\right\}\right]^{1/3}$	$\text{Max }\sigma_C = 0.918\left[\left\{\frac{P}{D^2\left\{\frac{1-\nu_1^2}{E_1}+\frac{1-\nu_2^2}{E_2}\right\}^2}\right\}\right]^{1/3}$
Sphere on a sphere (P, D_2, a, D_1)	$a = 0.721\left[P\left\{\frac{D_1D_2}{D_1+D_2}\right\}\left\{\frac{1-\nu_1^2}{E_1}+\frac{1-\nu_2^2}{E_2}\right\}\right]^{1/3}$	$\text{Max }\sigma_C = 0.918\left[P\frac{\left(\frac{D_1+D_2}{D_1D_2}\right)^2}{\left\{\frac{1-\nu_1^2}{E_1}+\frac{1-\nu_2^2}{E_2}\right\}^2}\right]^{1/3}$

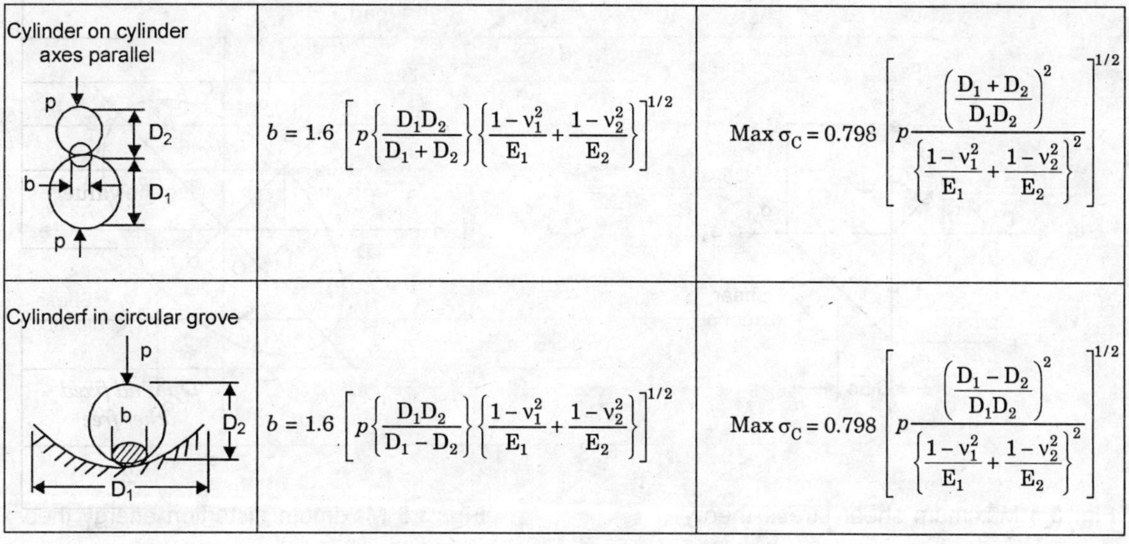

Cylinder on cylinder axes parallel	$b = 1.6\left[p\left\{\dfrac{D_1D_2}{D_1+D_2}\right\}\left\{\dfrac{1-\nu_1^2}{E_1}+\dfrac{1-\nu_2^2}{E_2}\right\}\right]^{1/2}$	$\text{Max}\ \sigma_C = 0.798\left[p\dfrac{\left(\dfrac{D_1+D_2}{D_1D_2}\right)^2}{\left\{\dfrac{1-\nu_1^2}{E_1}+\dfrac{1-\nu_2^2}{E_2}\right\}^2}\right]^{1/2}$
Cylinderf in circular grove	$b = 1.6\left[p\left\{\dfrac{D_1D_2}{D_1-D_2}\right\}\left\{\dfrac{1-\nu_1^2}{E_1}+\dfrac{1-\nu_2^2}{E_2}\right\}\right]^{1/2}$	$\text{Max}\ \sigma_C = 0.798\left[p\dfrac{\left(\dfrac{D_1-D_2}{D_1D_2}\right)^2}{\left\{\dfrac{1-\nu_1^2}{E_1}+\dfrac{1-\nu_2^2}{E_2}\right\}^2}\right]^{1/2}$

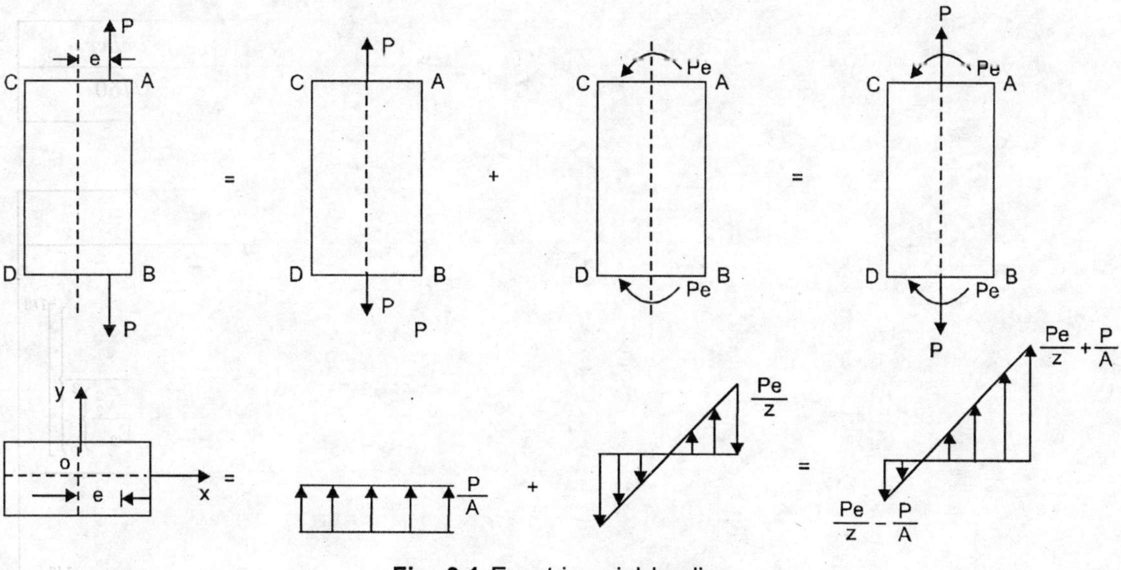

Fig. 3.1 Ecentric axial loading

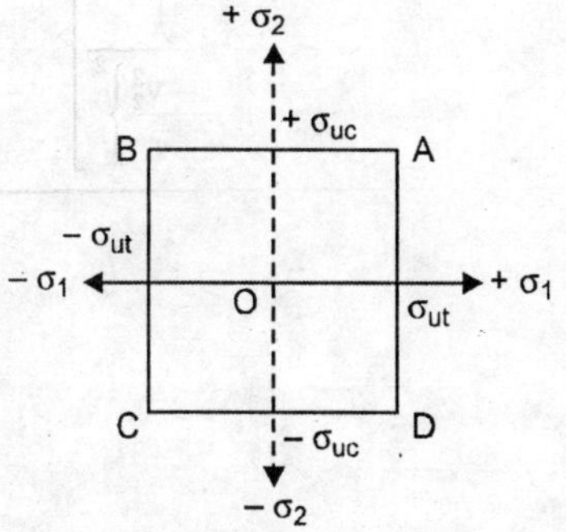

Fig. 3.2 Maximum normal stress theory

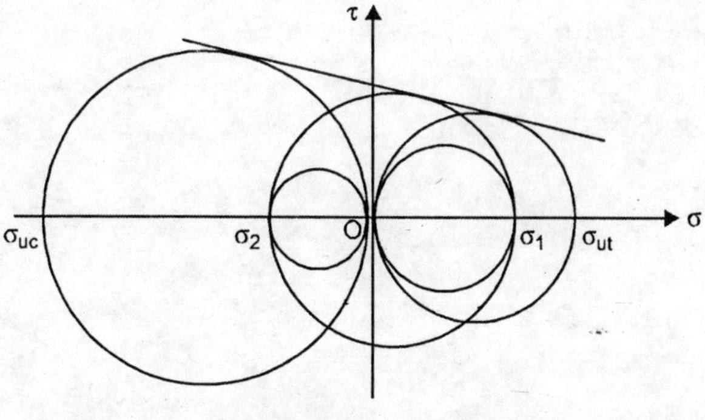

Fig. 3.3 Mohr's theory

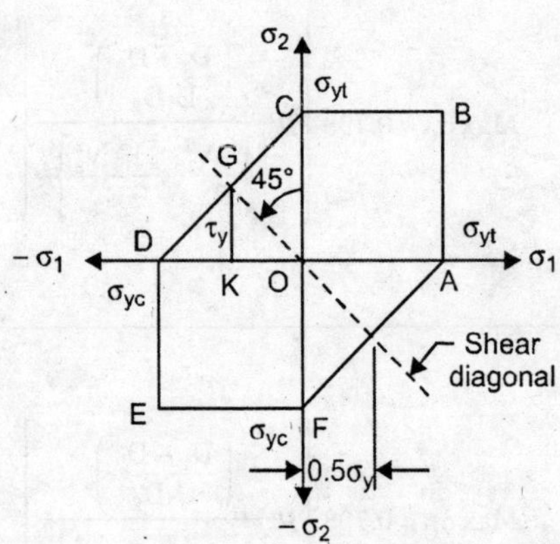

Fig. 3.4 Maximum shear stress theory

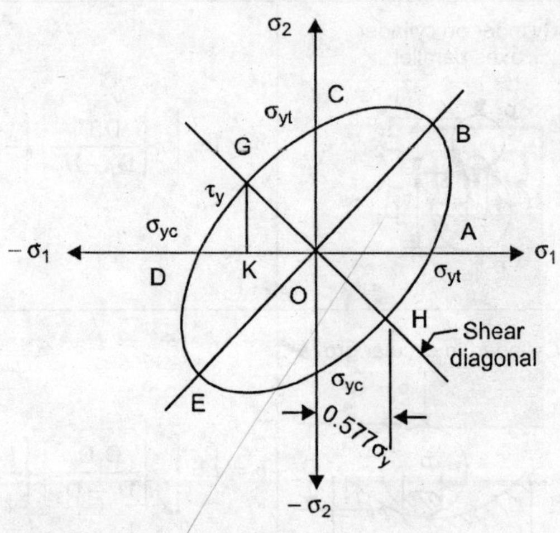

Fig. 3.5 Maximum distortion energy theory

CHAPTER 4

Design for Variable Loading

1. Stress – concentration factor,

$$K_t = \sigma_{max}/\sigma_{norm} \quad ...(4.1)$$

Refer to Table 4.1 for application of K_t for brittle and ductile materials subjected to static or fluctuating loading. The charts for the determination of stress – concentration factors are given in Figs. 4.17 to 4.29.

2. Completely Reversed and Fluctuating Stresses (Fig. 4.1)

Mean stress, $\sigma_m = (\sigma_{max} + \sigma_{min})/2$ ...(4.2)

Alternating stress, $\sigma_a = (\sigma_{max} - \sigma_{min})/2$ ...(4.3)

Stress range, $\sigma_r = \sigma_{max} - \sigma_{min}$ ...(4.4)

Where σ_{max} = maximum stress

σ_{min} = minimum stress

3. Fatigue Stress – Concentration Factor

$$K_f = \frac{\text{Fatigue limit of un-notched specimen (axial or bending)}}{\text{Fatigue limit of notched specimen (axial or bending)}} \quad ...(4.5)$$

4. Notch Sensitivity (Fig. 4.2 and 4.3)

For axial or bending loads,

$$q = (K_f - 1)/(K_t - 1) \quad ...(4.6)$$

$$K_f = 1 + q(K_t - 1) \quad ...(4.7)$$

For torsional loading,

$$q = (K_{fs} - 1)/(K_{ts} - 1) \quad ...(4.8)$$

Where $K_{fs} = \frac{\text{Fatigue limit of un-notched specimen (torsion)}}{\text{Fatigue limit of notched specimen (torsion)}}$ ...(4.9)

K_{ts} = Shear stress – concentration factor

$= 1 + q(K_{ts} - 1)$ ...(4.10)

5. Endurance Limit

$S' = k_a\, k_b\, k_c\, k_d\, k_e\, k_f\, k_g\, S'_e$ ...(4.11)

Where S'_e = rotating beam (unconnected) theoretical endurance limit of a specimen

k_a = surface finish factor (Fig. 4.4 Table 4.2)

k_b = size factor (Table 4.3)

k_c = reliability factor (Table 4.4)

k_d = temperature factor

$= 344/(273 + t),\ t = C°$ ...(4.12)

k_e = load factor (Table 4.5)

k_f = fatigue strength reduction factor

$= 1/K_f$ ...(4.13)

k_g = miscellaneous correction factor for press – fitted collar, residual stress, corrosion, impact, etc.

For steel, S_{ut} = 3.5 BHN MPa ...(4.14)

$S'_e = 0.5\, S_{ut}$ ...(4.15)

Refer to Table 4.6 for ratio of S'_e/S_{ut} values and Fig. 4.5.

For fluctuating torsional shear stresses, endurance limit.

$S'_e = 0.5\, S'_e$, based on maximum shear stress theory ...(4.16)

$= 0.577\, S'_e$, based on distortion energy theory ...(4.17)

6. Design for Finite (or Limited) and Infinite (or unlimited) Life

(*a*) Completely Reversed Stresses

Here $\sigma_m = 0$ and $\tau_m = 0$ and only σ_a and τ_a are present.

(*i*) Design for Infinite life ($N > 10^6$)

Induced stress in bending,

$\sigma_a = S_e/n$ ...(4.18)

Fig. 4.6 shows the S – N (or Woher's) curve on log-log scale.

Induced stress in torsion,

$\tau_a = S_{es}/n$ ...(4.19)

S_e, S_{es} = corrected endurance strengths in bending and torsion respectively.

σ_a, τ_a = amplitude (or alternating) normal and shear stresses respectively.

n = factor of safety

(*ii*) Design for finite life ($N < 10^6$), the $\log_{10} S_f - \log_{10} N$ curve (Fig. 4.7) for steels can be used. It consists of a straight line AB drawn from $\log_{10}(0.9\, S_{ut})$ at $\log_{10}(10^3)$ =

3 cycles to $\log_{10}(S_e)$ at $\log_{10}(10^6) = 6$ cycles. The design procedure for such problems is as follows:

1. Locate point A with coordinates [3, $\log_{10}(0.9\ S_{ut})$].
2. Locate point B with coordinated [6, $\log_{10}(S_e)$].
3. Join AB, which is used as a criterion of failure for finite life problems.
4. For the desired life N cycles of the machine component, draw a vertical line passing through $\log_{10}(N)$ on the abscissa, intersecting line AB at C.
5. Draw a line CD parallel to the abscissa. The ordinate at point D gives the endurance strength corresponding to N cycles. This value of fatigue strength S_f is used for the design calculations.

For finite life,

$$S_f = 10^b/N^m \quad ...(4.20)$$

For $10^3 \le N \le 10^6$, the values of b and m are given by:

$$b = \log [(0.9\ S_{ut})^2/S_e] \quad ...(4.21)$$

$$m = (1/3) \log [0.9S_{ut}/S_e] \quad ...(4.22)$$

For $N \ge 10^6$; $S_f = S_e$

(*b*) Fluctuating Stresses

1. Ductile Materials under Axial or Bending Loads (Fig. 4.8)

(*i*) Soderberg line

$$\sigma_{m/}\ S_{yt} + s_a/S_e = 1 \quad ...(4.23)$$

(*ii*) Goodman line

$$\sigma_{m/}\ S_{ut} + \sigma_a/S_e = 1 \quad ...(4.24)$$

(*iii*) Gerber parabola

$$(\sigma_{m/}\ S_{ut})^2 + \sigma_a/S_e = 1 \quad ...(4.25)$$

(*iv*) Modified Goodman Diagram for Axial or Bending Loads (Fig. 4.9)

$$\sigma_m + \sigma_a = S_{yt} \quad ...(4.26)$$

$$\sigma_a/\sigma_m = (P_a/A)/(P_m/A) = P_a/P_m \quad ...(4.27)$$

$$\tan\theta = P_a/P_m \quad ...(4.28)$$

or

$$\tan\theta = (M_a/M_m) \quad ...(4.29)$$

$$\sigma_{da} = S_{m/n} \quad ...(4.30a)$$

$$\sigma_{da} = S_a/n \quad ...(4.30b)$$

Complete Goodman diagram is shown in Fig. 4.10.

2. Ductile Materials under Torsional Loads (Fig. 4.11)

(*i*) Soderberg line

$$\tau_m/S_{ys} + \tau_m/S_{es} = 1 \quad ...(4.31)$$

(*ii*) Goodman line

$$\tau_m/S_{us} + \tau_a/S_{es} = 1 \quad ...(4.32)$$

(*iii*) Gerber parabola

$$(\tau_m/S_{us})^2 + \tau_a/S_{es} = 1 \quad ...(4.33)$$

(*iv*) Modified Goodman Diagram for Torsional Loads (Fig. 4.12)

$$\tau_{da} = S_{as}/n \quad ...(4.34a)$$

$$\tau_{dm} = S_{ms}/n \quad ...(4.34b)$$

3. $\sigma_m - \sigma_a$ Diagram for Brittle Materials

$$\tau_m / S_{us} + \tau_a/S_{es} = 1 \quad ...(4.35)$$

7. Working Stresses

(*i*) Ductile Materials

(*a*) Axial or Bending Stresses

For the Soderberg failure line,

$$\sigma_m/Syt + K_f\sigma_a/S_e = 1/n \quad ...(4.36)$$

For the Goodman failure line,

$$\sigma_m/S_{ut} + K_f\sigma_a/S_e = 1/n \quad ...(4.37)$$

(*b*) Torstonal Stresses

Soderberg line: $\tau_m/S_{ys} + K_{fs}\tau_a/S_{es} = 1/n$ ...(4.38)

Goodman line: $\tau_m/S_{ut} + K_{ft}\,\tau_a/S_{es} = 1/n$ ...(4.39)

(*ii*) Brittle Materials

(*a*) Axial or Bending Stresses

Soderberg line: $K_t[\sigma_m/S_{yt} + \sigma_a/S_e] = 1/n$ ...(4.40)

Goodman line: $K_t[\sigma_m/S_{ut} + \sigma_a/S_e] = 1/n$ ...(4.41)

(*b*) Torsional Stresses

Soderberg line: $K_{ts}[\tau_m/S_{ys} + \tau_a/S_{es}] = 1/n$ ...(4.42)

Goodman line: $K_{ts}[\tau_m/S_{ut} + \sigma_a/S_{es}] = 1/n$ ...(4.43)

Where $S_{us} = 0.8S_{ut+}$ ...(4.44)

$S_{es} = 0.8\ S_{us}$ ...(4.45)

(*iii*) Equivalent Stresses

(*a*) Axial or Bending Stresses

(*i*) Soderberg line

Equivalent mean stress,

$$\sigma_{me} = \sigma_m + (S_{yt}/S_e)\sigma_a \quad ...(4.46)$$

Equivalent alternating stress,

$$\sigma_{ae} = (S_e/S_{yt})\sigma_m + \sigma_a \quad ...(4.47)$$

(*ii*) Goodman line

Equivalent mean stress,

$$\sigma_{me} = \sigma_m + (S_{ut}/S_e)\sigma_e \quad ...(4.48)$$

Equivalent alternating stress,

$$\sigma_{ae} = \sigma_m + (S_e/S_{ut})\sigma_m + \sigma_a \quad ...(4.49)$$

(*b*) Torsionae Stresses

Equivalent mean shear stress,

$$\tau_{me} = \tau_m + (S_{ys}/S_{es})\tau_a \quad ...(4.50)$$

Equivalent alternating shear stress,

$$\tau_{ae} = (S_{es}/S_{ys})\tau_m + \tau_a \quad ...(4.51)$$

8. Combined Fluctuating Stresses

(*a*) Soderberg Equation:

(*i*) Axial and bending loading (Fig. 4.14)

$$P_{eq}/A + M_{eq}/z = S_{yt}/n \quad ...(4.52)$$

where $$P_{eq} = P_m + (S_{yt}/S_{ae})P_a \quad ...(4.53)$$

(*ii*) Two – dimensional normal stresses

$$\sigma_{xm} = [(\sigma_x)_{max} + (\sigma_x)_{min}]/2 \quad ...(4.55)$$

$$\sigma_{xa} = [(\sigma_x)_{max} - (\sigma_x)_{min}]/2 \quad ...(4.56)$$

$$\sigma_{ym} = [(\sigma_y)_{max} + (\sigma_y)_{min}]/2 \quad ...(4.57)$$

$$\sigma_{ya} = [(\sigma_y)_{max} + (\sigma_y)_{min}]/2 \quad ...(4.58)$$

$$\sigma_m = [(\sigma^2{}_{xm} + \sigma^2{}_{ym} - \sigma_{xm}\,\sigma_{ym})^{1/2} \quad ...(4.59)$$

$$\sigma_a = [(\sigma^2{}_{xa} + \sigma^2{}_{ya} - \sigma_{xa}\,\sigma_{ya})^{1/2} \quad ...(4.60)$$

Soderberg equation using maximum distortion energy theory gives,

$$\sigma_m + (S_{yt}/S_e)\,\sigma_a = S_{yt}/n \quad ...(4.61)$$

(*iii*) Two – dimensional normal stresses and shear stress (Fig. 4.16)

$$\tau_{xym} = (\tau_{max} + \tau_{min})/2$$

$$\tau_{xya} = (\tau_{max} - \tau_{min})/2$$

$$\sigma_m = [(\sigma^2{}_{xm} + \sigma^2{}_{ym} - \sigma_{xm}\,\sigma_{ym} + 3\tau^2{}_{xym})^{1/2} \quad ...(4.62)$$

$$\sigma_a = [(\sigma^2{}_{xa} + \sigma^2{}_{ya} - \sigma_{xa}\,\sigma_{ya} + 3\tau^2{}_{xya})^{1/2} \quad ...(4.63)$$

Soderberg equation using maximum distortion energy theory gives,

$$\sigma_m + (S_{yt}/S_e)\sigma_a = \sigma_a = S_{yt}/n \quad ...(4.64)$$

If stress – concentration factors are given in bending and torsion, then for ductile materials, the stress – concentration factors are applied only to the alternating components of stress.

$$\sigma_m = [(\sigma^2{}_{xm} + \sigma^2{}_{ym} - \sigma_{xm}\,\sigma_{ym} + 3\tau^2{}_{xym})^{1/2} \quad ...(4.65)$$

$$\sigma_a = [(K_{tb}\,\sigma_{xa})^2 + (K_{tb}\,\sigma_{ya})^2 - K^2{}_{tb}\sigma_{xa}\,\sigma_{ya} + 3(K_{ts}\,\tau_{xya})^2]^{1/2} \quad ...(4.66)$$

Table 4.1: Application of stress – concentration factor

Loading	*Brittle material*		*Ductile material*	
	Normal	*Shear*	*Normal*	*Shear*
Static	K_t	K_{ts}	Ignore	Ignore
Fluctuating	K_t	K_{ts}	K_t	K_{ts}

Table 4.2: Surface finish factor, k_a

Ultimate tensile strength S_{ut} *MPa*	*Machined surface (Cold drawn)* k_a	*Hot rolled ruface* k_a
410	0.91	0.72
480	9.90	0.68
550	0.88	0.62
620	0.86	0.58
690	0.85	0.55
760	0.84	0.52
820	0.82	0.48
1030	0.78	0.38
1370	0.72	0.30

Table 4.3: Size factor, k_b

Type of loading	*Size factor,* k_b
1. Reserved bending and torsional loads:	
(*i*) $d \leq 0.75$ mm	1.00
(*ii*) $14.5 < d \leq 50$ mm	0.85
(*iii*) $d > 50$ mm	0.75
2. Reversed axial loads for $d < 50$ mm	1.00

Table 4.4: Reliability factor

Realibility, %	*Realibility factor,* k_c
50.00	1.000
90.00	0.897
95.00	0.868
99.00	0.814
99.90	0.753
99.99	0.702

Tabte 4.5: Load factor

Types of loading	*Load factor,* k_e
1. Reversed for rotating bending	1.0
2. Reversed axial loading with no bending	0.9
3. Reversed axial loading with indeterminate bending	0.6 – 0.85
4. Reversed torsion:	
(*i*) Ductile materials	0.577
(*ii*) Brittle materials	0.80

Table 4.6: Ratios of S'_e/S_{ut}

Material	S'_e / S_{ut}
1. Wrought iron at 10^6 cycle	0.4
2. Cast steel iron at 10^6 cycle	0.5
3. Cast iron at 10^6 cycle	0.4 – 0.5
4. Wrought aluminium alloys at 5×10^8 cycle	0.4
5. Copper alloys at 10^8 cycle	0.25 – 0.50

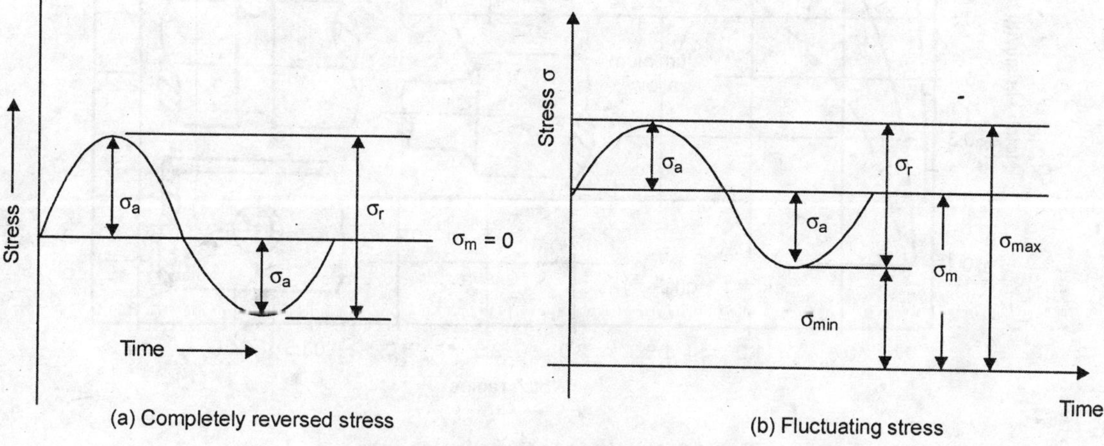

Fig. 4.1 Types of fatigue stresses

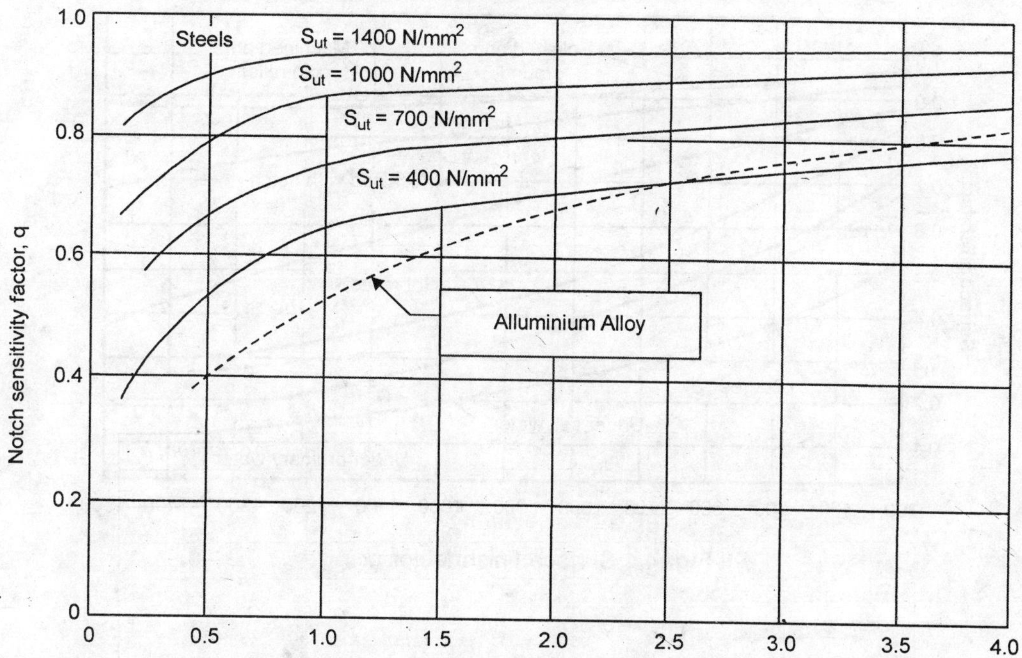

Fig. 4.2 Notch sensitivity charts for reversed bending and reversed axial stresses

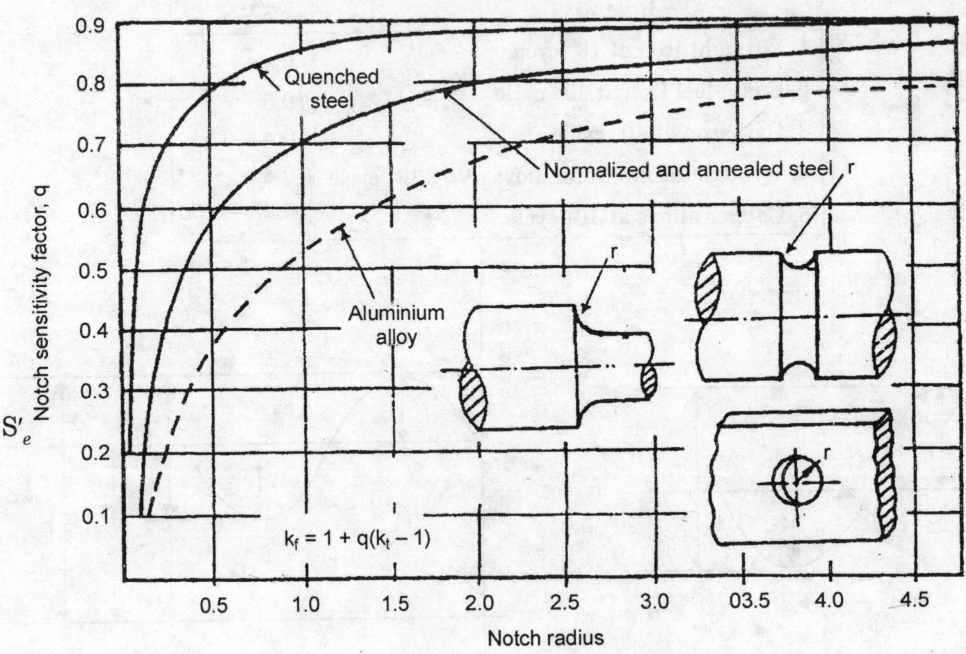

Fig. 4.3 Notch sensitivity factor (q)

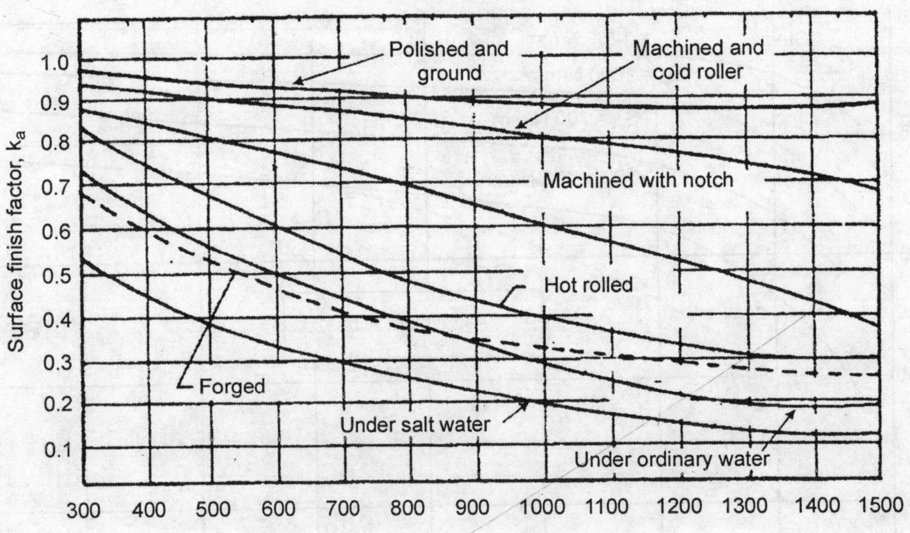

Fig. 4.4 Surface finish factor (k_a)

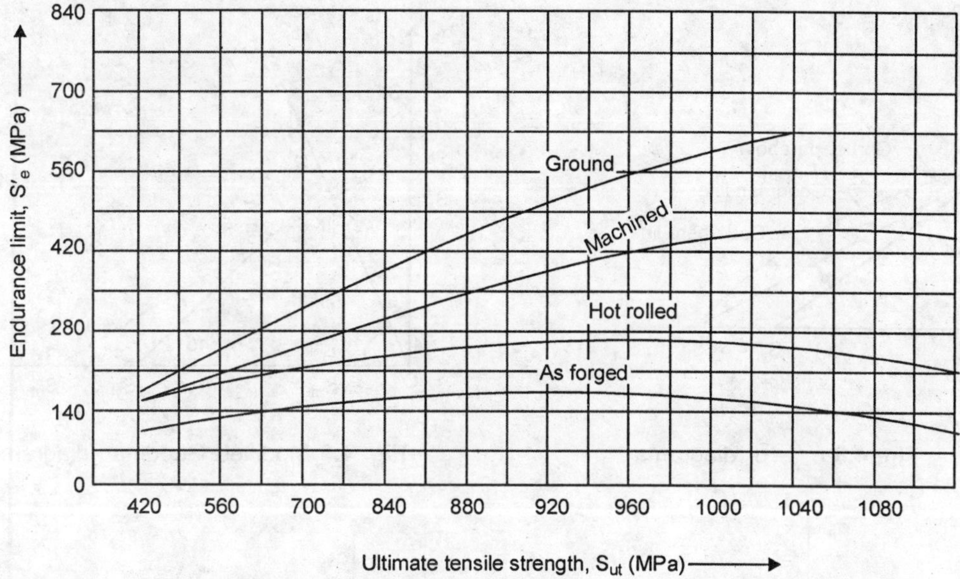

Fig. 4.5 Endurance limit of steel corresponding to ultimate tensile strength

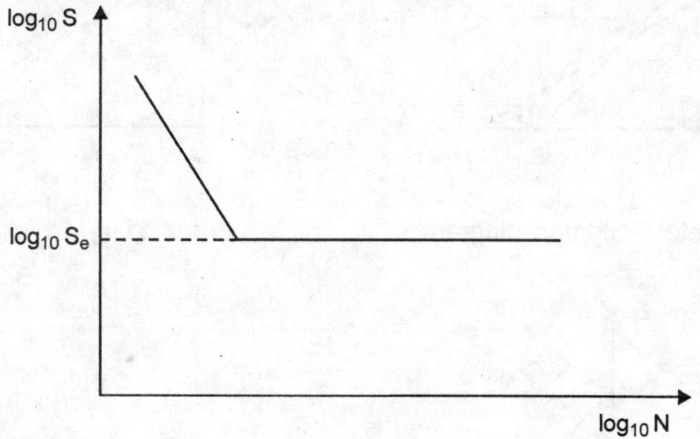

Fig. 4.6 S-N (or Wohler) Curve

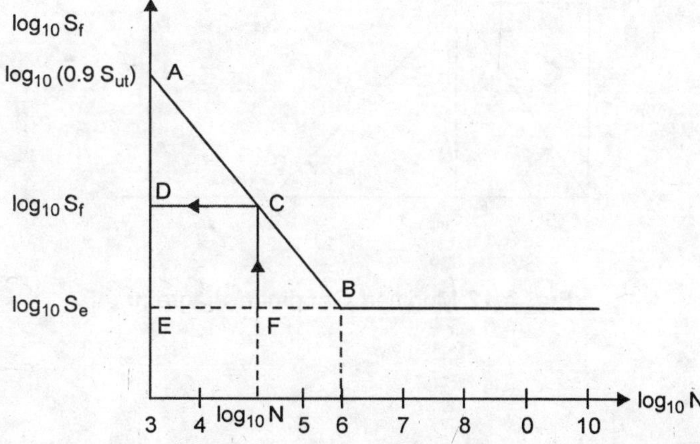

Fig. 4.7 S-N Curve for finite life

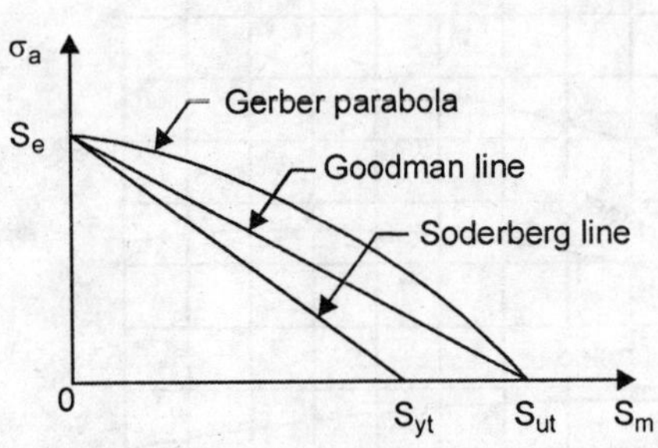

Fig. 4.8 $\sigma_m - \sigma_a$ diagrams

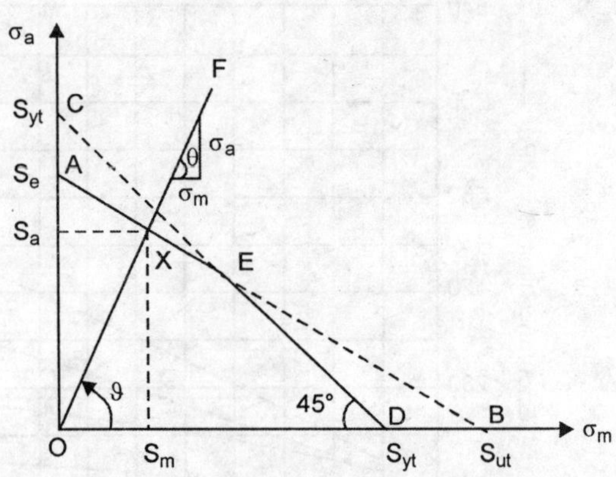

Fig. 4.9 Modified Goodman diagram

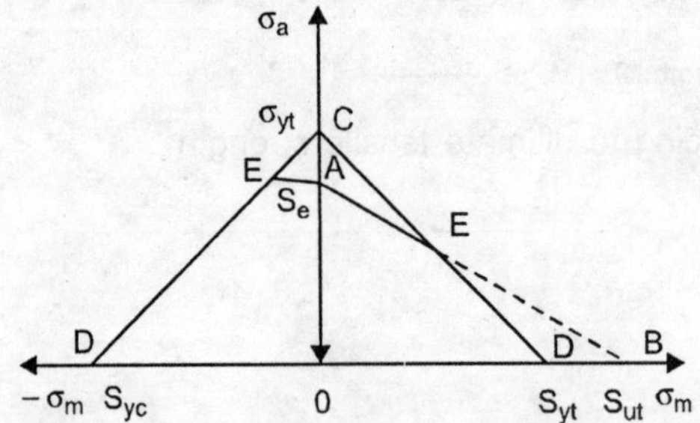

Fig. 4.10 Complete Goodman diagram

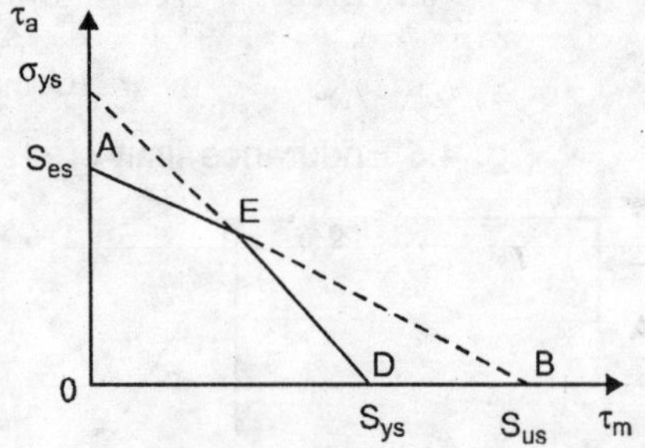

Fig. 4.11 $\tau_m - \tau_a$ diagram for torsion

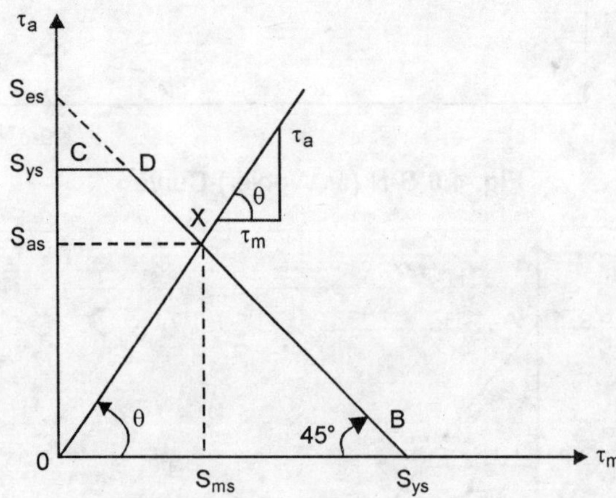

Fig. 4.12 Modified Goodman diagram

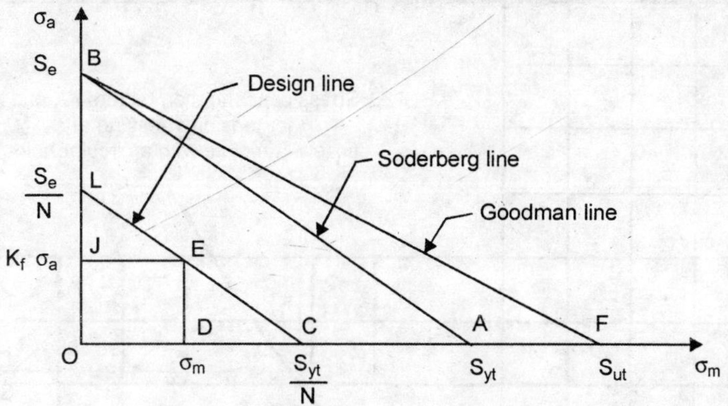

Fig. 4.13 Design line determination for ductile materials

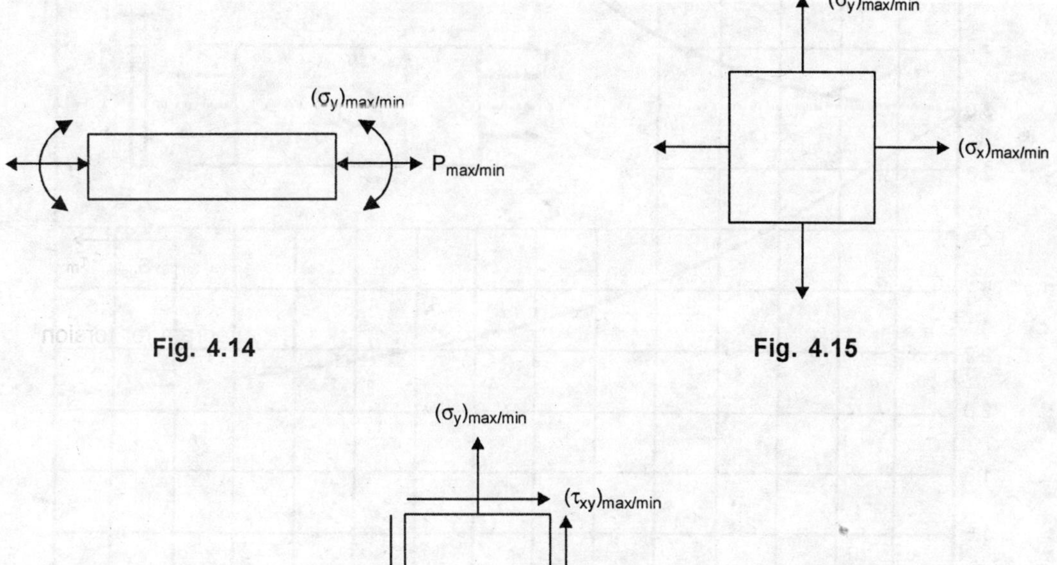

Fig. 4.14

Fig. 4.15

(σy)max/min

(τxy)max/min

(σx)max/min

Fig. 4.16

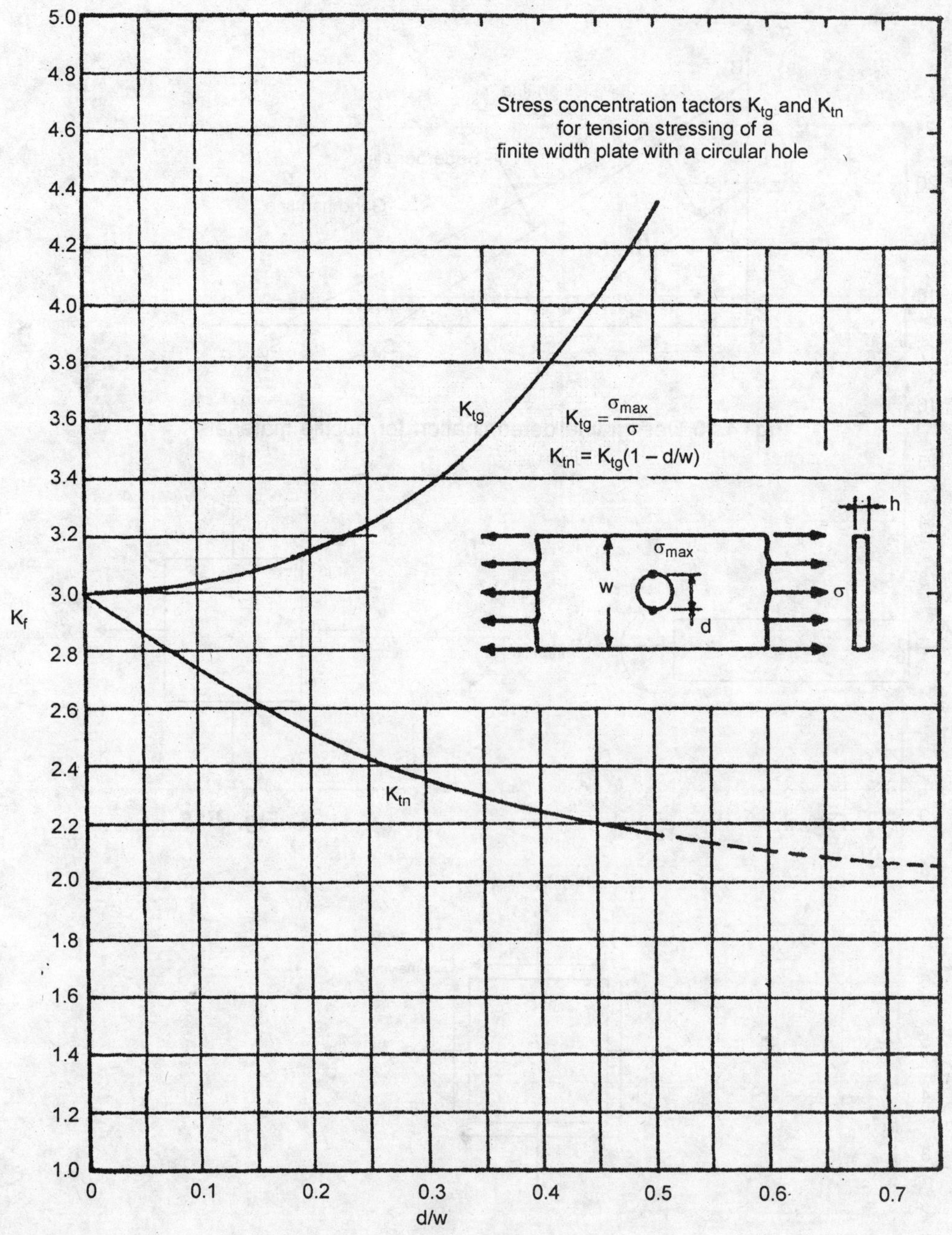

Fig. 4.17

$\frac{d}{w}$	0.05	0.1	0.15	0.20	0.25	0.30	0.35	0.40	0.45	0.50	0.55
K_{tn}	2.83	2.69	2.59	2.50	2.43	2.37	2.32	2.26	2.22	2.17	2.13

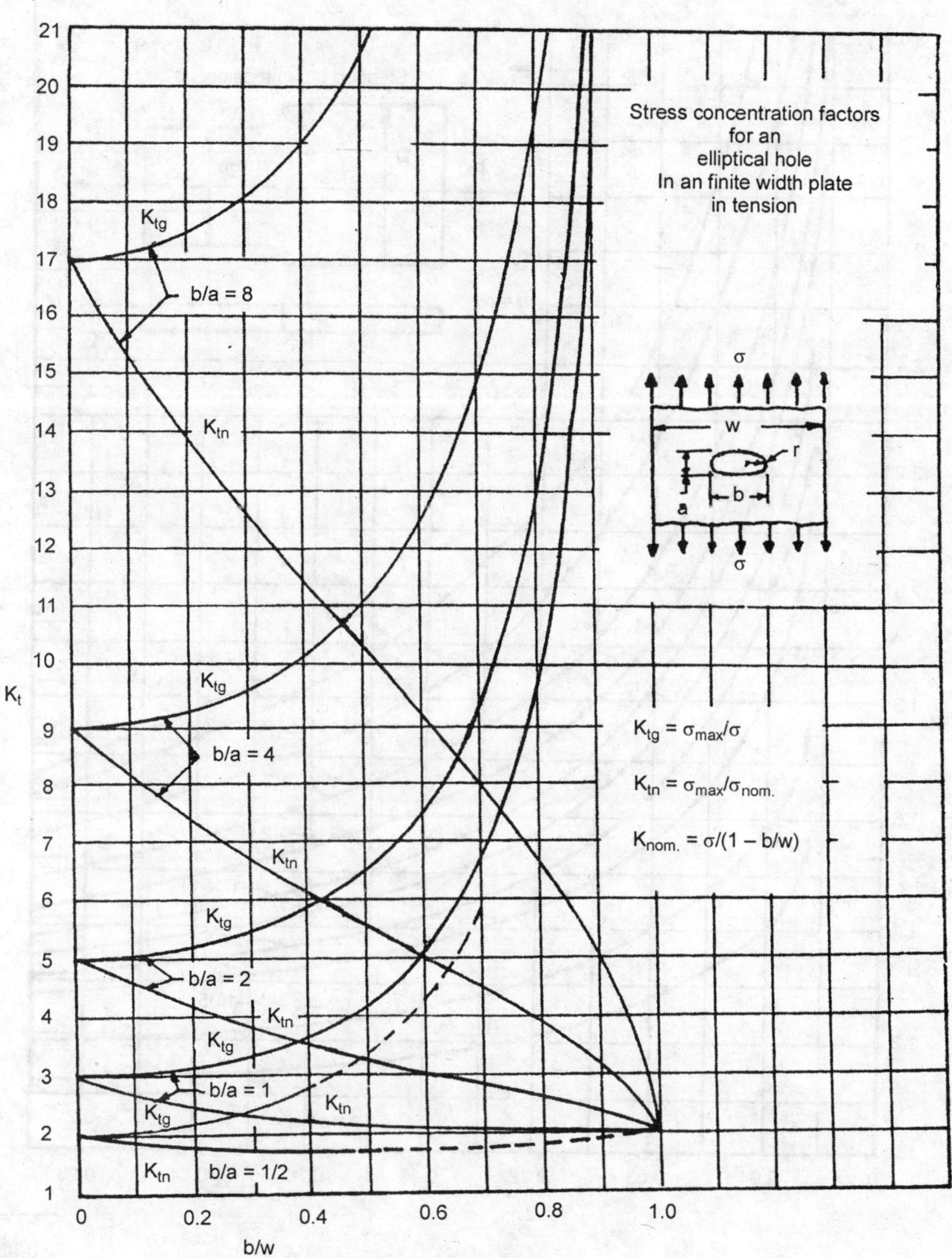
Stress concentration factors
for an
elliptical hole
In an finite width plate
in tension
σ
w
r
b
a
σ
$K_{tg} = \sigma_{max}/\sigma$
$K_{tn} = \sigma_{max}/\sigma_{nom.}$
$K_{nom.} = \sigma/(1 - b/w)$
K_{tg}
b/a = 8
K_{tn}
K_{tg}
b/a = 4
K_{tn}
K_{tg}
b/a = 2
K_{tn}
K_{tg}
b/a = 1
K_{tg}
K_{tn}
K_{tn}
b/a = 1/2
K_t
b/w
0
0.2
0.4
0.6
0.8
1.0
1
2
3
4
5
6
7
8
9
10
11
12
13
14
15
16
17
18
19
20
21

Fig. 4.18

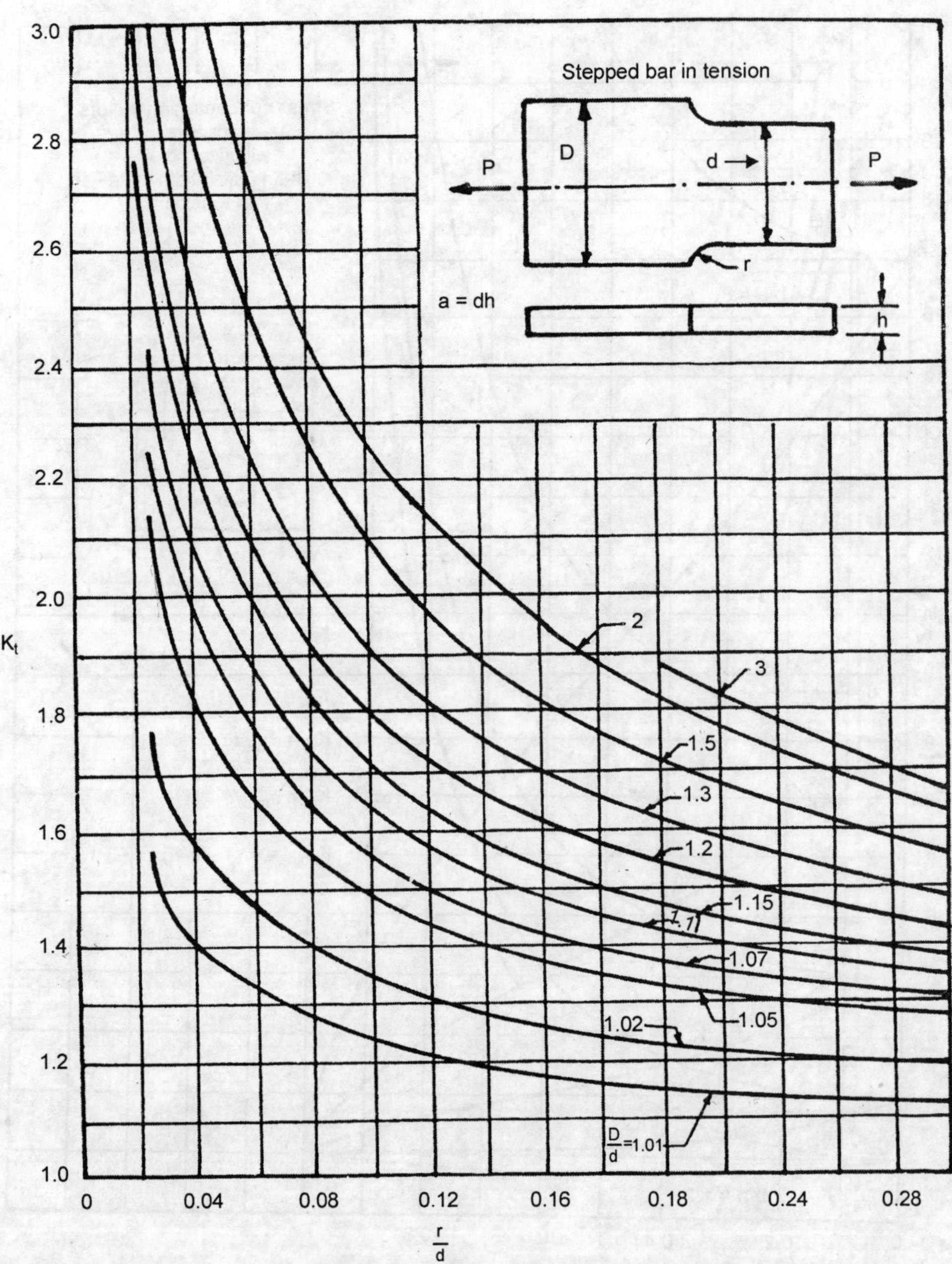

Stepped bar in tension
D
d
P
P
r
a = dh
h
K_t
3.0
2.8
2.6
2.4
2.2
2.0
1.8
1.6
1.4
1.2
1.0
0
0.04
0.08
0.12
0.16
0.18
0.24
0.28
$\frac{r}{d}$
2
3
1.5
1.3
1.2
1.15
1.1
1.07
1.05
1.02
$\frac{D}{d}$=1.01

Fig. 4.19

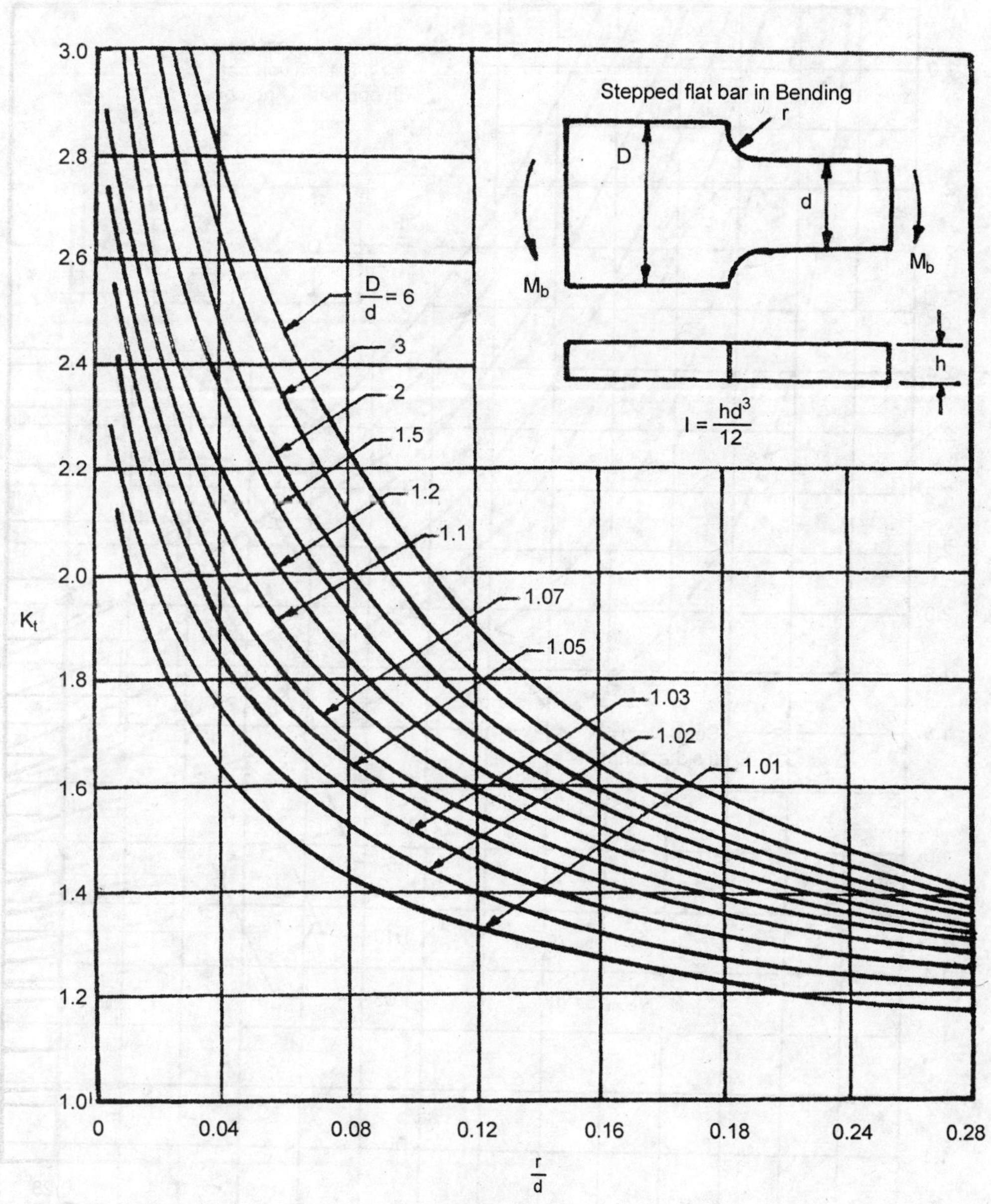
Stepped flat bar in Bending
r
D
d
M_b
M_b
h
$I = \frac{hd^3}{12}$
$\frac{D}{d} = 6$
3
2
1.5
1.2
1.1
1.07
1.05
1.03
1.02
1.01
K_t
3.0
2.8
2.6
2.4
2.2
2.0
1.8
1.6
1.4
1.2
1.0
0
0.04
0.08
0.12
0.16
0.18
0.24
0.28
$\frac{r}{d}$

Fig. 4.20

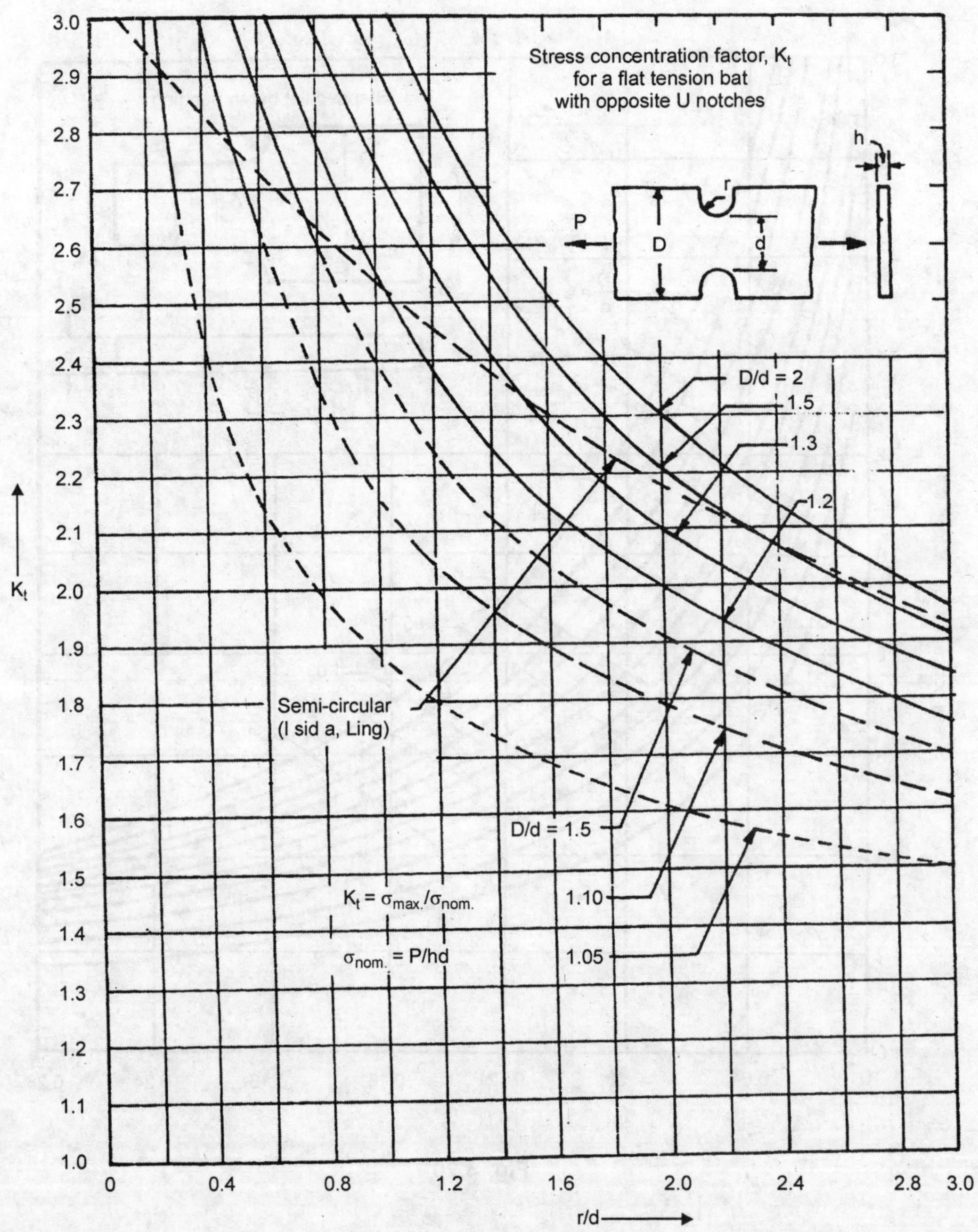

Stress concentration factor, K_t for a flat tension bat with opposite U notches
h
r
P
D
d
D/d = 2
1.5
1.3
1.2
Semi-circular
(I sid a, Ling)
D/d = 1.5
1.10
1.05
$K_t = \sigma_{max.}/\sigma_{nom.}$
$\sigma_{nom.} = P/hd$
K_t
r/d
3.0
2.9
2.8
2.7
2.6
2.5
2.4
2.3
2.2
2.1
2.0
1.9
1.8
1.7
1.6
1.5
1.4
1.3
1.2
1.1
1.0
0
0.4
0.8
1.2
1.6
2.0
2.4
2.8
3.0

Fig. 4.21

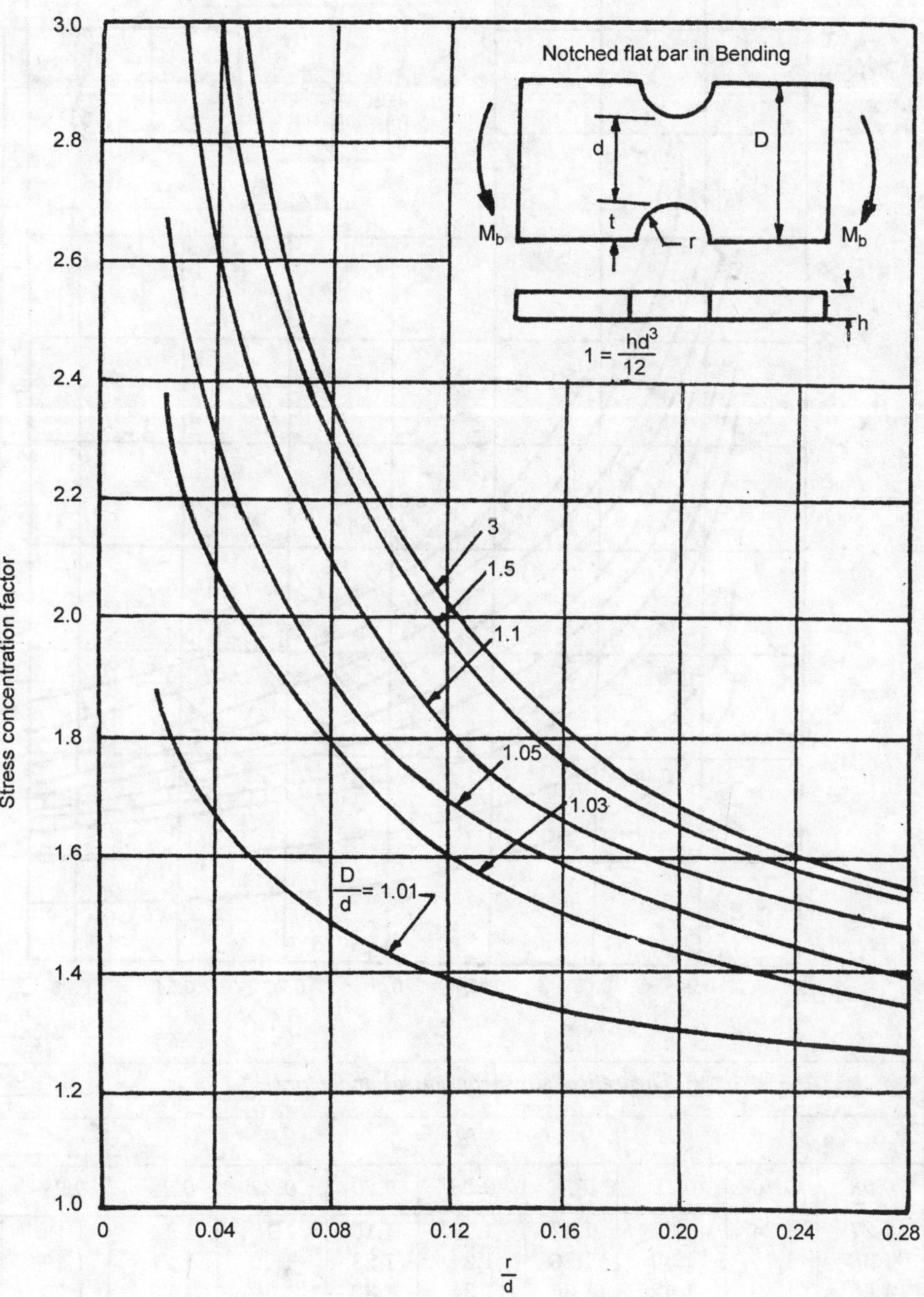
Notched flat bar in Bending
d
D
t
r
M_b
M_b
h
$1 = \frac{hd^3}{12}$
Stress concentration factor
3.0
2.8
2.6
2.4
2.2
2.0
1.8
1.6
1.4
1.2
1.0
0
0.04
0.08
0.12
0.16
0.20
0.24
0.28
$\frac{r}{d}$
3
1.5
1.1
1.05
1.03
$\frac{D}{d}$ = 1.01

Fig. 4.22

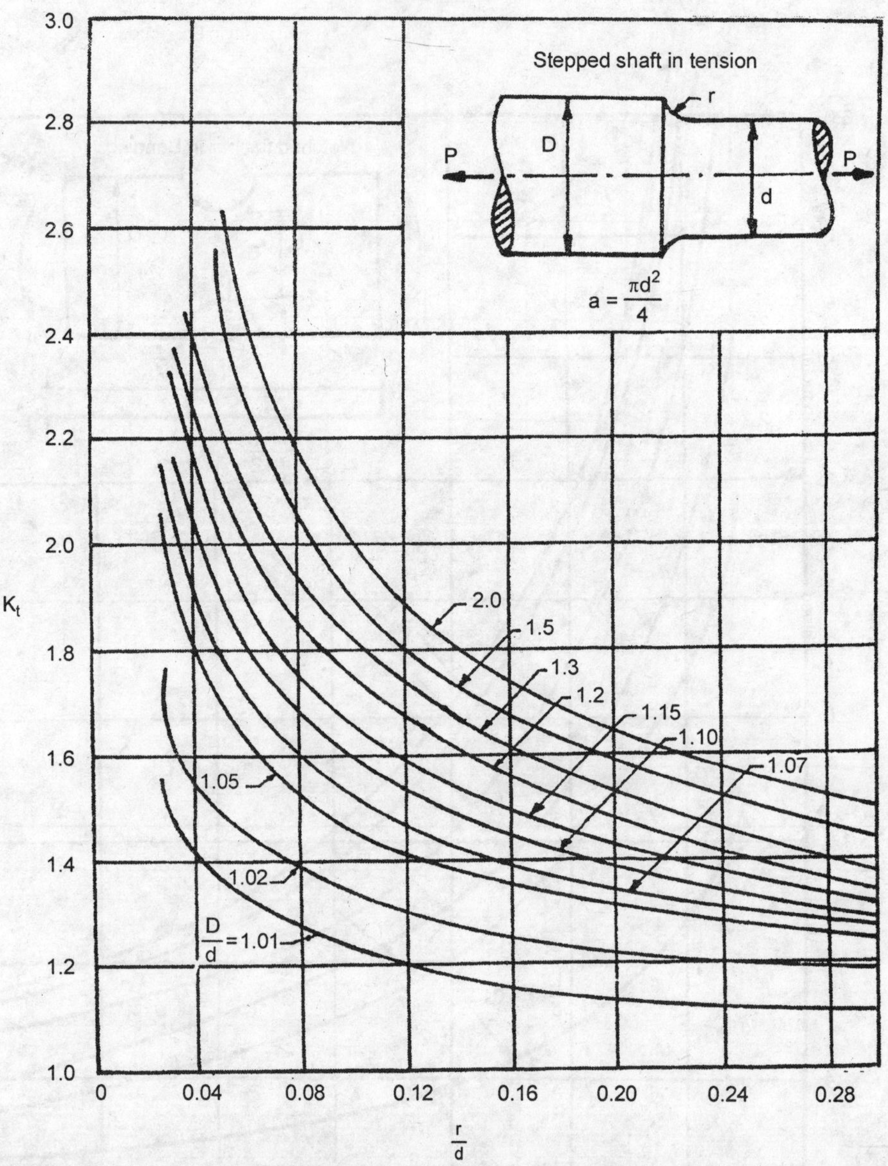

$\frac{D}{d}$	Theoretical stress concentration factor (K_t)									
	r/d									
	0.08	*0.10*	*0.12*	*0.16*	*0.18*	*0.20*	*0.22*	*0.24*	*0.28*	*0.30*
1.01	1.27	1.24	1.21	1.17	1.16	1.15	1.15	1.14	1.13	1.13
1.02	1.38	1.34	1.30	1.26	1.24	1.23	1.22	1.21	1.19	1.19
1.05	1.53	1.46	1.42	1.36	1.34	1.32	1.30	1.28	1.26	1.25
1.10	1.65	1.56	1.50	1.43	1.39	1.37	1.34	1.33	1.30	1.28
1.13	1.73	1.63	1.56	1.46	1.43	1.40	1.37	1.35	1.32	1.31
1.20	1.82	1.68	1.62	1.51	1.47	1.44	1.41	1.38	1.35	1.34
1.50	2.03	1.84	1.80	1.66	1.60	1.56	1.53	1.50	1.46	1.44
2.00	2.14	1.94	1.89	1.74	1.68	1.64	1.59	1.56	1.50	1.47

Fig. 4.23

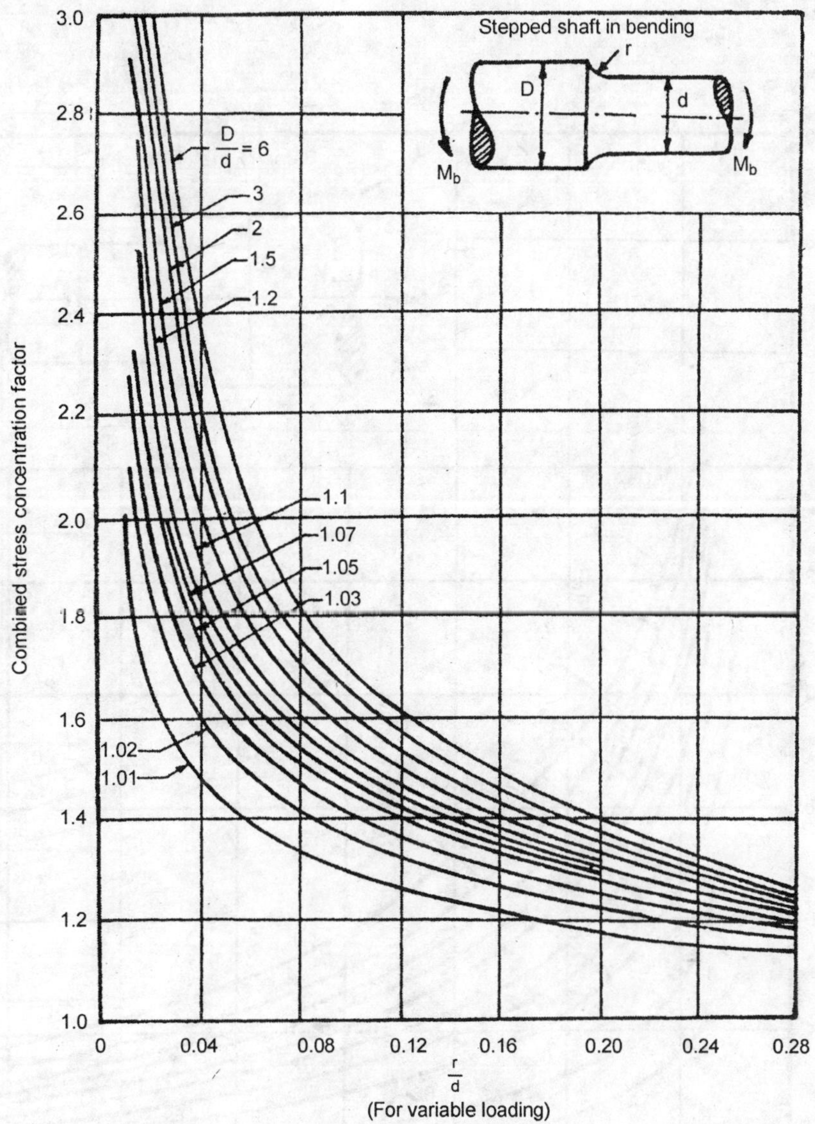

$\frac{D}{d}$	*Theoretical stress concentration factor (K_t)*									
	r/d									
	0.02	*0.04*	*0.05*	*0.10*	*0.12*	*0.16*	*0.20*	*0.24*	*0.28*	*0.30*
1.01	1.85	1.61	1.42	1.36	1.32	1.24	1.20	1.17	1.15	1.14
1.02	1.97	1.72	1.50	1.44	1.40	1.32	1.27	1.23	1.21	1.20
1.05	2.20	1.88	1.60	1.53	1.48	1.40	1.34	1.30	1.27	1.25
1.10	2.36	1.99	1.66	1.58	1.53	1.44	1.38	1.33	1.28	1.27
1.20	2.52	2.10	1.72	1.62	1.56	1.46	1.39	1.34	1.29	1.28
1.50	2.75	2.20	1.78	1.68	1.60	1.50	1.42	1.36	1.31	1.29
2.00	2.86	2.32	1.87	1.74	1.64	1.53	1.43	1.37	1.32	1.30
3.00	3.00	2.45	1.95	1.80	1.69	1.56	1.46	1.38	1.34	1.32
6.00	3.04	2.58	2.04	1.87	1.76	1.60	1.49	1.41	1.35	1.33

Fig. 4.24

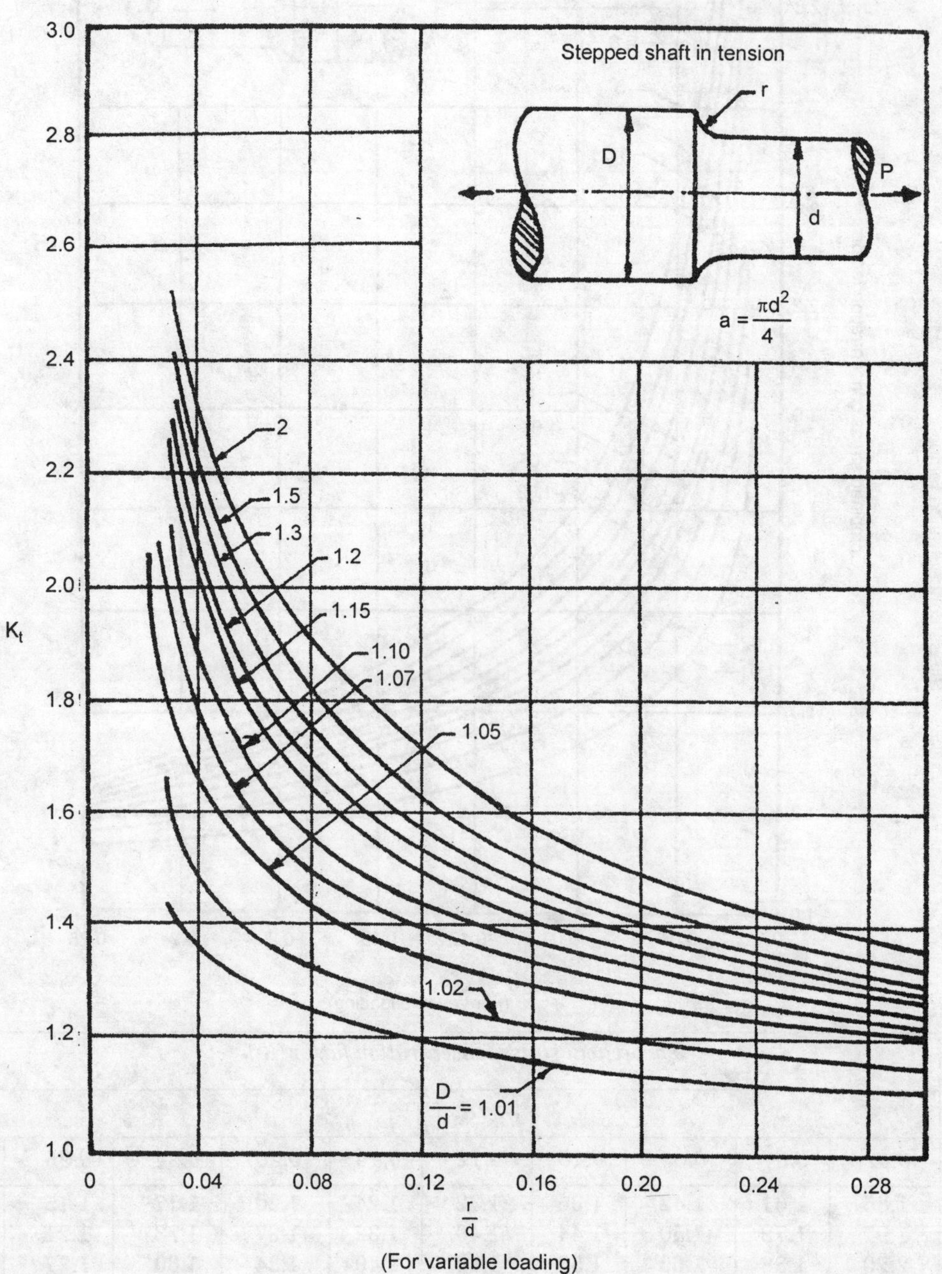

Stepped shaft in tension
r
D
P
d
a = $\frac{\pi d^2}{4}$
3.0
2.8
2.6
2.4
2.2
2.0
1.8
1.6
1.4
1.2
1.0
K_t
2
1.5
1.3
1.2
1.15
1.10
1.07
1.05
1.02
$\frac{D}{d}$ = 1.01
0
0.04
0.08
0.12
0.16
0.20
0.24
0.28
$\frac{r}{d}$
(For variable loading)

Fig. 4.25

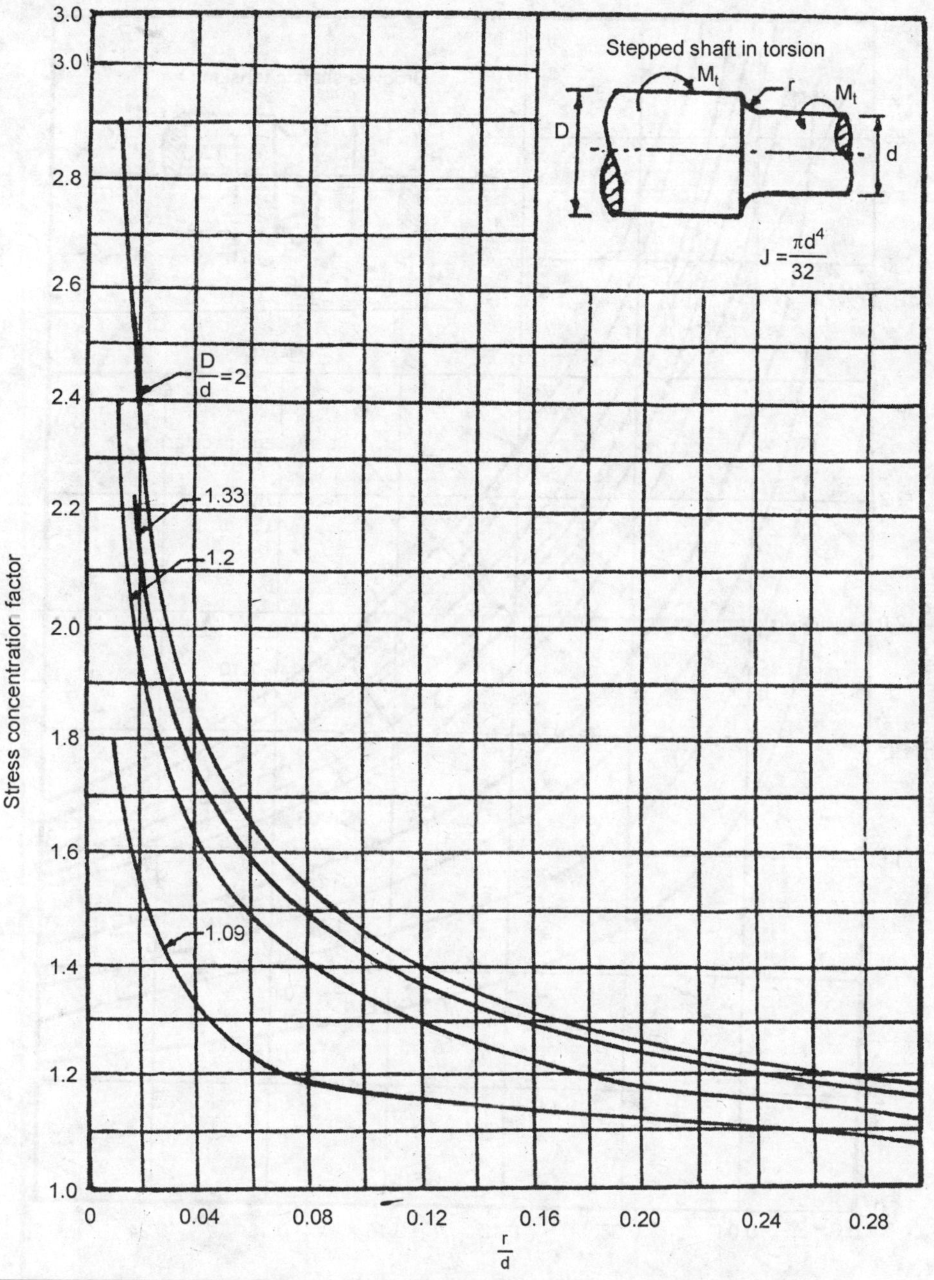

$\frac{D}{d}$	r/d									
	0.02	*0.04*	*0.08*	*0.10*	*0.12*	*0.16*	*0.20*	*0.24*	*0.28*	*0.30*
1.09	1.54	1.32	1.19	1.16	1.15	1.12	1.11	1.10	1.09	1.09
1.20	1.98	1.67	1.40	1.33	1.28	1.22	1.18	1.15	1.13	1.13
1.33	2.14	1.79	1.48	1.41	1.35	1.28	1.22	1.19	1.17	1.16
2.00	2.27	1.84	1.53	1.46	1.40	1.32	1.26	1.22	1.19	1.18

Fig. 4.26

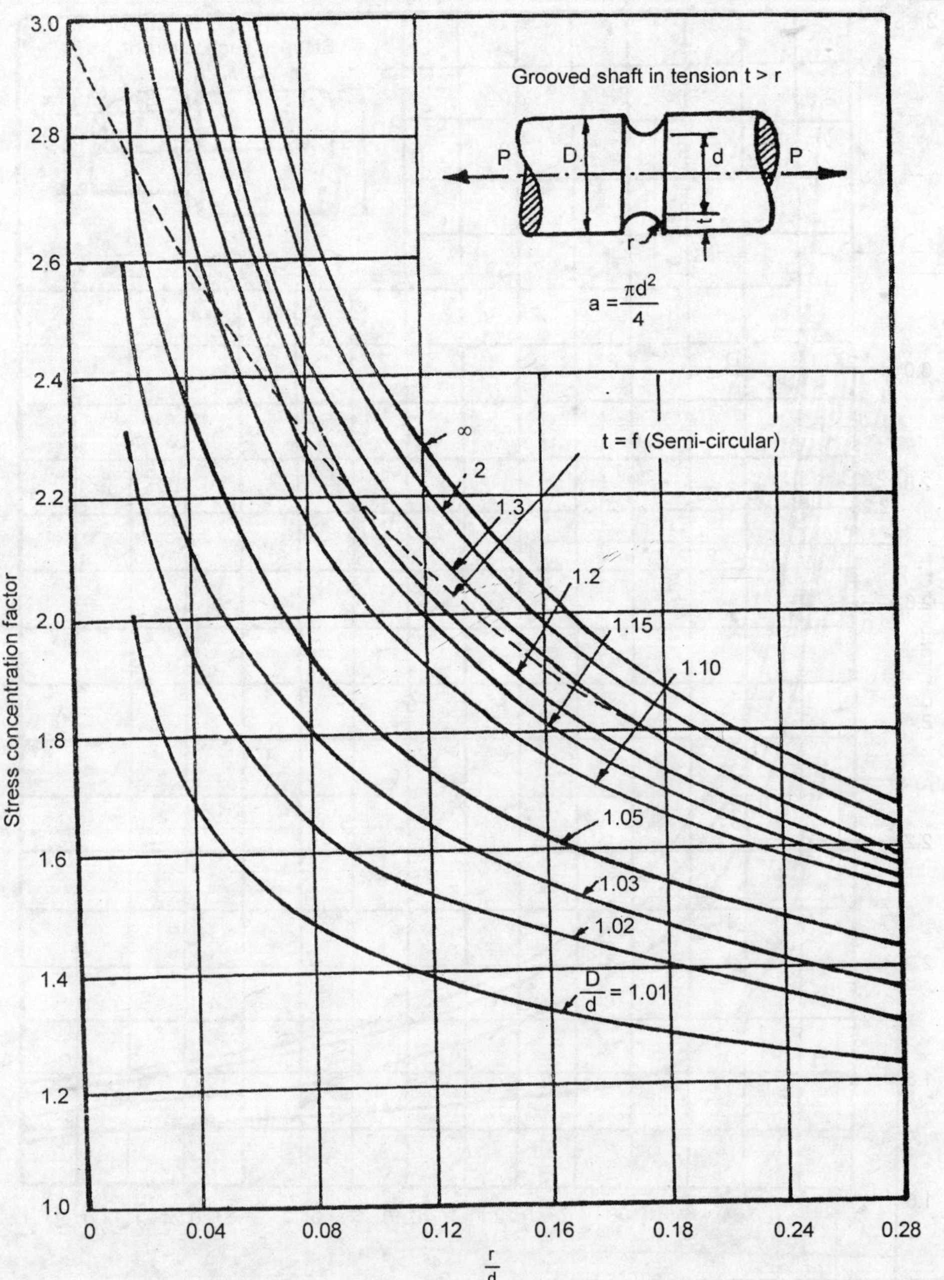

$\frac{D}{d}$	Theoretical stress concentration factor (K_t)									
	r/d									
	0.02	*0.04*	*0.08*	*0.10*	*0.72*	*0.16*	*0.20*	*0.24*	*0.28*	*0.30*
1.01	1.98	1.71	1.47	1.42	1.38	1.33	1.28	1.25	1.23	1.22
1.02	2.30	1.94	1.66	1.59	1.54	1.45	1.40	1.36	1.33	1.31
1.03	2.60	2.14	1.77	1.69	1.63	1.53	1.46	1.41	1.37	1.36

1.05	2.85	2.36	1.94	1.81	1.73	1.61	1.54	1.47	1.43	1.41
1.10	–	2.70	2.16	2.01	1.90	1.75	1.70	1.57	1.50	1.47
1.20	–	2.90	2.36	2.17	2.04	1.86	1.74	1.64	1.56	1.54
1.30	–	–	2.46	2.26	2.11	1.91	1.77	1.67	1.59	1.56
1.50	–	–	2.54	2.33	2.16	1.94	1.79	1.69	1.61	1.57
2.00	–	–	2.61	2.38	2.22	1.98	1.83	1.72	1.63	1.59
∞	–	–	2.69	2.44	2.26	2.03	1.86	1.74	1.65	1.61

Fig. 4.27

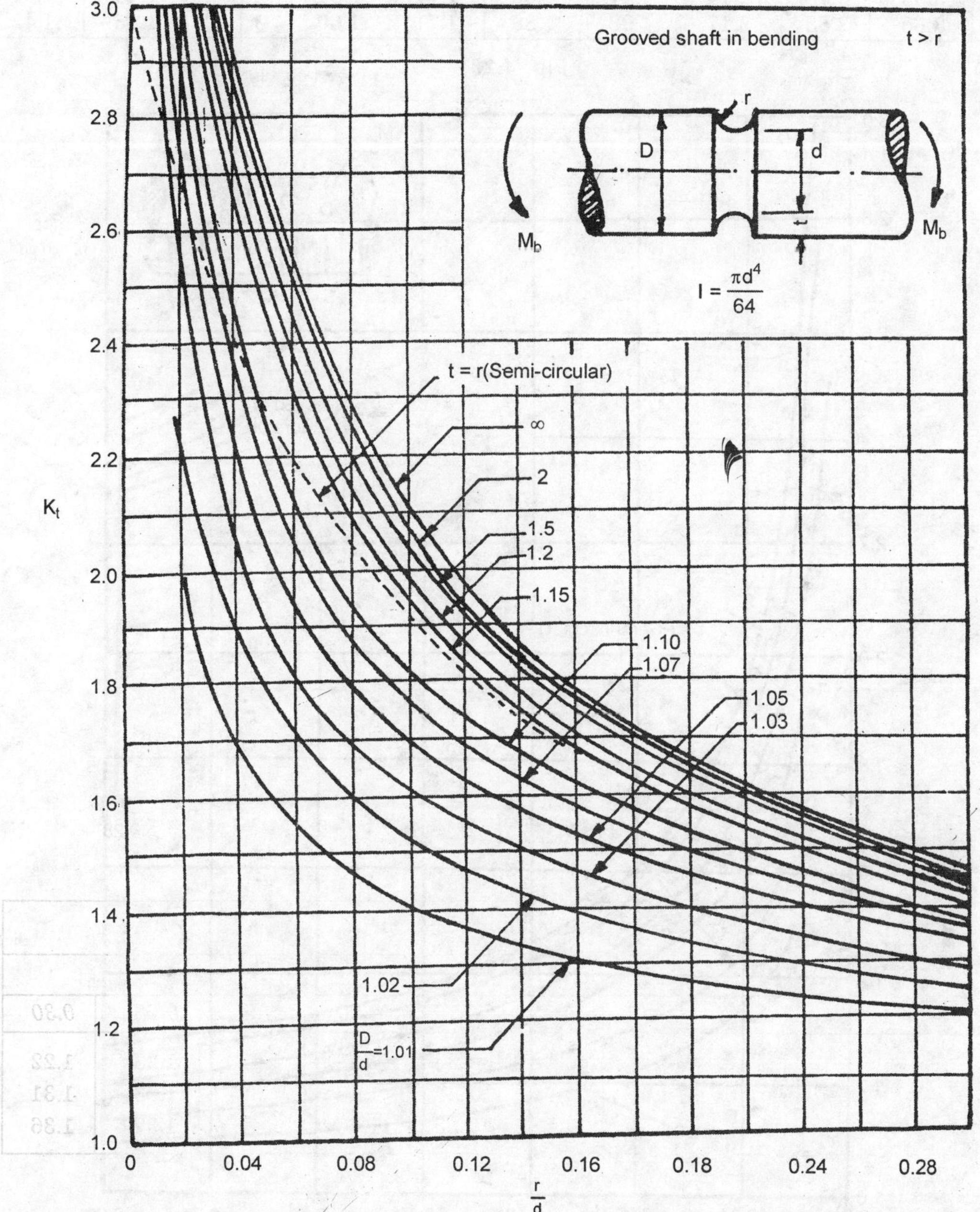

$\frac{D}{d}$	*Theoretical stress concentration factor* (K_t)									
	r/d									
	0.02	*0.04*	*0.08*	*0.10*	*0.12*	*0.16*	*0.20*	*0.24*	*0.28*	*0.30*
1.01	1.74	1.68	1.47	1.41	1.38	1.32	1.27	1.23	1.22	1.20
1.02	2.28	1.89	1.64	1.53	1.48	1.40	1.34	1.30	1.26	1.25
1.03	2.46	2.04	1.68	1.61	1.55	1.47	1.40	1.35	1.31	1.28
1.05	2.75	2.22	1.80	1.70	1.63	1.53	1.46	1.40	1.35	1.33
1.12	3.20	2.50	1.97	1.83	1.75	1.62	1.52	1.45	1.38	1.34
1.30	3.40	2.70	2.04	1.91	1.82	1.67	1.57	1.48	1.42	1.38
1.50	3.48	2.74	2.11	1.95	1.84	1.69	1.58	1.49	1.43	1.40
2.00	3.55	2.78	2.14	1.97	1.86	1.71	1.59	1.55	1.44	1.41
∞	3.60	2.85	2.17	1.98	1.88	1.71	1.60	1.51	1.45	1.42

Fig. 4.28

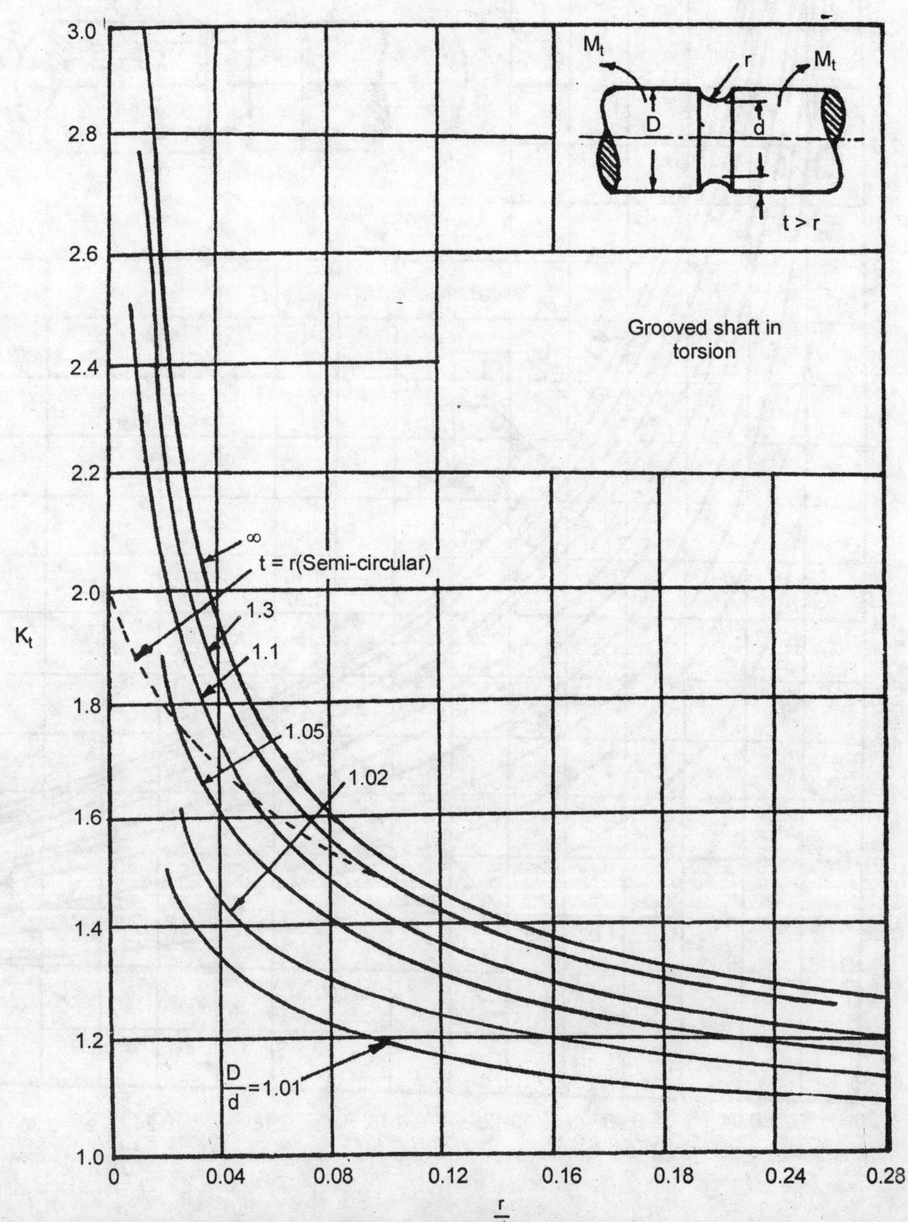

$\frac{D}{d}$	r/d									
	0.02	*0.04*	*0.08*	*0.10*	*0.12*	*0.16*	*0.20*	*0.24*	*0.28*	*0.30*
1.01	1.50	1.30	1.22	1.20	1.18	1.16	1.13	1.12	1.12	1.12
1.02	1.62	1.45	1.31	1.27	1.23	1.20	1.18	1.16	1.15	1.16
1.05	1.88	1.61	1.40	1.35	1.32	1.26	1.22	1.20	1.18	1.17
1.10	2.05	1.73	1.47	1.41	1.37	1.31	1.26	1.24	1.21	1.20
1.20	2.26	1.83	1.53	1.46	1.41	1.34	1.27	1.25	1.22	1.21
1.30	2.32	1.89	1.55	1.48	1.43	1.35	1.30	1.26	1.22	1.22
2.00	2.40	1.93	1.58	1.50	1.45	1.36	1.31	1.26	1.23	1.22
∞	2.50	1.96	1.60	1.51	1.46	1.38	1.32	1.27	1.24	1.23

Fig. 4.29

Riveted Joints

1. Modes of Failure of a Riveted Joint

(*i*) Margin, $m = 1.5d_0$ ...(5.1)

where d_0 = diameter of rivet hole.

(*ii*) Tearing strength of plate per pitch length,

$$P_t = A_t \times \sigma_t = (p - d_0)\, \sigma_t \quad ...(5.2)$$

where σ_t = allowable tearing strength for plate material.

$$P_s = A_s \times \tau$$

(*iii*) Shearing strength of rivets,

$$A_s = n \times z \times \pi/4 \times d^2{}_0 \quad ...(5.3)$$

n = number of rivets per pitch length

τ = allowable shear stress for rivets material

z = number of shears

= 1 for single shear

= 2 for double shear

= 1.875 for double shear as per Indian Boiler Regulations (I.B.R.)

(*iv*) Crushing strength of rivets,

$$P_c = A_c \times \sigma_c \quad ...(5.4)$$

Crushing or projected area of n rivets, A_c = ndt

σ_c = crushing strength for rivets.

(*v*) Strength of joint is the smallest value of P_t, P_s and P_c.

(*vi*) Strength of solid plate, $P = p \times t \times \sigma_t$ ...(5.5)

(*vii*) Efficiency, η = Least value of P_t, P_s and P_c/P ...(5.6)

2. Design of Boiler Joints

(*a*) Longitudinal Butt Joint

(*i*) Plate thickness,

$$t = p_s D/(2 \sigma_t \eta_1) + c \qquad ...(5.7)$$

where p_s = steam pressure, MPa gauge

D = shell diameter, mm

σ_t = permissible tensile stress for shell plates material at operating temperature, MPa (see Table 5.1 for tensile strength of boiler plates and rivets.

= efficiency of longitudinal joint

c = corrosion allowance $\approx$ 1 mm

The boiler plate thickness should not be lass than 7 mm. For standard sizes of plates, refer to Table 5.5.

(*ii*) Rivet hole diameter. Rivet hole diameter is given by Unwin's formula,

$$d_0 = 6\sqrt{t} \quad \text{for} \quad t \geq 8 \text{ mm} \qquad ...(5.8)$$

For $t \leq 8$ mm, calculate rivet hole diameter as follows:

$$d_0 = \sqrt{50t} - 4 \text{ mm for lap joints} \qquad ...(5.9a)$$

$$d_0 = \sqrt{50t} - 5 \text{ mm for single riveted butt joints} \qquad ...(5.9b)$$

$$d_0 = \sqrt{50t} - 6 \text{ mm for double riveted butt joints} \qquad ...(5.9c)$$

(*iii*) Diameter of rivets.

$$\text{Diameter of rivet, } d = d_0 - 1 \text{ mm for } 12 \text{ mm} \leq d \leq 24 \text{ mm} \qquad ...(5.10a)$$

$$= d_0 - 1.5 \text{ mm for } 27 \text{ mm} \leq d \leq 36 \text{ mm} \qquad ...(5.10b)$$

$$= d_0 - 2 \text{ mm for } 39 \text{ mm} \leq d \leq 48 \text{ mm} \qquad ...(5.10c)$$

(*iv*) Pitch of rivets,

$$p = C \times t + 40 \text{ mm} \qquad ...(5.11)$$

where C = a coefficient (Table 5.6).

(*v*) In joints where the pitch in the outermost row is double the pitch of inner rows and the inner rows are chain riveted:

Transverse pitch between the outermost row and the inner rows,

$$p_t = 2d_0 \text{ for chain riveting} \qquad ...(5.12a)$$

$$= 0.33p + 0.67\, d_0 \text{ for zig - zag riveting} \qquad ...(5.12b)$$

Transverse pitch between the successive inner rows,

$$p_t > 2d_0 \qquad ...(5.13)$$

(*vi*) In joints where the pitch in the outermost row is double the pitch of inner rows and the inner rows are zig – zag riveted:

Transverse pitch between the outermost row and the inner rows,

$$p_t \geq (0.2p + 1.15\, d_0) \qquad ...(5.14a)$$

Transverse pitch between the successive inner rows,

$$p_t \geq (0.165p + 0.67d_0) \quad ...(5.14b)$$

(*vii*) Minimum thickness of butt strap,

$$t_1 = 1.12t, \text{ for ordinary chain riveting with single butt strap.} \quad ...(5.15a)$$

$$= 1.125t\ [(p - d_0) / (p - 2d_0)] \quad ...(5.15b)$$

for single butt straps, every alternate rivet in outer rows being omitted.

$$= 0.625t \text{ for} \quad ...(5.15c)$$

double butt straps of equal widths having chain riveting.

$$= 0.625t\ [(p - d_0) / (p - 2d_0)] \quad ...(5.15d)$$

for double butt straps of equal width having every alternate rivet in outer rows being omitted.

For unequal width of butt straps, the thickness of butt strap will be;

$$t_1 = 0.75t, \text{ for wide trap on the inside, and} \quad ....(5.16a)$$

$$= 0.625\ t, \text{ for narrow strap on the outside.} \quad ....(5.16b)$$

(*viii*) Margin.

$$m = 1.5d_0 \quad ...(5.17)$$

(*b*) Circumferential Lap Joint

(*i*) Thickness of boiler shell and diameter of rivets shad be same as for the longitudinal joint

(*ii*) Number of rivets.

$$n = (D/d_0)\ (p_s/t) \quad ...(5.18)$$

(*iii*) Pitch of rivets.

Let η_c = defficiency of circumferential joint

$$= 0.5\eta_1$$

Then

$$\eta_c = (p_c - d_0)/p_c \quad ...(5.19)$$

(*iv*) Number of rows.

Number of rows of rivets,

$$N = n\ (D + t)/p_c \quad ...(5.20)$$

(*v*) Overlap.

$$\text{Overlap} = (N - 1)p_b + m \quad ...(5.21)$$

3. Structural Riveted Joint – Lozenge Joint (Fig. 5.3)

(*i*) Diameter of rivet hole,

$$d_0 = 6\sqrt{t} \quad ...(5.22)$$

Diameter of rivets,

$$d = d_0 - \text{clearance} \quad ...(5.23)$$

Diameter of rivet hole,

$$d_0 = d + 1.5 \text{ mm for } 12 \text{ mm} \leq d \leq 24 \text{ mm} \quad ...(5.24a)$$

$$= d + 2 \text{ mm for } 27 \text{ mm} \leq d \leq 50 \text{ mm} \quad ...(5.24b)$$

(*ii*) Number of rivets.

Tearing resistance of the plate in the outermost row,

$$P_t = (w - d_0)\, t\sigma_t$$

Shearing resistance of rivets,

$$P_s = 1.75 \times \pi/4 \times d^2{}_0 \times \tau$$

Crushing resistance of one rivet,

$$P_c = dt\sigma_c$$

Number of rivets, $n = P_t/\text{Least of } P_s \text{ and } P_c$ ...(5.25)

(*iii*) Thickness of butt straps.

$t_1 = 1.25t$ for single cover strap ...(5.26*a*)

$= 0.75$ for double cover strap ...(5.26*b*)

(*iv*) Efficiency of joint.

At section 1 – 1, there is only one rivet hole.

Resistance of the joint against tearing along 1 – 1,

$P_1 = (w - d_0)t\sigma_t$ ...(5.27*a*)

Resistance of joint along section 2 – 2,

P_2 = tearing resistance along 2 – 2 + shearing resistance of one rivet at 1 – 1

$= (w - 2d_0)\, t\sigma_t + 1.75 \times \pi/4 \times d^2{}_0\tau$ ...(5.27*b*)

Resistance of joint along 3 – 3,

P_3 = tearing resistance along 3 – 3 + shearing resistance of two rivets at 2 – 2 and one rivet at 1 – 1.

$= (w - 3d_0)t\sigma_t + 3 \times 1.75 \times \pi/4 \times d^2{}_0 \times \tau$ ...(5.27*c*)

Strength of solid plate, $P = wt\sigma_t$ ...(5.27*d*)

Shear resistance of all rivets,

$P_s = n \times 1.75 \times \pi/4 \times d^2{}_0 \times t$ ...(5.27*e*)

Crushing resistance of all rivets,

$P_c = nd\tau\sigma_c$ ...(5.27*f*)

Efficiency of joint, η = Least of P_1, P_2, P_3, and P_c/P ...(5.28)

(*v*) Pitch of rivets, $p = 3d_0 + 5$ mm ...(5.29)

(*vi*) Margin, $m = 1.5d_0$ ...(5.30)

(*vii*) Distance between rows of rivets $= 2.5d_0$ ...(5.31)

4. Eccentrically Loaded Riveted Joint (Fig. 5.4)

(*i*) Locate the centre of gravity of the rivet group.

Let

A = area of cross – section of each rivet

x_i = distance of ith rivet from y – axis

y_i = distance of ith rivet from x – axis

$x = A\,\Sigma\, x_i/(nA)$

$y = A\,\Sigma\, y_i/(nA)$

(*ii*) Direct primary shear force in each rivet,

$$P_d = P/n \quad ...(5.32)$$

(*iii*) Couple about C.G., $C = P \times e$ ...(5.33)

(*iv*) Let F_i = secondary shear force in each rivet

l_i = perpendicular distance of F_i from C.G.

The secondary shear force in each rivet is proportional to the perpendicular distance from the C.G.

(*v*) Equate the couple to the resisting moment of all the rivets.

$$C = \Sigma F_i l_i = (F_1/l_1)\, \Sigma l^2_i \quad ...(5.34)$$

(*vi*) Calculate the resultant shear force in each rivet due to the primary shear and secondary shear.

$$R_j = [P^2_d + F^2_i + 2P_d\, F_i \cos \theta_i]^{1/2} \quad ...(5.35)$$

Where θ_i = angle between P_d and F_i

(*vii*) Find the maximum resultant shear force in a rivet. Then, as the rivets are in double shear,

$$R_{max} = 1.75 \times \pi/4 \times d^2_0 \times \tau \quad ...(5.36)$$

Note: The rivet nearest to the load line shall be subjected to maximum resultant shear force, which may be designed.

Table 5.1: Mechanical properties of boiler plates and rivets.

Grade	*Tensile strength, MPa*	*Yield strength as percentage of minimum tensile strength*
1	410 – 490	55
2	510 – 610	50

Table 5.2: Recommended type of joint for pressure vessels

Diameter of shell, mm	*Thickness of shell, mm*	*Type of joint*
600 – 1800	6 – 13	Double riveted
900 – 2100	13 – 25	Triple riveted
1500 – 2700	16 – 40	Quadruple riveted

Table 5.3: Range of expected efficiencies for riveted joints.

Type of joint	*Efficiency, %*	*Type of joint*	*Efficiency, %*
Lap joints:		Butt joints:	
Single	45 – 65	Single	60 – 70
Double	63 – 77	Double	75 – 85
Triple	75 – 85	Triple	80-90
		Quadruple	85-95

Table 5.4: Recommended factor of safety for boiler joints

Type of joint	*Factor of safety based on ultimate strength*	
	Hand riveting	*Machine riveting*
Lap joint	4.75	4.50
Single strap butt joint	4.75	4.50
Single riveted butt joint with two equal cover straps	4.75	4.50
Double riveted butt joint with two equal cover straps	4.25	4.00

Table 5.5: Standard nominal size of steel plates

Thickness, mm	5, 6, 8, 12, 14, 16, 18, 20, 22, 25, 28, 32, 36, 40, 45, 50, 56 and 63
Width, mm	900, 1000, 1100, 1200, 1250, 1400, 1500, 1600, 1800, 2000, 2200 and 2500
Length, mm	220, 2500, 2800, 3200, 3600, 4000, 4500, 5000, 5600, 6300, 7100, 8000, 9000, 10000, 11 000 and 12000

Table 5.6: Values of coefficient C

Number of rivets per pitch	*Lap joint*	*Single strap*	*Double strap butt joint butt joint*
1	1.31	1.53	1.75
2	2.62	3.06	3.50
3	3.47	4.05	4.63
4	4.17	–	5.52
5	–	–	6.00

Table 5.7: Nominal diameter of rivets, mm

d, mm	1, 1.2, 1.6, 2, 2.5, 3, 4, 5, 6, 8, 10, 12, 14, 16, 18, 20, 22, 24, 27, 30, 33, 36, 39, 42 and 45

Table 5.8: Allowable stress values for carbon and low alloy steel plates at elevated temperature

Grade or designation	Tensile strength MPa	Yield strength MPa	Design temperature, °C								
			50	100	150	200	250	300	350	375	400
2A	363	199.7	121	121	113	102	93	85	76	73	71
2B	412	206	137	126	117	106	96	88	79	75	72
20Mo6	510	255	170	156	144	130	119	109	98	81	81
20Mn6	471	275	157	157	157	150	140	129	121	117	113
15Cr4Mn6	510	294	170	170	167	151	137	125	114	108	81
15C8	490	294	163	163	163	163	157	149	141	135	131
	412	226	137	137	127	116	105	96	87	82	79

Table 5.9: Allowable stresses in steel for structural joints

Type of stress	*Shop assembly*	*Field assembly*
Tensile stress of plate, MPa	115	115
Shear stress of rivet, MPa	86	70
Crushing stress of rivet, MPa	210	140

Table 5.10: Rivet Hole Diameters and Strap Thickness

All dimensions in mm

Plate thickness	*Strap thickness Min.*	*Hole diameter*
5		
6	6	17
8		
10	8	21
12		
14	10	25
16		
18	12	31.5
20		
22	16	34.5
25	18	37.5
28	20	
32	22	
36		
40		
45	25	41
50		
56		
63		

Table 5.11: Recommended Practice of Leakproof Joints for Pressure Vesseles

All dimensions in mm

Type of joint	*Rivet hole dia d_0*	*Margin m*	*Pitch p*	*Transverse pitch p_t*	*Rivets in single shear*	*Rivets in double shear*	*Efficiency*
Lap joint :							
Single riveted	$t + 8$	$1.5\ d_0$	$2\ d_0 + 8$	$2\ d_0$	1	–	45–60
Double riveted	$t + 8$	$1.5\ d_0$	$2.6\ d_0 + 15$	$2\ d_0$	2	–	63–60
Triple riveted	$t + 6$ to 8	$1.5\ d_0$	$3\ d_0 + 22$	$2\ d_0$	3	–	72–80
Butt joint :							
Double riveted	$t + 5$ to 6	$1.5\ d_0$	$3.5\ d_0 + 15$	$2\ d_0$	2	–	76–84

Triple riveted (unequal strap width)	$t + 5$	$1.5\ d_0$	$6\ d_0 + 20$	$2\ d_0$ to $2.75\ d_0$	1	4	80–88
Quadruple riveted (unequal strap width)	$t + 5$	$1.5\ d_0$	$6\ d_0 + 20$	$2\ d_0$ to $2.75\ d_0$	3	8	86-94

Table 5.15: Recommended Dimensions of Riveted Joints for Structural Work

Types of joint	*Rivet dia d*	*Marge in m*	*Pitch p*	*Transverse pitch*	*Rivets in p_t*	*Rivets in single shear*	*Efficiency % double shear*
Lap Joint :							
Single riveted	2*t*	1.5*d*	3*d*	(2-2.5)*d*	1	–	50-60
Double riveted	2*t*	1.5*d*	3*d*	(2-2.5)*d*	2	–	60-72
Butt Joint :							
Single riveted, single strap	2*t*	1.5*d*	3*d*	(2-2.5)*d*	1	–	53-60
Double riveted, single strap	2*t*	1.5*d*	4*d*	(2-2.5)*d*	1	–	70-83
Single riveted, double strap	1.5*t*	1.5*d*	3.5*d*	(2-2.5)*d*	–	1	55-60
Double riveted, double strap	1.5*t*	1.5*d*	6*d*	(2-2.5)*d*	–	2	76-84
Triple riveted, double strap (unequal width)	1.5*t*	1.5*d*	8*d*	2*d* and 2.75*d*	1	4	80-88
Quadruple riveted, double strap (unequal width)	1.5*t*	1.5*d*	20*d*	2*d* and 2.75*d*	3	8	86-94

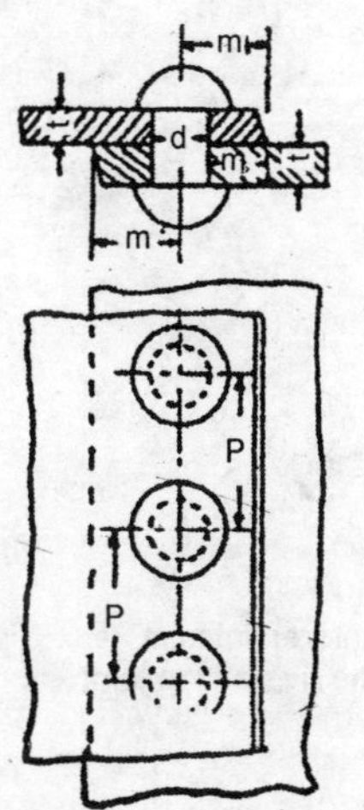

(a) Single riveted lap joint

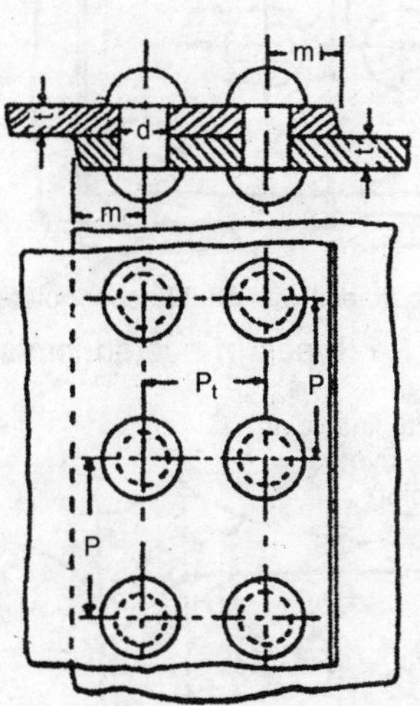

(b) Double riveted lap joint (chain)

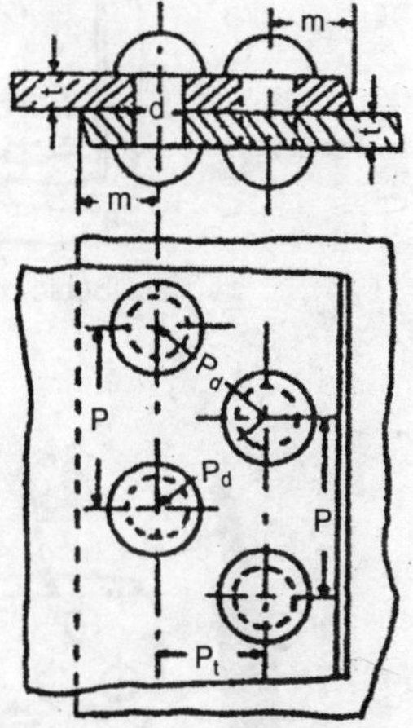

(c) Double riveted lap joint (zig-zag)

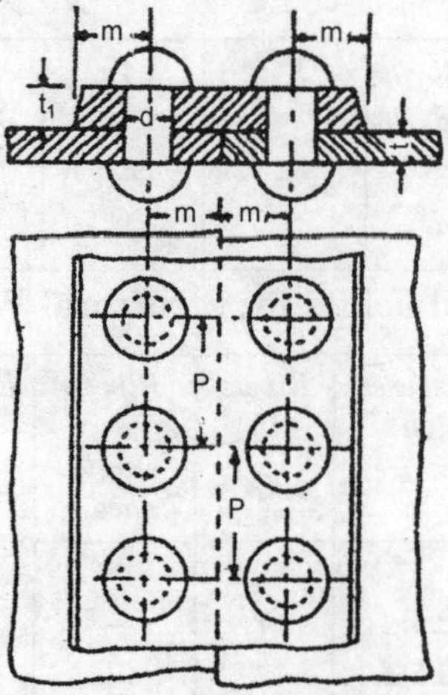

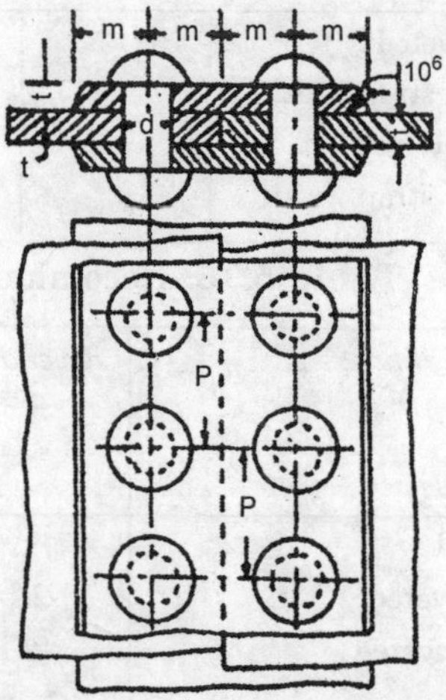

(d) Single riveted but joint with double cover plate

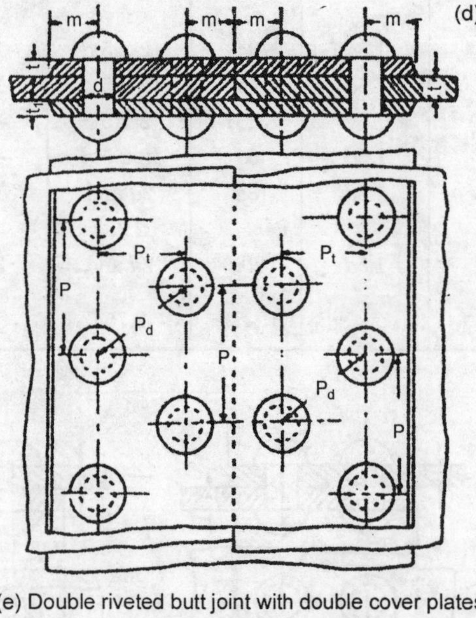

(e) Double riveted butt joint with double cover plates

Fig. 5.1 Types of riveted joints

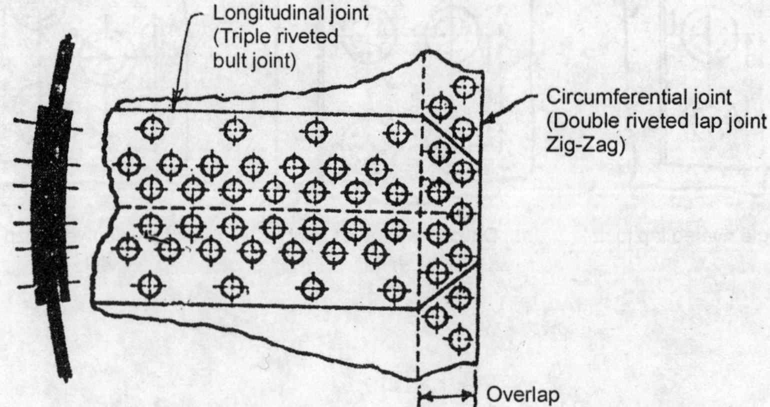

Fig. 5.2 Circumferential and longitudinal joints of a boiler

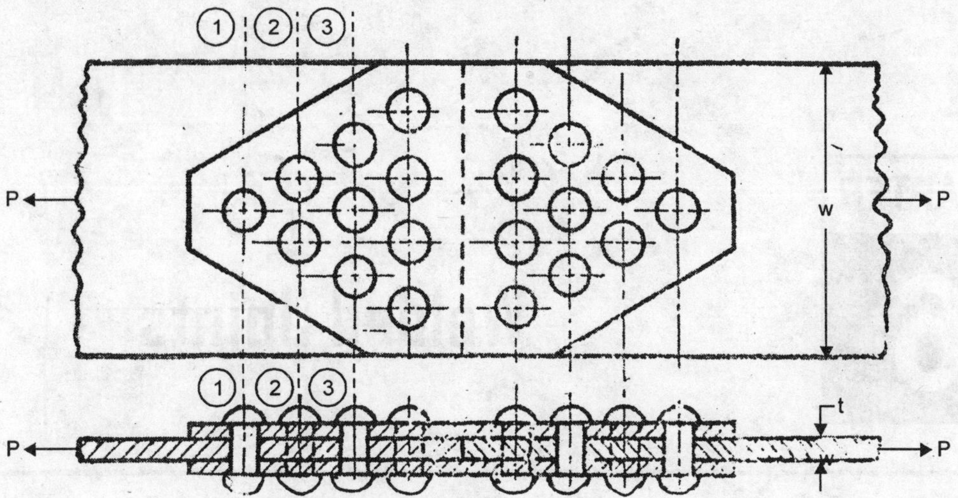

Fig. 5.3 Lozenge joint

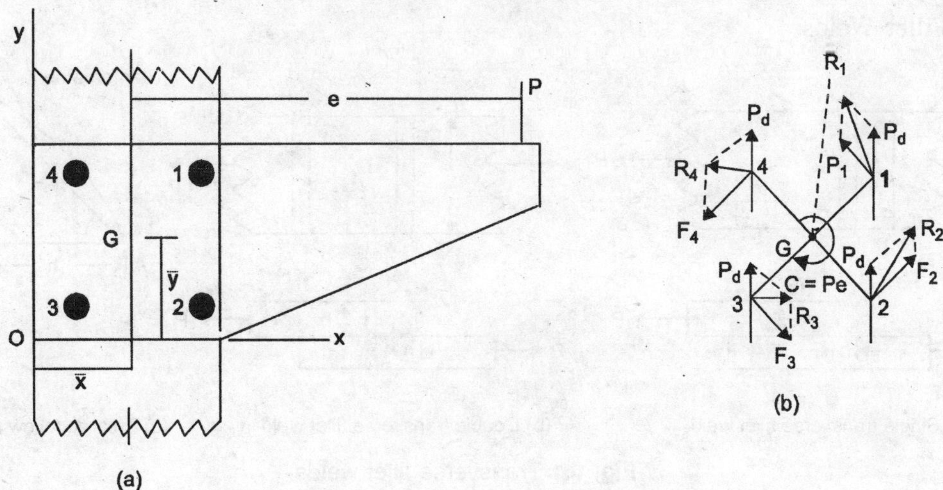

Fig. 5.4 Eccentrically loaded riveted joint

CHAPTER

6

Welded Joints

1. Design of Welded Joints

(*a*) Fillet Welds:

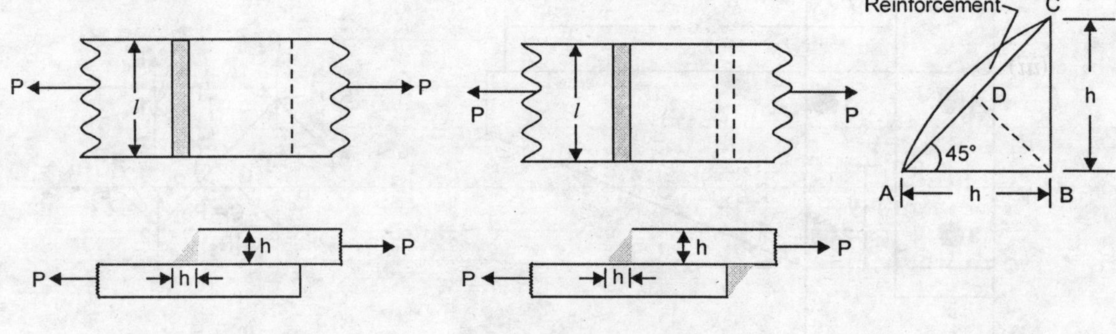

Fig. 6.1 Transverse fillet welds

(*i*) Transverse Fillet Welds

h = leg or size of the weld = thickness of plate

t = throat thickness = $h \sin 45° = 0.707h$

l = length of the weld

σ_t = allowable tensile stress in weld

P = applied tensile force

Minimum area of weld, $A = t \times l = 0.707\ hl$ ...(6.1)

Tensile strength of single fillet weld,

$$P = 0.707\ h \times l \times \sigma_t \quad ...(6.2)$$

Tensile strength of double fillet weld,

$$P = 1.414\ h \times l \times \sigma_t \quad ...(6.3)$$

(*ii*) ParallelFiltetWelds

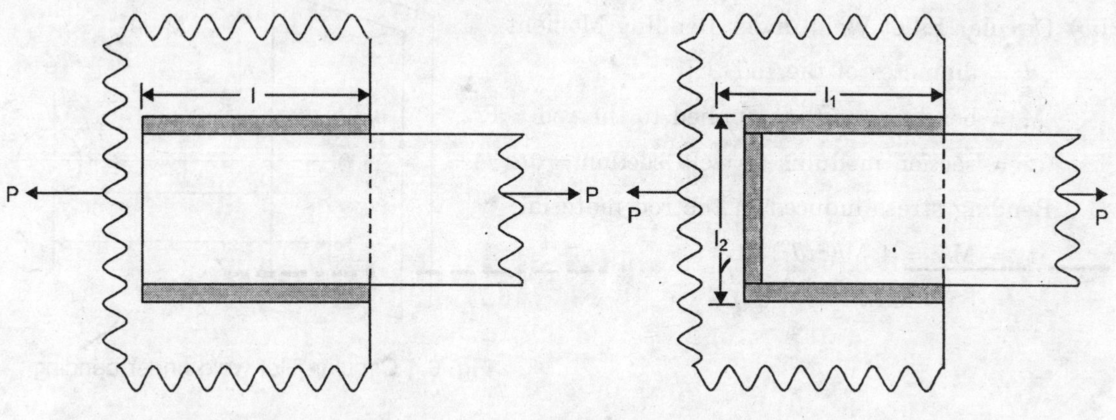

(a) Double parallel fillet weld

(b) Combination of transverse and parallel fillet welds

Fig. 6.2

$$P = 2A\tau = 1.414\ h \times l \times \tau \qquad ...(6.4)$$

For the combined single transverse and double parallel fillet welds,

$$P = 0.707\ h \times l_1 \times \sigma_t + 1.414\ h \times l_2 \times \tau \qquad ...(6.5)$$

(*iii*) Circular Fillet Weld under Torsion

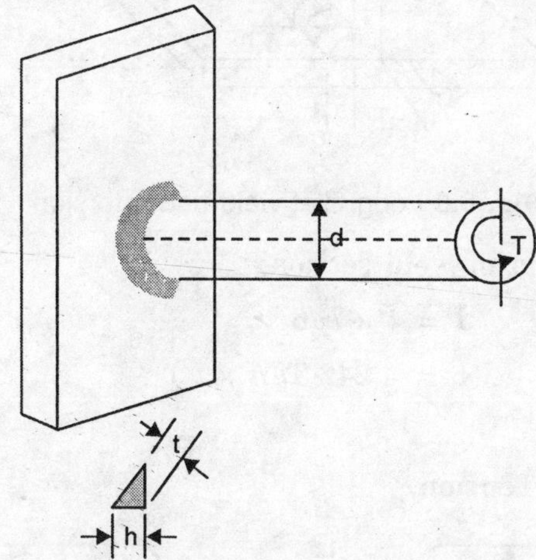

Fig. 6.3 Circular fillet weld under torsion

d = diameter of the rod

T = torque applied to the rod

J = polar moment of inertia of weld section $\pi td^3/4$

$z_p = \pi td^2/2$

Shear stress induced in the rod material,

$$\tau = 2.83\ T/(\pi \times h \times d^2) \qquad ...(6.6)$$

(*iv*) Circular Fillet Weld under Bending Moment

d = diameter of the rod

M = bending moment applied to the rod = Pl

z = section modulus of weld section = $\pi t d^2/4$

Bending stress induced in the rod material,

$$\sigma_b = M/z = 4\ M/(\pi t d^2)$$

$$= 5.66\ M/(\pi \times h \times d^2) \qquad ...(6.7)$$

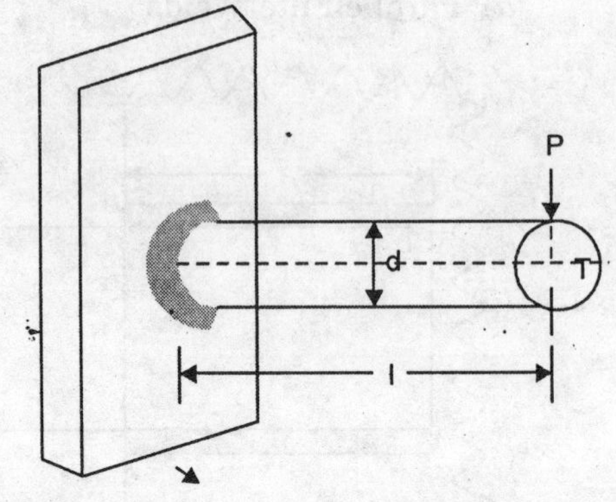

Fig. 6.4 Circular fillet weld under bending

(*v*) Long Fillet Weld under Torsion

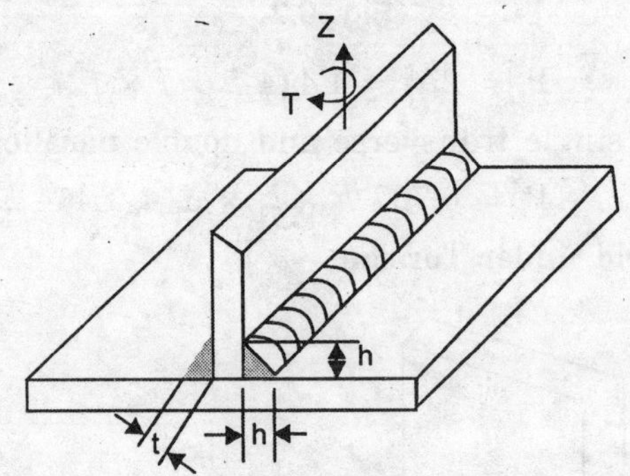

Fig. 6.5 Long fillet weld under torsion

Polar moment of inertia of weld section,

$$J = t \times l^3/6$$

Shear stress in weld, $\tau = 4.242\ T/(h \times l^2)$ ...(6.8)

(*b*) Butt Welds

(*i*) Butt Welds under Tension

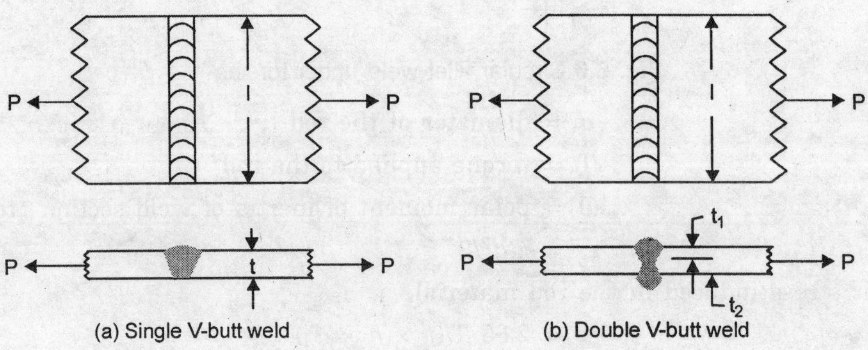

(a) Single V-butt weld (b) Double V-butt weld

Fig. 6.6 Butt joints under tension

Tensile strength of single V – butt weld,

$$P = t \times l \times \sigma_t \quad ...(6.9)$$

Where l = length of weld = width of plate

Tensile strength of double V – butt weld, $P = (t_1 + t_2)\, l \times \sigma_t$, ...(6.10)

Where t_1, t_2 = throat thicknesses at top and bottom respectively,

(*ii*) Butt Weld under Shear

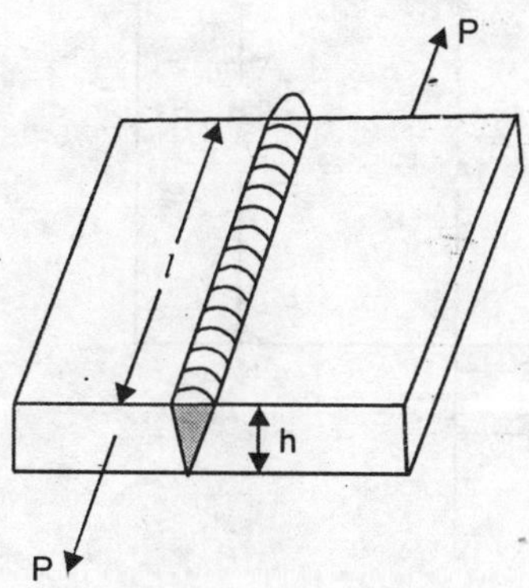

Fig. 6.7 Butt weld in shear

Shear stress in weld, $\tau = P/(hl)$ ...(6.11)

2. Axially Loaded Unsymmetrical Welded Sections

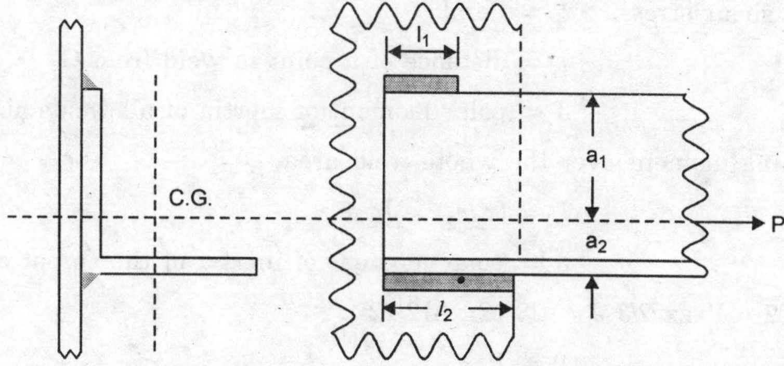

Fig. 6.8 Axially loaded unsymmetrical welded section

l_1 = length of weld at top

l_2 = length of weld at the bottom

$l = l_1 + l_2$ = total length of weld

a_1 = distance of top weld from C.G. axis

a_2 = distance of bottom weld from C.G. axis

P = axial load

$$l_1 = la_2/(a_1 + a_2) \quad ....(6.12a)$$

$$l_2 = la_1/(a_1 + a_2) \quad ....(6.12b)$$

3. Eccentrically Loaded Welded Joints

(*i*) Welded Joint under Shear

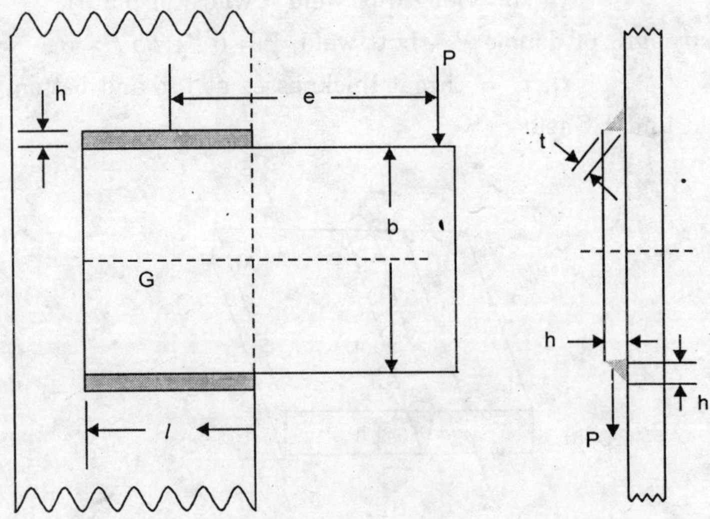

Fig. 6.9 Eccentrically loaded weld in shear

Primary shear stress, $\tau_1 = P/A$ ...(6.13)

Where A = throat area of weld throat

Secondary shear stress, $\tau_2 = Per/J$ ...(6.14)

Where r = distance of a point in weld from G

J = polar moment of inertia of all welds about G.

Total turning moment over the whote weld area,

$$T = (\tau_2/r_2) \times J$$

Where J = polar moment of inertia of the throat area about G.

T = Pxe T2 = Pxexr2/J J = tl3/12= AI2/12

Now $T = P \times e$

$$\tau_2 = P \times e \times r_2/J \quad ...(6.15)$$

$$J = tl^3/12 = Al^2/12$$

$$J_G = J + Ay^2{}_1 = A[l^2/12 + y^2{}_1]$$

$$J = J_1 + J_2 + J_3 + ...$$

Resultant shear stress, $\tau = [\tau^2{}_1 + \tau^2{}_2 + 2\tau_1\tau_2 \cos\theta]^{1/2}$ ...(6.16)

Where θ = angle between τ_1 and τ_2.

(*ii*) Welded Joint under Bending

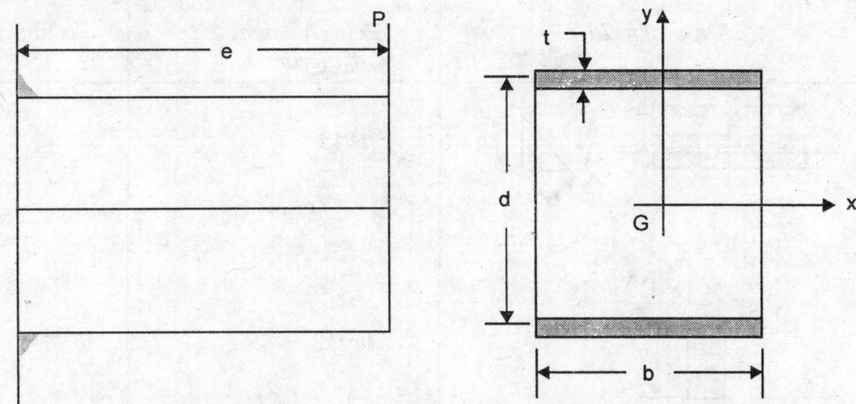

Fig. 6.10 Eccentrically loaded weld under bending

Primary shear stress, $\tau_1 = P/A$ ...(6.17)

Where A = throat area of welds

$$\sigma_b = Pe/(btd) \quad ...(6.18)$$

Resultant shear stress in the welds,

$$T = [(\sigma_b/2)^2 + \tau^2_1] \quad ...(6.19)$$

4. Welded Joint Subjected to Fluctuating Forces

The design procedure as explained in chapter 4 may be adopted to solve a problem.

Table 6.1: Stresses for welded joints.

Type of weld	*Bare electrode*		*Coated electrode*	
	Steady load MPa	*Fatigue load MPa*	*Steady load MPa*	*Fatigue load MPa*
Fillet weld (all types)	80	21	98	75
Butt welds:				
Tension	90	35	110	55
Compression	100	35	125	55
Shear	55	21	70	35

Table 6.2: Fatigue stress – concentration factors for welded joints.

Types of joint	*Fatigue stress – concentrationfactor, K_f*
Reinforced butt welds	1.2
Toe of transverse fillet weld	1.5
End of parallel fillet weld	2.7
T – butt joint with sharp corner	2.0

Table 6.3: Plate thickness and minimum size of fillet weld

Plate thickness, mm	3 – 5	6 – 8	10 – 16	18 – 24	26 – 35	Over 38
Minimum weld size, mm	3	5	6	10	15	20

Table 6.4: Polar moment of inertia and section modulus for welds

Sl. No.	Types of weld	Polar moment of inertia, J	Section modulus, z
1.	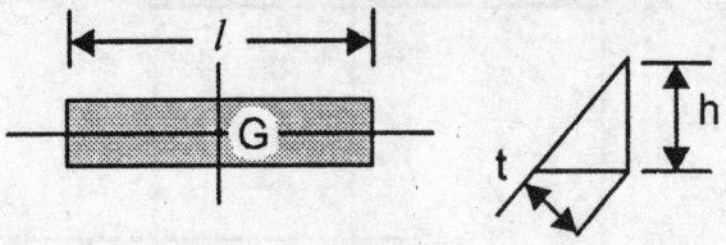	$tl^3/12$	–
2.	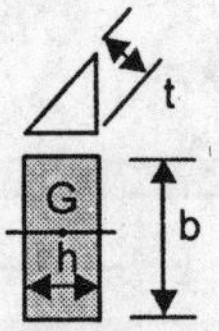	$tb^3/12$	$tb^2/6$
3.	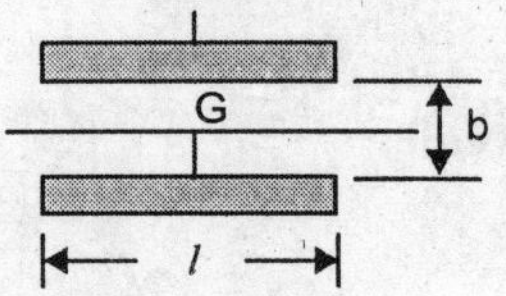	$tl(3b^2 + l^2)/6$	tb_1
4.	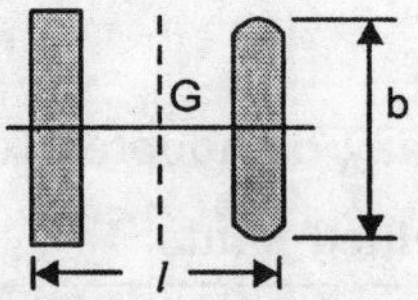	$tb(b^2 + 3l^2)/6$	$tb^2/3$
5.	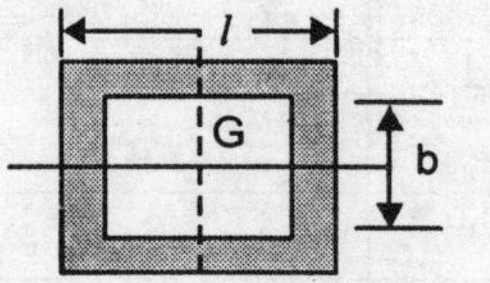	$t(b + l)^3/6$	$t(bl + b^2/3)$
6.	$\bar{x} = \dfrac{l^2}{2(l+b)}$ $\bar{y} = \dfrac{b^2}{2(l+b)}$ 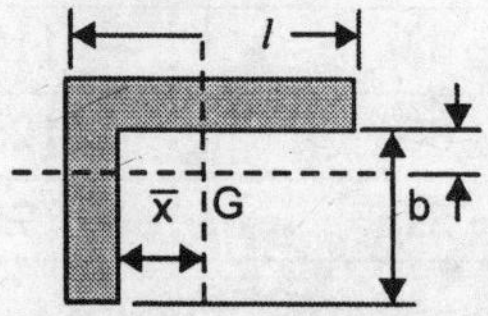	$t[\{(b + l)^4 - 6b^2t^2\}$ $/\{12(t + b)\}]$	$t[(4lb + b^2)/6]$ top $t[\{b^2(4lb + b)\}]/$ $\{6(2l + b)\}]$ (bottom)
7.	$\bar{x} = \dfrac{l^2}{2l+b}$ 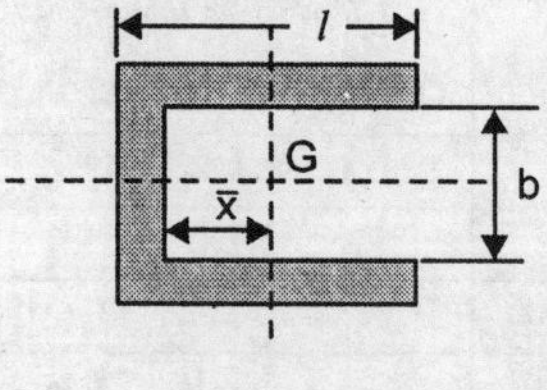	$t[(b + 2l)^3/12 -$ $l^2(b + l)^2/(b + 2l)]$	$t[tb + b^2/6]$
8.	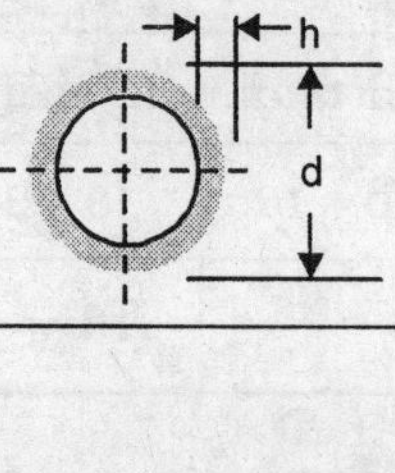	$\pi td^3/4$	$\pi td^2/4$

CHAPTER 7

Screwed Joints

1. Stresses in Screwed Joints under Static Loading

(*a*) Initial stresses

(*i*) Torsional shear stress caused by the frictional resistance of threads during tightening—

Helix angle, $\alpha = \tan^{-1} [p/(\pi d_m)]$ ...(7.1)

Where p = pitch of threads

d_m = mean pitch diameter of threads.

μ = coefficient of friction

μ_c = coefficient of friction between the collar (nut or bolt head) and the surface on which the collar bears.

d_c = mean collar diameter (nut or bolt head)

P_i = initial load

Total frictional force,

$$H = P_i \, [(\tan \alpha + \mu \sec \beta)/(1 - \mu \tan \alpha \sec \beta) + \mu_c \, d_c/d_m] \quad ...(7.2)$$

Frictional torque, $T_f = Hd_m/2$ ...(7.3)

(*ii*) Initial tensile stresses due to screwing up forces

P_i = 2805d N, for a leak proof joint ...(7.4*a*)

= 1402d N, for an ordinary joint ...(7.4*b*)

Where d = nominal diameter of bolt, mm

Initial tensile stress, $\sigma_i = P_i/A_s$ ...(7.5)

Where A_s = stress area of bolt.

(*iii*) Torsional stress, $\tau_i = 16T_f/(\pi d^3)$ ...(7.6) (*iv*) Shear stress across the threads,

$$\tau_s = P_i/(\pi d_c bn) \qquad ...(7.7)$$

Where b = width of thread section at the root

n = number of threads in engagement

(*v*) Crushing stress on threads,

$$\sigma_c = P_i/[\pi(d^2 - d^2_c)\ n] \qquad ...(7.8)$$

Where d = major diameter

(*vi*) Bending stress, $\sigma_b = aE/(2l)$ ...(7.9)

Where a = difference in height between the extreme corners of the nut or head

l = length of shank of the bolt

E = Young's modulus of elasticity of bolt material

(*b*) Stress due to external forces

(*i*) Tensile stress, $\sigma_t = P / A_s$ ...(7.10)

(*ii*) Shear stress, $\tau = 4P/(\pi d^2 n)$ ...(7.11)

(*iii*) Combined tension and shear stresses,

Maximum shear stress,

$$\tau_{max} = [(\sigma_t/2\ )^2 + \tau^2]^{1/2} \qquad ...(7.12)$$

Maximum principal stress, $\sigma_{max} = o_t/2 + [(\sigma_t/2)^2 + \tau^2]^{1/2}$ ...(7.13)

2. Fluctuating Loading

The surface finish of the bolt is usually considered as equivalent to a machined surface. The fatigue stress – concentration factor K_f varies from 2 to 3. The modified Goodman diagram may be used to determine the bolt size.

3. External Load with Preload

(*a*) Axial loads with a flexible gasket

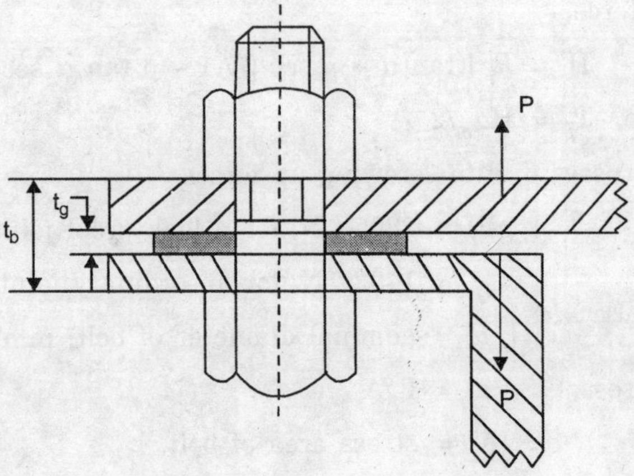

Fig. 7.1 Bolt joint with flexible gasket

Let P_i = initial load due to tightening of the bolt.

P = external load

P_g = compression load on the gasket

P_b = combined load on the bolt

k_b, k_g = stiffness of bolt and gasket respectively, N/mm

δ_b = elongation of bolt, mm/N = $1/k_b$

δ_g = compression of gasket, mm/N = 1/kg

n = total number of bolts

Initial compression of gasket = $P_i\delta_g$

Initial elongation of bolt = $P_i\delta_b$

Combined load on bolt, $P_b = P_i + CP$ ...(7.14)

Where $C = \delta_g/(\delta_g + \delta_b) = k_b/(k_b + k_g)$ ...(7.15)

$P_g = P_i - [\delta_g/(\delta_g + \delta_b]P$ ...(7.16)

Now $\delta_b = l_b/(A_bE_b) = 1/k_b$ ...(7.17)

Where l_b, A_b = length and area of bolt respectively

For the gasket and flange,

$$\delta_g = t_g/(A_gE_g) + t_f/(A_fE_f) = 1/k_g \quad ...(7.18)$$

For a thick gasket, the second term is very small. Hence

$$C = t_gA_bE_b/(t_gA_bE_b + l_bA_gE_g) \quad ...(7.19)$$

The value of P should be such that P_g remains compressive.

(*b*) Metal – to - metal Joint – For this case, when $E_b = E_f$, $t_f = l_b$ and $A_f >> A_b$, then $C \to 0$, and the external load does not affect the stress in the bolt as long as $P < P_i$. For a tight joint, $P < P_i$ by about 20%. If $P > P_i$, the joint will separate and the stress will be due to the external load alone.

4. Eccentric Loading on Bolts

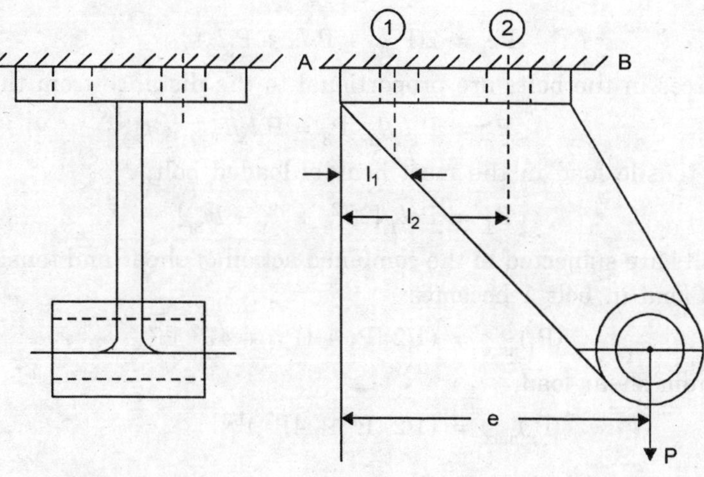

Fig. 7.2

(*i*) Eccentric Load Parallel to the Axis of Bolts

Direct load in each bolt P_d = P/n ...(7.20)

where n = number of bolts.

Moment due to load P about the tilting edge = Pe

The most heavily loaded bolts are those which are located at the greatest distance from the tilting edge. Thus

$$P_2 = Pel_2/[2(l^2_1 + l^2_2)]$$

Total tensile load on the heavily loaded bolt,

$$P_t = P_d + P_2$$

$$= P/n + Pel_2/[2(l^2_1 + l^2_2)] \quad ...(7.21)$$

If d_c is the core diameter of the bolt, then

$$(\pi/4)\, d_c^2\ \sigma_t = P_t \quad ...(7.22)$$

(*ii*) Eccentric Load Acting Perpendicular to the Axis of Bolts.

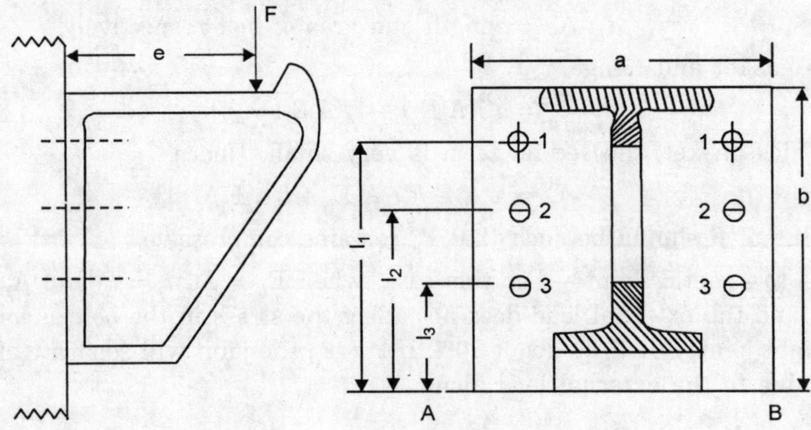

Fig. 7.3

$$Pe = 2(P_1l_1 + P_2l_2 + P_3l_3) \quad ...(7.23)$$

The forces in the bolts are proportional to the distance from the edge AB. Hence

$$P_2 = P_1l_2/l_1,\ P_3 = P_1l_3/l_1 \quad ...(7.24)$$

Hence tensile load in the most heavily loaded bolt,

$$P_1 = Pel_1/[2(l^2_1 + l^2_2 + l^2_3)] \quad ...(7.25)$$

The bolts are subjected to the combined action of shear and tension. The maximum tensile load in bolt 1 becomes:

$$(P_1)_{max} = (1/2)[P_1 + (P^2_1 + 4P^2_s)^{1/2} \quad ...(7.26)$$

Maximum shear load,

$$(P_s)_{max} = (1/2)(P^2_1 + 4P^2_s)^{1/2} \quad ...(7.27)$$

$$(\pi/4)\ d_c^2\ \sigma_t = P_t \quad ...(7.28a)$$

(*iii*) Eccentric Load on a Bracket with Circular Base

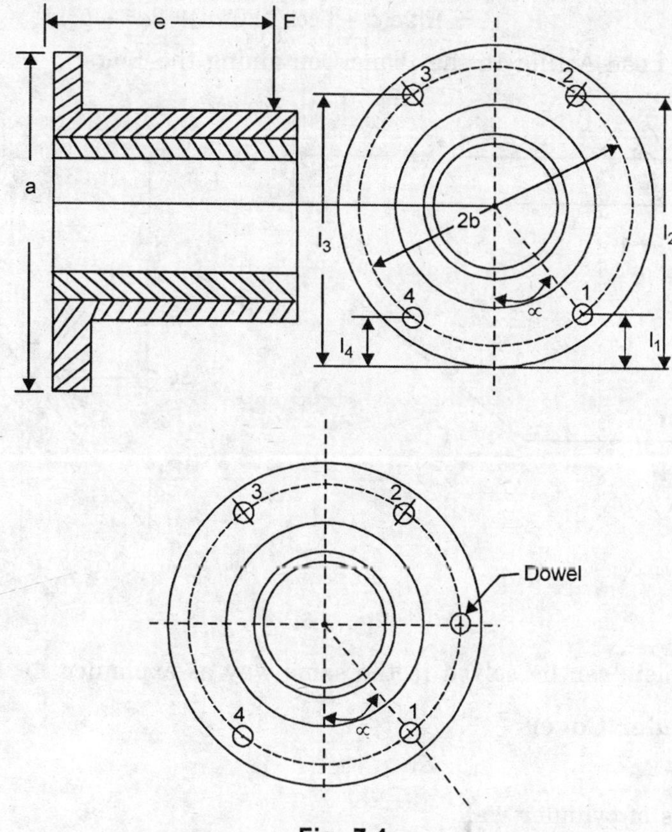

Fig. 7.4

a = radius of the column flange

b = radius of the bolt pitch circle.

e = distance of the load from the tilting edge

l_1, l_2, l_3, l_4 = distances of bolt centres from the tilting edge.

$$P_1 = Pel_1/(l_1^2 + l_2^2 + l_3^2 + l_4^2)$$

$$l_1 = a - b \cos \alpha$$

$$l_2 = a + b \sin \alpha$$

$$l_3 = a + b \cos \alpha$$

$$l_4 = a - b \sin \alpha$$

$$P_1 = P[e(a - b \cos \alpha)/(4a^2 + 2b^2)] \quad ...(7.29)$$

For n bolts, equally spaced, the general-expression is,

$$P_1 = P[2e\ (a - b \cos \alpha)/(2a^2 + b^2)n] \quad ...(7.30)$$

Where $\alpha = 180°/n$

For P_1 to be maximum, cos α = – 1 or α = 180°

$$P_{max} = P[2e(a + b)/(2a^2 + b^2)n] \quad ...(7.31)$$

$$\alpha = 180° - 180°/(2n) \quad ...(7.32)$$

$$P_{max} = P[2e(a + b\cos(180°/n))/(2a^2 + b^2)n] \quad ...(7.33)$$

(*iv*) Eccentric Load Acting In the Plane Containing the Bolts

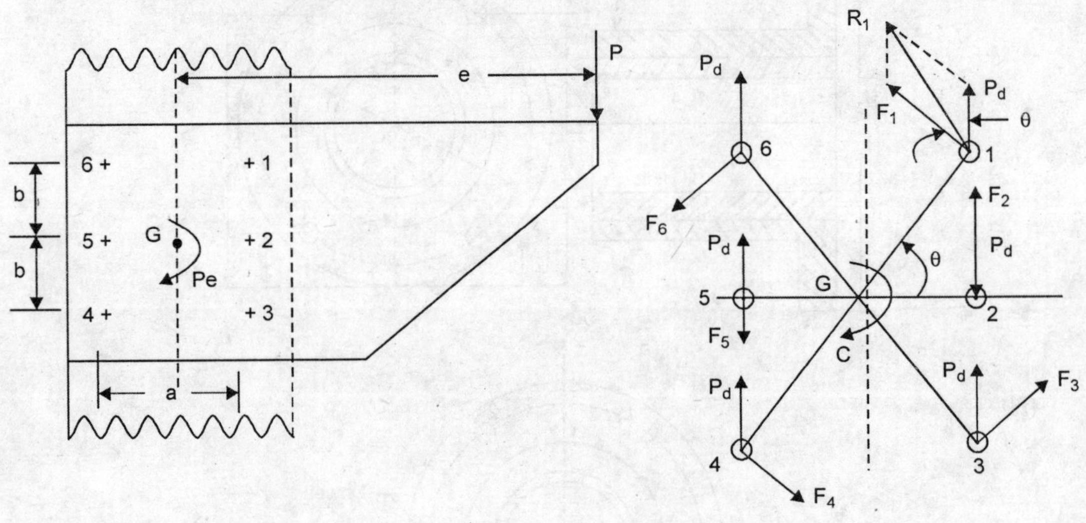

Fig. 7.5

This problem can be solved in the same way as explained for rivets.

5. Design of Cylinder Cover

(*i*) Cylinder

Thickness of cylinder wall,

$$t = r[\{(\sigma_\theta + p)/(\sigma_\theta - p)\}^{1/2} - 1] \quad ...(7.34)$$

where r = inner radius of cylinder

σ_θ = permissible hoop stress in cylinder wall

p = internal pressure

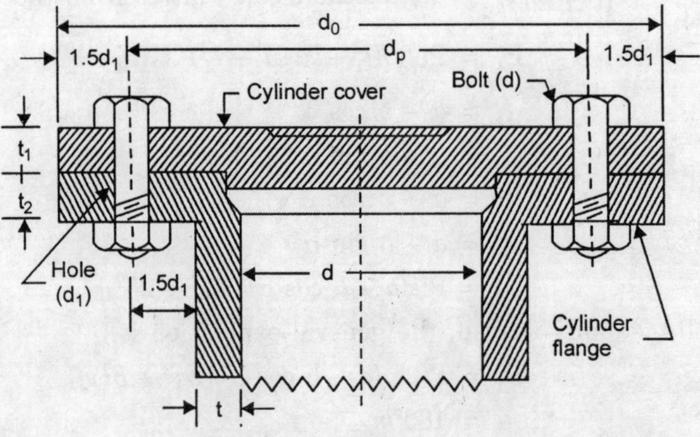

(a) Arrangement of securing the cylinder cover with bolts.

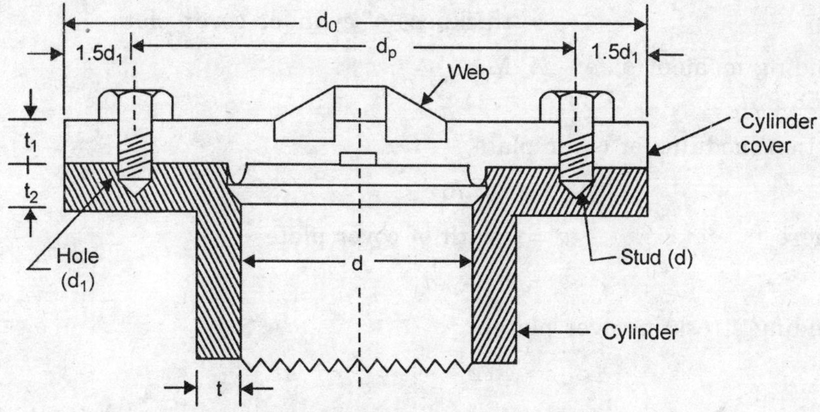

(b) Arrangement of securing the cylinder cover with studs.

Fig. 7.6 Cylinder head designs

(*ii*) Bolts or Studs

Let d = inner diameter of cylinder

d_c = core diameter of bolts or studs

n = number of bolts or studs

σ_t = permissible tensile stress in bolts or studs

$$d_c = d[p/(n\,\sigma_t)]^{1/2} \quad \text{...(7.35)}$$

Assume some even number of bolts or studs and determine their core diameter. Otherwise assume $d_c > 16$ mm and calculate the number of bolts, which should be even. The effect of screwing up force on gasket may be accounted for to obtain a leak proof joint.

The circumferential pitch of bolts or studs for leak proof joint,

$$p = 20\sqrt{d_1} \text{ to } 30\sqrt{d_1} \quad \text{...(7.36)}$$

where d_1 = diameter of hole in mm for bolt or stud.

Pitch circle diameter of bolts or studs,

$$d_p = d + 2t + 3d_1 \quad \text{...(7.37)}$$

Outside diameter of cover,

$$d_0 = d_p + 3d_1 = d + 2t + 6d_1 \quad \text{...(7.38)}$$

(iii) Cylinder Cover Plate

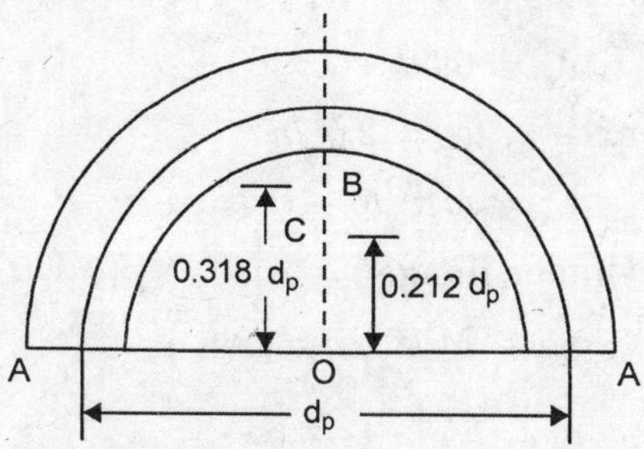

Fig. 7.7 Semi-cover plate of a cylinder

Let t_1 = thickness of cylinder cover plate ...(7.39)

Bending moment at A – A, M = 0.053 $P_b d_p$ ... (7.39)

Section modulus of cover plate,

$$z_c = wt^2{}_1/6 \quad ...(7.40)$$

Where w = width of cover plate

$$= d_0 - 2d_1 \quad ...(7.41)$$

Bending stress in cover plate,

$$\sigma_{bc} = M/z_c \quad ...(7.42)$$

Knowing the allowable bending stress in cover plate material, its thickness t_1 can be determined.

(*iv*) Cylinder Flange

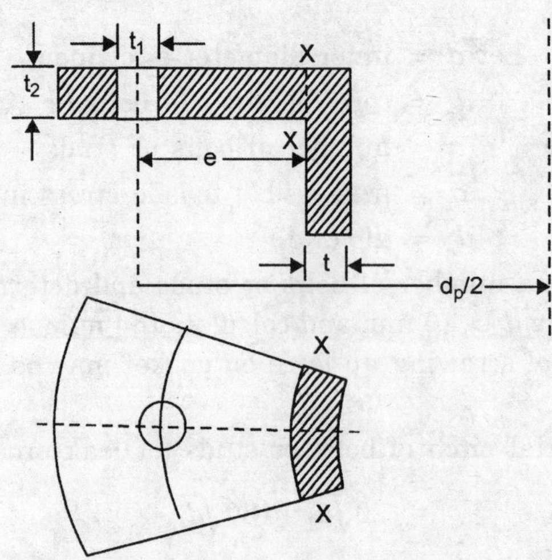

Fig. 7.8 A portion of the cylinder flange

Eccentricity of force from section $x - x$ is,

$$e = 0.5dp - (0.5d_1 + t) \quad ...(7.43)$$

Bending moment, $M_x = P_b \times e/n$ ...(7.44)

Radius of the section

$$x - x,\ r_x = 0.5d + 1 \quad ...(7.45)$$

Width of the sectionx – x, $w_x = 2\pi r_x/n$ ...(7.46)

Section modulus, $z_x = w_x \tau^2{}_2/6$ ...(7.47)

Bending stress in cylinder flange,

$$\sigma_{bf} = M_x/z_x \quad ...(7.48)$$

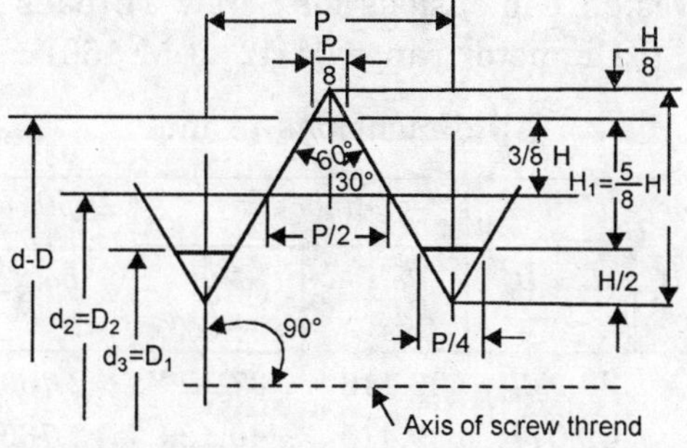

Basic profit for ISO metric screw thereads

Table 7.1: Dimensions for Screw Threads for Bolts and Nuts (Diameter range M2.5 to M36)

All dimensions in mm.

	Size	*Pitch* P	*Major Dia* $d = D$	*Pitch dia* $d_1 = D_2$	*Minor dia*		*Depth of Thread* h_3	*Maximum depth of engagement* H_1	*Stress area** A_s mm^2
					Bolt d_3,	*Nui* d_1,			
Coarse Series	M2.5	0.45	2.5	2.208	1.948	2.013	0.276	0.244	3.39
	M3	0.5	3	2.675	2.387	2.459	0.307	0.271	5.03
	M4	0.7	4	3.545	3.141	3.242	0.429	0.379	8.78
	M5	0.8	5	4.480	4.019	4.134	0.491	0.433	14.2
	M6	1	6	5.350	4.773	4.918	0.613	0.541	20.1
	M8	1.25	8	7.188	6.466	6.647	0.767	0.677	36.6
	M10	1.5	10	9.026	8.160	8.376	0.920	0.812	58.0
	M12	1.75	12	10.863	9.853	10.106	1.074	0.947	84.3
	M16	2	16	14.701	13.546	13.835	1.227	1.083	157
	M20	2.5	20	18.376	16.933	17.294	1.534	1.353	245
	M24	3	24	22.051	20.320	20.752	1.840	1.624	353
	M30	3.5	30	27.727	25.706	26.21 1	2.147	1.894	561
	M33	3.5	33	30.727	28.706	29.211	2.147	.894	694
	M36	4	36	33.402	31.093	31.67	2.454	2.165	817
Fine Series	M8×1	1	8	7.350	6.773	6.918	0.613	0.541	39.2
	M10×1.25	1.25	10	9.188	8.466	8.647	0.767	0.677	61.2
	M12×1.25	1.25	12	11.188	10.466	10.647	0.767	0.677	92.1
	M16×1.25	1.5	16	15.026	14.16	14.376	0.920	0.812	167
	M20×1.5	20	19.026	18.16	18.376	0.920	0.812	272	
	M24×2	2	24	22.701	21.546	21.835	1.227	1.083	384
	M30×2	2	30	28.701	27.546	27.835	1.227	1.083	621
	M36×3	3	36	34.051	32.32	35.752	1.840	1.629	865

IS : 4218–1976

$$\cdot\text{Stress area} = \frac{\pi}{4}\left(\frac{d_2 + d_3}{2}\right)^2$$

Designation : A fine series screw of major diameter 8 mm and pitch 1 mm shall be designated as : M8×1.

A coarse series screw of major diameter 6 mm and pitch 1 mm shall be designated as : M6.

Table 7.2: Design Dimensions for Screw Threads for Bolts and Nuts (Diameter range M 42 to M 150)

All dimensions in mm

Size	*Pitch* P	*Allow-ance pmm*	*Major dia* $d = D$	*Pitch dia* $d_2 = D_2$	*Minor dia*		*Depth of thread bolt*	*Maximum depth of engage-* h_3	*Radius* r H_1	*Stres area* $A_s\ mm^2$
					Bolt d_3	*Nut* D_1				
M42	4.5	63	42	39.077	36.416	37.129	2.760	2.436	0.650	1120
(M45)	4.5	63	45	42.077	39.416	40.129	2.760	2.436	0.650	1300
M48	5	71	48	44.752	41.795	42.587	3.067	2.706	0.722	1470
(M52)	5	71	52	48.752	45.795	46.587	3.067	2.706	0.722	1760
MS6	5.5	75	56	52.428	49.177	50.046	3.374	2.977	0.794	2030
(M60)	5.5	75	60	56.428	53.177	54.046	3.374	2.977	0.794	2360
M64	6	80	64	60.103	56.559	57.505	3.681	3.248	0.866	2680
(M68)	6	80	68	64.103	60.559	61.505	3.681	3.248	0.866	3060
M92	6	80	72	68.103	64.559	65.505				3460
(M76)	6	80	76	72.103	68.559	69.505				3890
M80	6	80	80	76.103	72.559	73.505				4340
(M85)	6	80	85	81.103	77.559	78.505				4950
M90	6	80	90	86.103	62.559	83.505				5596
(M95)	6	80	95	91.103	87.559	88.505				6270
M100	6	80	100	96.103	92.559	93.505	3.681	3.248	0.866	7000
(M105)	6	80	105	101.103	97.559	98.505				7760
M 110	6	80	110	106.103	102.559	103.505				8560
(M 1 15)	6	80	US	111.103	107.559	108.505				9400

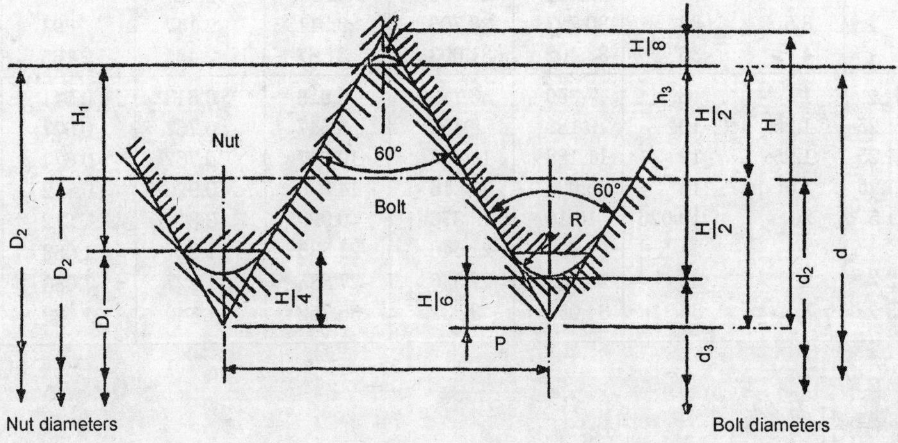

Table 7.3: Bask Dimensions for Design Profiles of ISO Metric Screw Threads

Basis Dia, mm	Pitch P, mm	Major Diameter d, mm	Pitch Diameter d_2, mm	Minor Diameter, mm		Lead Angle at Basis Pitch Stress		Tensile Stress Area, A_c mm^2
				External Threads, d_1	Internal Threads, D_1	deg	min	
1	0.25	1.0	0.837620	0.693283	0.729367	5	27	0.46
	0.20	1.0	0.870096	0.754626	0.783494	4	11	0.53
2	0.40	2.0	1.740192	1.509252	1566987	4	11	2.07
	0.25	2.0	1.837620	1.693283	1.729367	2	29	2.45
2.5	0.45	2.5	2.207716	1.947909	2.012861	3	43	3.39
	0.35	2.5	2.272668	2.070596	2.121114	2	20	3.70
3.0	0.50	3.0	2.675240	2.386565	2.458734	3	24	5.03
	0.35	3.0	2.772668	2.570596	2.621114	2	18	5.61
4.0	0.70	4.0	3.545337	3.141191	3.242228	3	36	8.78
	0.50	4.0	3.675240	3.386565	3.458734	2	29	9.79
5.0	0.80	5.0	4.480385	4.018505	4.133975	3	15	14.2
	0.50	5.0	4.675240	4.386565	4.458734	2	57	16.1
6.0	1.00	6.0	5.350481	4.773131	4.917468	3	24	20.1
	0.75	6.0	5.512861	5.079848	5.188101	2	29	22.0
7.0	1.00	7.0	6.350481	5.773131	5.917408	2	52	28.9
	0.75	7.0	6.512861	6.079848	6.188101	2	6	313
8.0	1.25	8.0	7.188101	6.466413	6.646835	3	10	36.6
	1.00	8.0	7.350481	6.773131	6.917468	2	29	39.2
10	1.50	10.0	9.025721	8.159696	8376202	3	2	58.0
	1.25	10.0	9.188101	8.466413	8.646835	2	29	61.2
	1.00	10.0	9.350481	8.773131	8.917468	1	57	645
12	1.75	12	10.863342	9.852979	10.105569	2	56	843
	1.50	12	11.025721	10.159686	10376202	2	29	88.1
	1.25	12	11.188101	10.466413	10.646835	2	2	92.1
	1.00	12	11350481	10.773131	10.917468	1	36	96.1
14	2.00	14	12.700962	11546261	11.834936	2	52	115
	1.50	14	13.025721	12.159696	12376202	2	6	125
	1.25	14	13.188101	12.466413	12.646835	1	44	129
16	2.00	16	14.700962	13546261	13.834936	2	29	157
	1.50	16	15.025721	14.159696	14376202	1	49	167
18	2.50	18	16.376202	14.932827	15.293671	2	47	192
	2.00	18	16.700962	15546261	15.834936	2	11	204
	1.50	18	17.025721	15.159696	16376202	1	36	216
20	2.50	20	18376202	16.932827	17.293671	2	29	245
	2.00	20	18.700962	17516261	17.834936	1	57	258
	1.50	20	19.025721	18.159696	18376202	1	26	272
22	2.50	22	20376202	18.932827	19.293671	2	14	303
	2.00	22	20.700962	19546261	19.834936	1	46	318
	1.50	22	21.025721	20.159696	20376202	1	18	333

Table 7.3: (Contd.)

Basis Dia, mm	*Pitch P, mm*	*Major Diameter d, mm*	*Pitch Diameter d_2, mm*	*Minor Diameter, mm*		*Lead Angle at Basis Pitch Stress*		*Tensile Stress Area, A_c*
				External Threads, d_1	*Internal Threads, D_1*	*deg*	*min*	
24	3.00	24	22.051443	20319392	20.752405	2	49	353
	2.00	24	22.700962	21556261	21.834936	1	39	384
	1.50	24	23.025721	22.159696	22376202	1	11	401
25	3.00	25	23.051443	21319392	21.752405	2	36	385
30	3.50	30	27.726683	25.705957	26.211139	2	18	561
	3.00	30	28.051443	26319392	26.752405	1	57	581
	2.00	30	28.700962	27546261	27.834936	1	16	621
	1.50	30	29.025721	28.159696	28376202	0	57	642
35	1.50	35	34.055721	33.159696	33376202	0	48	860
42	4.5	42	39.072114	36.479088	37.128607	2	6	1120
	4.0	42	39.401924	37.092523	37.669873	1	51	1150
	3.0	42	40.051443	38.319392	38.752405	1	22	1210
	2.0	42	40.700962	39546261	39.834936	0	52	1 60
	1.5	42	41.025771	40.159696	40376202	0	40	1290
45	4.5	45	42.077164	39.479088	40.128607	1	57	1300
	4.0	45	42.401924	40.092523	40.669873	1	43	1340
	3.0	45	43.051443	41.319392	41.752405	1	16	1400
	2.0	45	43.700962	42.546261	42.834936	0	50	1460
	1.5	45	44.025771	43.159696	43.376202	0	37	1490
52	5.0	52	48.752405	45.865653	46.587341	1	52	1760
	4.0	52	49.401924	47.092523	47.669873	1	29	1830
	3.0	52	50.051443	48.319392	48.752405	1	6	1900
	2.0	52	50.700962	49.546261	49.834936	0	43	1970
	1.5	52	51.025721	50.159696	50.376202	0	32	2010
60	5.5	60	56.427645	53.252219	54.046075	1	47	2360
	4.0	60	57.401924	55.092523	55.669873	1	16	2490
	3.0	60	58.051443	56.319392	56.752405	0	57	2570
	2.0	60	58.700962	57.546261	57.834936	0	37	2650
	1.5	60	59.025721	58.159696	58.376202	0	28	2700
72	6	72	68.102886	64.638784	66.504809	1	36	3460
	4	72	69.401924	67.092523	67.669873	1	3	3660
	3	72	70.051443	68.319392	68.752405	0	47	3760
	2	72	70.700962	69.546261	69.834936	0	31	3860
80	6	80	76.102886	72.638724	73.504809	1	26	4340
	4	80	77.401924	75.092523	75.669873	0	57	4570
	3	80	78.051443	76.319392	76.752405	0	42	4680
	2	80	78.700962	77.546261	77.834936	0	28	4790

Table 7.3: (Contd.)

Basis Dia, mm	Pitch P, mm	Major Diameter d, mm	Pitch Diameter d_2, mm	Minor Diameter, mm		Lead Angle at Basis Pitch Stress		Tensile Stress Area, A_c
				External Threads, d_1	Internal Threads, D_1	deg	min	
90	6	90	86.102886	82.638784	83.504809	1	16	5590
	4	90	87.401924	85.092523	85.669873	0	50	5840
	3	90	88.051449	86.319292	86.752405	0	37	5970
	2	90	88.700962	87.546261	87.834936	0	25	6100
100	6	100	96.102886	92.638784	93.504809	1	8	7000
	4	100	97.401924	95.092523	95.669873	0	45	7280
	3	100	98.051443	96.319392	96.752405	0	33	7420
	2	100	98.700962	97.546261	97.834936	0	22	7560
110	6	110	106.102886	102.638784	103.504809	1	2	8560
	4	110	107.401924	105.092523	105.669873	0	41	8870
	3	110	108.051443	106.319392	106.752405	0	30	9020
120	6	120	116.102886	112.638784	113.504819	0	57	10300
	4	120	117.401924	115.092523	115.669873	0	37	10600
	3	120	118.051443	116.319392	116.752405	0	28	10800
150	6	150	146.102886	142.538784	143.504809	0	45	16400
	4	150	147.401924	145.092523	145.669873	0	30	16800
	3	150	148.051443	146.319392	146.752405	0	22	17000
160	6	160	156.102886	152.638784	153.504809	0	42	18700
	4	160	157.401924	155.092523	155.669873	0	28	19200
	3	160	158.051443	156.319392	156.752405	0	21	19400
180	6	180	176.102886	172.638784	173.504809	0	37	23900
	4	180	177.401924	175.092523	175.669873	0	25	24400
	3	180	178.051443	176.319392	176.752405	0	18	24700
200	6	200	196.102886	192.638784	193.504809	0	33	29700
	4	200	197.401924	195.092523	195.669873	0	22	30200
	3	200	198.051453	196.319392	196.752405	0	17	30500
250	6	250	246.102886	242.638784	243.504809	0	27	46900
	4	250	247.401924	245.092523	245.669873	0	18	47600
	3	250	248.051443	246.319392	246.752405	0	13	48000
300	6	300	296.102886	295.638784	293.504809	0	22	68100
	4	300	297.401924	292.092523	295.669873	0	15	68900

IS : 4218—1967 (Part III)

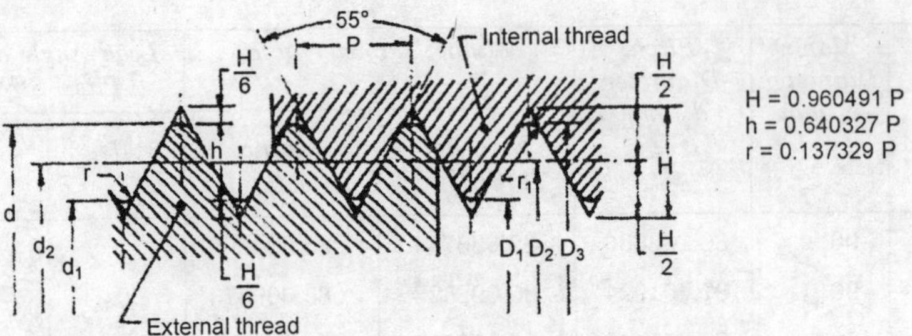

Pipe Threads for Fastening Purposes.

Table 7.4: Dimensions for Pipe Threads for Fastening Purposes
(All dimensions in mm)

Size	*Pitch*	*Major dia* $d = D$	*Pitch dia* $d_2 = D_2$	*Minor dia* $d_1 = D_1$	*Depth of thread* t	*Radius* r ≈
1/16	0.907	7.723	7.142	6.561	0.581	0.125
1/18	0.907	9.728	9.147	8.566	0.581	0.125
1/4	1.337	13.157	12.301	11.445	0.856	0.184
3/8	1.337	16.662	15.806	14.950	0.856	0.184
1/2	1.814	20.955	19.793	18631	1.162	0.249
5/8	1.814	22.911	21.749	20.587	1.62	0.249
3/4	1.814	26.441	25.279	24.117	1.162	0.249
7/0	1.814	30.201	29.039	27.877	1.162	0.249
1	2.309	33.249	31.291	30.291	1.479	0.317
1½		37.897	36.418	34.939		
1¼	2.309	41.910	40.431	38.952	1.479	0.371
1¼		47.803	46.324	44.845		0.317
1¾		53.746	52.267	50.788		
2	2.309	59.614	58.135	56.656	1.479	0.317
2¼		65.710	64.231	62.752		
2½		75.184	73.705	72.226		
2¾	2.309	81.534	80.055	78.576	1.479	0.317
3		87.884	86.405	84.926		
3½		100.330	98.851	97.372		
		113.030	111.551	100.072		
4	2.309				1.479	0.317
4½		125.730	124.251	122.772		
5		138.430	136.951	135.472		
5½	2.309	151.130	149.651	148.172	1.479	0.317
6		163.830	162.351	160.872		

IS : 2643 (Pwt-I)—1975

Table 7.5: Preferred pitches of screw threads in nun

Screw thread form	*Preferred Pitches*
For Bolts and Nuts	Coarse series—0.45, 0.5, 0.7, 0.8, 1, 1.25, 1.5, 1.75, 2, 2.5, 3, 3.5, and 4. Fine series — 1, 1.25. 1.5. 2 and 3.
Square	Fine series— 2, 3, 4 and 6. Normal scries—5, 6, 7, 8, 9, 10, 12, 14, and 16 Coarse series—8, 10, 12, 14, 16, 18, 20, 22 and 24.
Knuckle	2.54, 3.175, 4.233, and 6.35.
Saw tooth	2, 3, 4, 5, 6, 7, 8, 9, 10, 11, 12, 14, 16, 18, 20, 22, 24, 28, 32, 36, 40 and 44.
ISO Metric trapezoidal	1.5, 2, 3, 4, 5, 6, 7, 8, 9, 10, 12, 14, 16, 18, 20, 22, 24, 28, 32, 36, 40 and 44.

Table 7.6: Mechanical properties of Bolts, Screws and Studs

Mechanical property	4.6	4.8	5.6	5.9	6.8	8.8		9.8	10.9	12.9
						< M16	> M16			
Tensile strength, Nom.	400		500		600	800	800	900	1000	1200
S_{ut}, MPa Min.	400	420	500	520	600	800	830	900	1040	1220
Yield strength, Nom. S_{yt}, MPa Min.	240 240	320 340	300 300	400 420	480 480			–		
Stress at Nom. Permanent Min. set limit, $S_{0.2}$, MPa			–			640 640	640 640	720 720	900 940	1080 1100

Table 7.7: Values of Factor of Safety for Bolts, Screws and Studs

Material	*Steady load*	*Variable load*				
		d_o, *mm*				
		6 – 16	16 – 30	30 – 60	6 – 16	16 – 30
Carbon steels	4 – 5	1.5 – 4	1.5 – 2.5	8 – 12	8	
Alloy steels	5 – 7	3 – 5	3	6 – 10	7	

Table 7.8: Fatigue Stress – concentration Factors K_f for Screw Threads

Material	*Turned*	*Rolled*
Cast steel	3 – 4.5	1.5 – 4
Alloy steel	5.5 – 6	3 – 5.5

Table 7.9: Values of C

Type of joint	*Factor C*
1. Soft, elastic gasket with studs	1.00
2. Soft gasket with through bolts	0.90
3. Copper - asbestos gasket	0.60
4. Soft copper gasket with long through bolts	0.50
5. Soft copper corrugated gasket	0.40
6. Lead gasket with studs	0.10
7. Narrow copper ring	0.01
8. Metal -to -metal joint	0.00

Table 7.10: Design stress and permissible load for metric screw threads

Major diameter *mm*	*Stress area,* A_S mm^2	*Design stress* *MPa*	*Permissible load* *N*
16	157	19	2970
20	245	23	5590
24	353	27.2	9590
30	561	32.2	18040
36	817	37.8	30890
42	1120	43.2	48350
48	1472	48.4	71100
56	2030	55.2	111800

CHAPTER 8

Power Screws

1. Force Analysis of Square Threads

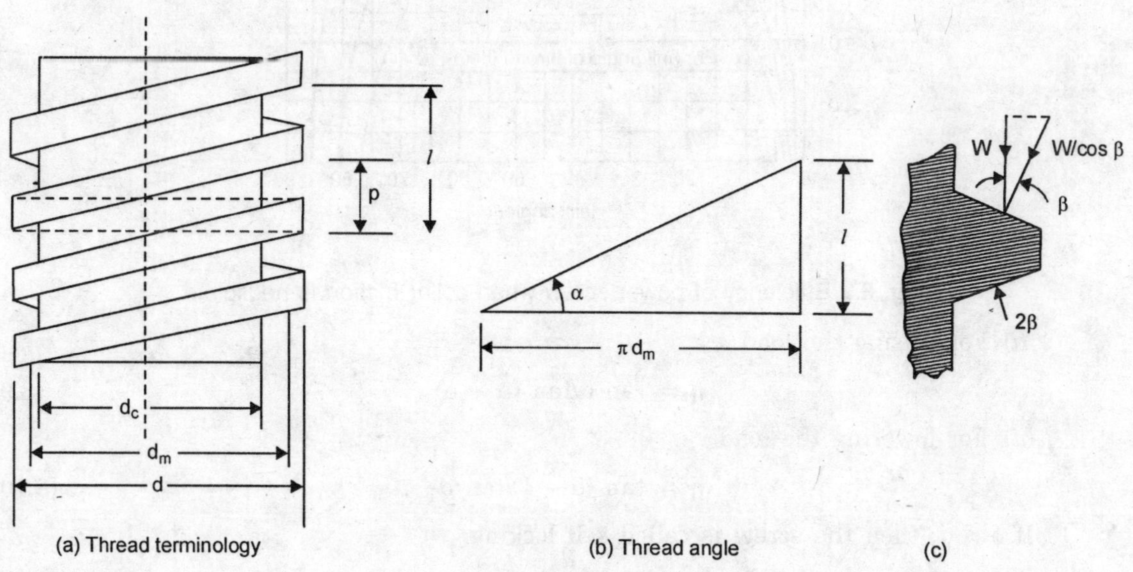

(a) Thread terminology (b) Thread angle (c)

Fig. 8.1

$$\text{Lead, } l = \text{number of starts} \times \text{pitch, } p \qquad ...(8.1)$$

$$\text{Helix angle, } \alpha = \tan^{-1} [p/(\pi d_m)] \text{ or } \tan^{-1} [l/(\pi d_m)] \qquad ...(8.2)$$

where d_m = mean diameter of screw

(*a*) Raising Road

$$P = W \tan (\alpha + \phi) \qquad ...(8.3a)$$

$$= W (\sin \alpha + \mu \cos \alpha)/(\cos \alpha - \mu \sin \alpha) \qquad ...(8.3b)$$

Where $\phi = \tan^{-1} \mu$ = angle of firction ...(8.4)

Frictional torque, $T_f = (Wd_m/2) \tan (\alpha + \phi)$ ...(8.5)

(b) Lowering Load

$$P = W(\mu \cos \alpha - \sin \alpha)/(\cos \alpha + \mu \sin \alpha) \quad ...(8.6a)$$

$$= W \tan (\phi - \alpha) \quad ...(8.6b)$$

Frictional torque, $$T_f = (Wd_m/2) \tan (\phi - \alpha) \quad ...(8.7)$$

(c) Efficiency of Screw

$$\eta = P_o/P \quad ...(8.8)$$

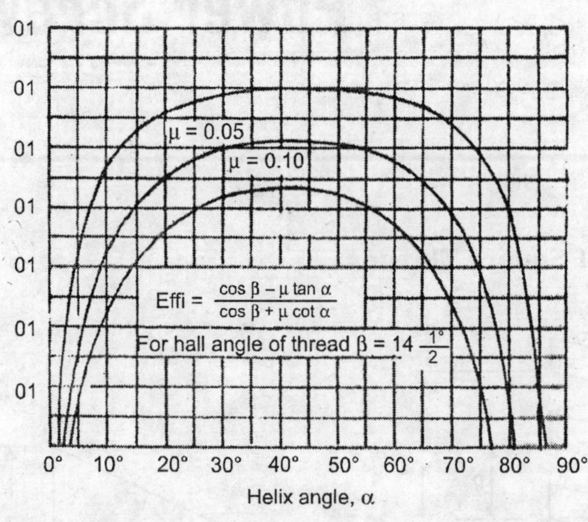

Fig. 8.2 Efficiency of power screw when collar friction is negligible.

(i) For raising the load

$$\eta = \tan \alpha/\tan (\alpha + \phi) \quad ...(8.9)$$

(ii) For lowering the load,

$$\eta = \tan (\alpha - \phi)/\tan \alpha \quad ...(8.10)$$

If $\phi > \alpha$, then the screw is called self locking.

If $\phi < \alpha$, then the screw is called over hauling.

For $\alpha = \phi$, the efficiency of the screw becomes,

$$\eta = \tan \phi/\tan (2\phi) \quad ...(8.11)$$

For a self locking screw,

$$\eta \geq [1/2 - \tan^2 (\phi/2)] \leq 50\% \quad ...(8.12)$$

2. Force Analysis of Trapezoidal Threads

Let 2β = angle of the threads.

Then equivalent or virtual coefficient of friction,

$$\mu_e = \mu/\cos \beta \quad ...(8.13)$$

3. For efficiency to be maximum

$$\alpha = 45° - \phi/2 \quad \text{...(8.14)}$$

$$\eta_{max} = [1 - \sin \phi]/[1 + \sin \phi] \quad \text{...(8.15)}$$

4. Collar Torque

$$T_c = (\mu_c W/3)\,[(D^3_o - D^3_i)/(D^2_o - D^2_i)], \text{ for uniform pressure} \quad \text{...(8.16a)}$$

$$= (\mu_c W/4)\,(D_o + D_i), \text{ for uniform wear} \quad \text{...(8.16b)}$$

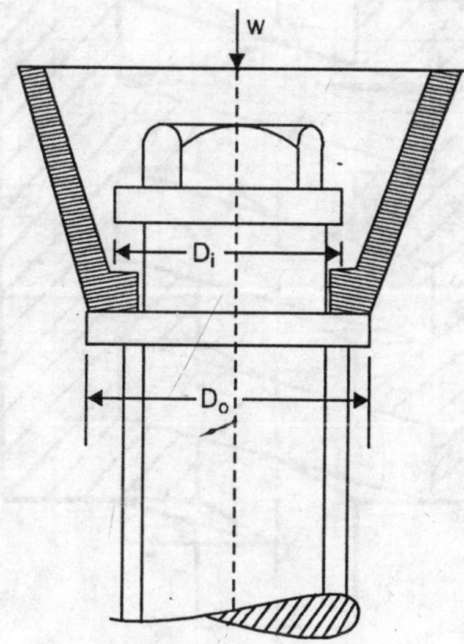

Fig. 8.3 Collar friction.

Where μ_c = collar coefficient of friction

Total torque, $T = T_f + T_c$ ...(8.17)

5. Stresses in Screw

Let d_c = core diameter of screw

W = axial load on the screw

T = total torque on the screw

(*i*) Direct compressive stress,

$$\sigma_c = 4\,W/(\pi d^2_c) \quad \text{...(8.18)}$$

(*ii*) Torsional shear stress,

$$\tau = 16\,T/(\pi d^3_c) \quad \text{...(8.19)}$$

(*iii*) Maximum normal stress,

$$\sigma_1 = 0.5[\sigma_c + \{\sigma^2_c + 4\tau^2\}^{1/2}] \quad \text{...(8.20)}$$

(*iv*) Maximum shear stress,

$$\tau_{max} = [(\sigma_c/2)^2 + \tau^2]^{1/2} \quad \text{...(8.21)}$$

(*v*) Transverse shear stress in screw

$$\tau_s = W/(\pi d_c t n) \quad \text{...(8.22)}$$

Where t = thickness of threads

n = number of threads in engagement

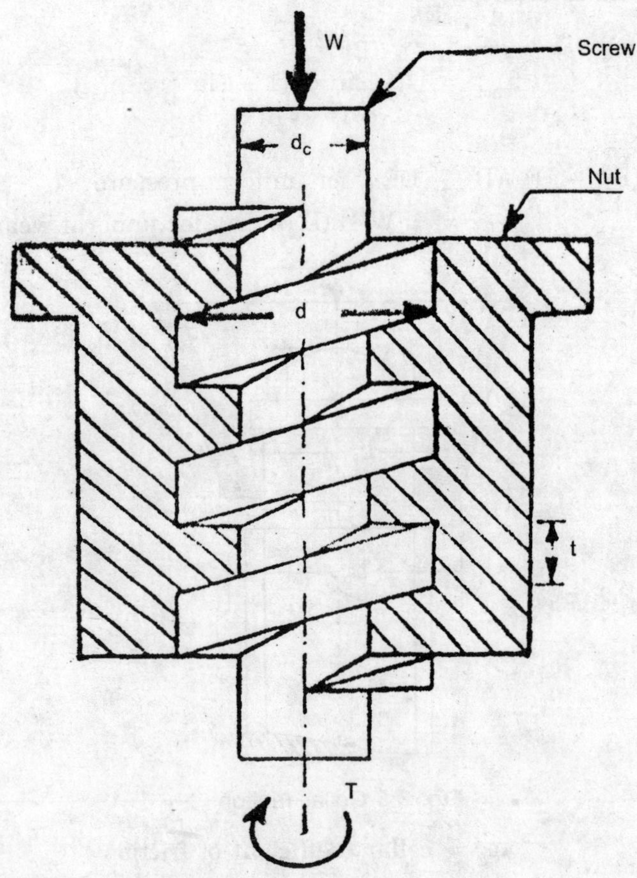

Fig. 8.4 Stresses in screw.

6. Stresses in nut

(*i*) Transverse shear stress in nut,

$$\tau_n = W/(\pi dtn) \quad \text{...(8.23)}$$

Where d = outside or major diameter of nut or screw

(*ii*) Bearing pressure between screw and nut.

$$p_b = 4W/[\pi(d^2 - d^2_c)n] \quad \text{...(8.24a)}$$

$$= W/(\pi d_m tn) \quad \text{...(8.24b)}$$

7. Compressive stress in screw considering column action

Johnson's formula for buckling of columns.

Critical load $W_{cr} = \sigma_y \times A_c[1 - \sigma_y(L/K)^2/(4C\pi^2E)]$ ...(8.25)

Where σ_y = yield stress of screw material

L = unsupported screw length

K = least radius of gyration = $d_c/4$

C = end condition coefficient = 0.25

E = Modulus of elasticity of screw material

A_c = core area of screw = $(\pi/4)\, d^2_c$

8. Design of Screw Jack

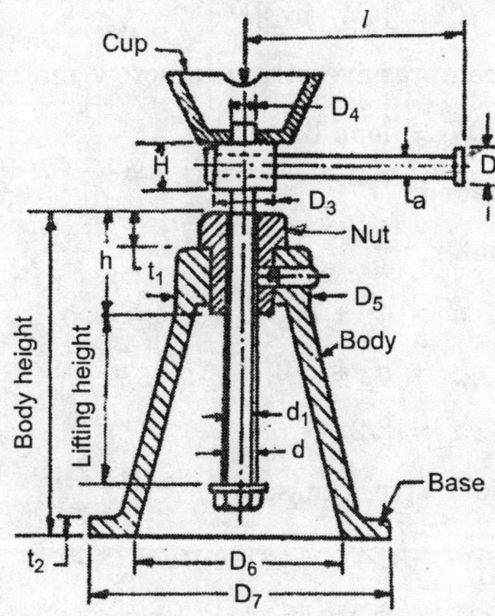

Fig. 8.5 Screw jack

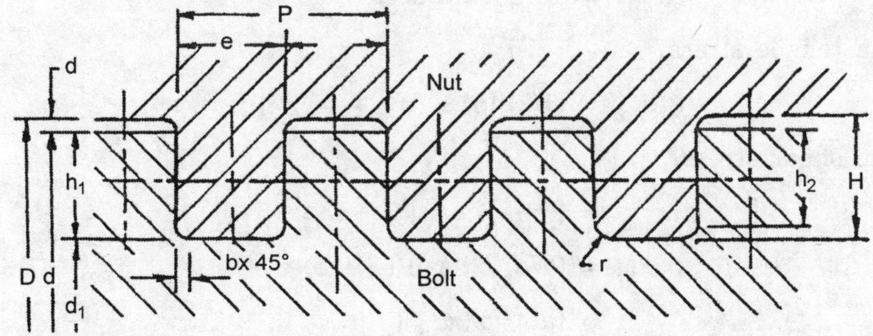

Fig. 8.6 Basic dimensions for square threads-fine series (dimensions in mm).

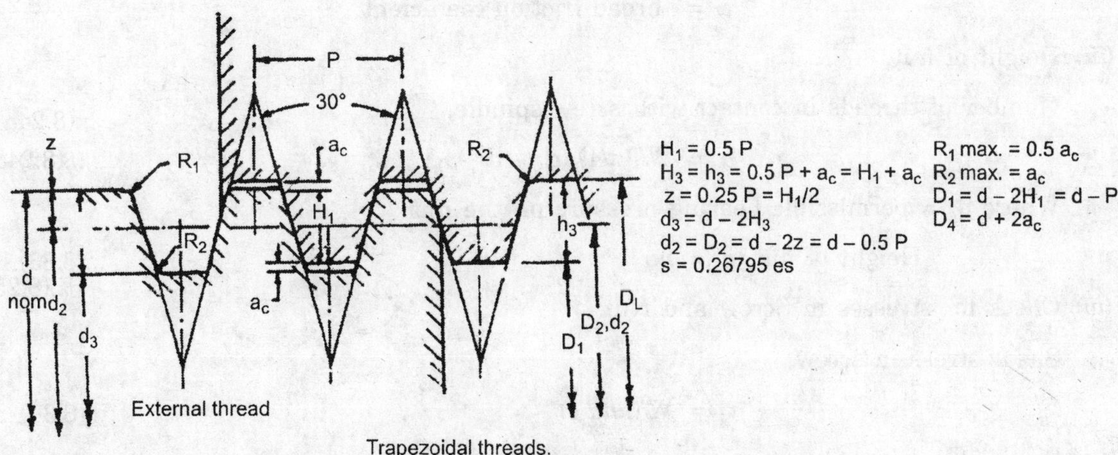

Fig. 8.7 ISO Metric trapezoidal screw threads.

(*i*) Screw diameter

$$d_c = [4W/(\pi\sigma_c)]^{1/2} \quad ...(8.26)$$

Where σ_c = permissible compressive stress in screw material

$$W = \text{load lifted}$$

The screw is square threaded

Nominal or outside diameter,

$$d = d_c + p$$

Mean diameter, $d_m = d_c + p/2$

Thread thickness, $t = p/2$

(*ii*) Maximum tensile and shear stress in screw.

Direct compressive stress,

$$\sigma_c = 4\ W/(\pi d^2{}_c) \quad ...(8.27)$$

Torsional shear stress,

$$\tau = 16\ T_t/(\pi \times d^3{}_c) \quad ...(8.28)$$

Maximum tensile stress,

$$\sigma_{1,2} = (1/2)\ [\sigma_c \pm (\sigma^2{}_c + 4\tau^2)^{1/2}] \quad ...(8.29)$$

Maximum shear stress,

$$\tau_{max} = [(\sigma_c/2)^2 + \tau^2]^{1/2} \quad ...(8.30)$$

Where $T_f = (Wd_m/2)\tan(\alpha + \phi)$ ...(8.31)

$$\alpha = \tan^{-1}[p/(\pi d_m)]$$

$$\phi = \tan^{-1}\mu$$

μ = thread friction coefficient

(*iii*) Height of nut.

Number of threads in contact with screw spindle,

$$n = W/[(\pi/4)\ (d^2 - d^2{}_c)p_b] \quad ...(8.32)$$

Where p_b = permissible bearing pressure on the nut

Height of nut, $n = np$ ...(8.33)

(*iv*) Check for stresses in Screw and Nut.

Shear stress in screw,

$$\tau_s = W/(\pi n d_c t) \quad ...(8.34)$$

Shear stress in nut,

$$\tau_n = W/(\pi n d t) \quad ...(8.35)$$

(*v*) Nut Collar size.

Let D_2 = outside diameter of collar

D_1 = inside diameter of collar

The inside diameter of nut collar is found by considering the tearing strength of nut.

$$(\pi/4)\,[D^2_1 - d^2]\,\sigma_t = W \quad \text{...(8.36)}$$

Where σ_t = permissible tensile strength of nut material

The outside diameter of nut collar is found by considering the crushing strength of nut collar.

$$(\pi/4)\,[D^2_2 - D^2_1]\,\sigma_c = W \quad \text{...(8.37)}$$

Where σ_c = permissible compressive strength of nut collar material.

The thickness of nut collar is found by considering the shearing strength of nut collar.

$$\pi D_1 t_c \tau_c = W \quad \text{...(8.38)}$$

Where t_c = nut collar thickness

τ = permissible shear stress in nut collar material

(*vi*) Head dimensions.

Diameter of head on the top of screw

$$D_3 = 1.75\text{ d} \quad \text{...(8.39)}$$

Pin diameter in the cup,

$$D_4 = D_3/4 \quad \text{...(8.40)}$$

(*vii*) Torque required between Cup and Head.

$$T_c = (\mu_o W/3)\,[(D^3_3 - D^3_4)/(D^2_3 - D^2_4)], \text{ according to uniform pressure theory} \quad \text{...(8.41a)}$$

$$= (\mu_c\, W/4)\,(D_4 + D_3), \text{ according to uniform wear theory} \quad \text{...(8.41b)}$$

Where μ_c = collar coefficient of friction

Total torque, $T = T_f + T_c$ ...(8.42)

(*viii*) Length of lever

A normal person can apply a force,

$$F = 100 \text{ to } 300 \text{ N} \quad \text{...(8.43)}$$

Length of lever, $l = T/F$

(*ix*) Diameter of lever.

Bending moment, $M = Fl$

Diameter of lever, $d_l = [32\, M/(\pi\, \sigma_b)]^{1/3}$ ...(8.44)

Where σ_b = permissible bending stress in lever.

(*x*) Height of Head

$$H = 2d_1 \quad ...(8.45)$$

(*xi*) Check of Screw in Buckling

Unsupported length of screw,

$$L = \text{lift of screw} + 0.5 \times \text{height of nut}$$

Buckling or critical load for screw,

$$W_{cr} = A_c \sigma_y [1 - \{\sigma_y/(4C\pi^2 E)\}\,(L/K)^2] \quad ...(8.46)$$

Where σ_y = yield strength of screw material

C = end condition coefficient

= 0.25, assuming the screw with lower end fixed and upper end free

K = least radius of gyration of screw

= $d_c/4$

For a safe design $W_{cr} > W$

(*xii*) Efficiency of screw.

Efficiency of screw, $\eta = T_0/T$ ...(8.47)

Where $T_o = (Wd_m/2)\tan\alpha$ ...(8.48)

(*xiii*) Body dimensions

Diameter of body at top,

$$D_5 = 1.5\ D_2 \quad ...(8.49)$$

Thickness of base,

$$t_2 = 2t_1 \quad ...(8.50)$$

Thickness of body,

$$t_3 = 0.25\ d \quad ...(8.51)$$

Inside diameter of bottom,

$$D_6 = 2.25\ D_2 \quad ...(8.52)$$

Outer diameter at the bottom,

$$D_7 = 1.75\ D_6 \quad ...(8.53)$$

Height of body = Maximum lift + height of nut + 100 mm ...(8.54)

Table 8.1: Basic dimensions for square threads fine series (dimensions in mm)

Nom. Dia.	*Major Dia* *Bolt* *D*	*Nut* *D*	*Minor Dia.* D_1	*Pitch* *P*	*e*	*R*	h_2	*b*	h_1	*a*	*h*	*Area of Core.* mm^2
10	10	10.5	8									50.3
12	12	12.5	10									78.5
14	14	14.5	12	2	1	0.12	0.75	0.25	1	0.25	1.25	113
16	16	16.5	14									154
18	18	18.5	16									201
20	20	20.5	18									254
22	22	22.5	19									284
24	24	24.5	21									346
26	26	26.5	23									415
28	28	28.5	25									491
30	30	30.5	27									573
32	32	32.5	29									661
(34)	34	34.5	31									755
36	36	36.5	33									855
(38)	38	38.5	35									962
40	40	40.5	37									1075
42	42	42.5	39	3	1.5	0.12	1.25	0.25	1.5	0.25	1.75	1195
44	44	44.5	41									1320
(46)	46	46.5	43									1452
48	48	48.5	45									1590
50	50	50.5	47									1735
52	52	52.5	49									1886
55	55	55.5	52									2124
(58)	58	58.5	55									2376
60	60	60.5	57									2552
(62)	62	62.5	59									2734
65	65	65.5	61									2922
(68)	68	68.5	64									3217
70	70	70.5	66									3421
(72)	72	72.5	68									3632
75	75	75.5	71									3959
(78)	78	78.5	74									4301
80	80	80.5	76									4536
(82)	82	82.5	78									4778
85	85	85.5	81	4	2	0.12	1.75	0.25	2	0.25	2.25	5153
(88)	88	88.5	84									5542
90	90	90.5	86									5802
(92)	92	92.5	88									6082
95	95	95.5	91									6504
(98)	98	98.5	94									6940
100	100	100.5	96									7238
(105)	105	105.5	101									8012
110	110	110.5	106									8825
(115)	115	115.5	109									9331
120	120	120.5	114									10207
(125)	125	125.5	119									11122
130	130	130.5	124	6	3	0.25	2.5	0.5	3	0.25	3.25	12076
(135)	135	135.5	129									13070
140	140	140.5	134									14103
(145)	145	145.5	139									15175
150	150	150.5	144									16286

Table 8.2: Basic dimensions for square threads–normal series (all dimensions in mm)

Nom. Dia.	*Major Dia* Bolt d	Nut D	*Minor Dia.* d_1	*Pitch* P	e	r	h_2	b	h_1	a	H	*Area of Core.* mm^2
22	22	22.5	17									227
24	24	24.5	19									284
26	26	26.5	21	5	2.5	0.25	2	0.5	2.5	0.25	2.75	346
28	28	28.5	23									415
30	30	30.5	24									452
32	32	32.5	26									531
(34)	34	34.5	28	6	3	0.25	2.5	0.5	3	0.25	3.25	616
36	36	36.5	30									707
(38)	38	38.5	31									755
40	40	40.5	33									855
(42)	42	42.5	35	7	3.5	0.25	3	0.5	3.5	0.25	3.75	962
44	44	44.5	37									1075
(46)	46	46.5	38									1134
48	48	48.5	40									1257
50	50	50.5	42	8	4	0.25	3.5	0.5	4	0.25	4.25	1385
52	52	52.5	44									1521
55	55	55.5	46									1662
(58)	58	58.5	49									1886
60	60	60.5	51	9	4.5	0.25	4	0.5	4.5	0.25	4.75	2043
(62)	62	62.5	53									2206
65	65	65.5	55									2376
(68)	68	68.5	58									2642
70	70	70.5	60									2827
(72)	72	72.5	62									3019
75	75	75.5	65	10	5	0.25	4.5	0.5	5	0.25	5.25	3318
(78)	78	78.5	68									3632
80	80	80.5	70									3848
(82)	82	82.5	72									4072
85	85	85.5	73									4185
(88)	88	88.5	76									4536
90	90	90.5	78									4778
(92)	92	92.5	80									5027
95	95	95.5	83	12	6	0.25	5.5	0.5	6	0.25	6.25	5411
(98)	98	98.5	86									5809
100	100	100.5	88									6082
(105)	105	105.5	93									6793
110	110	110.5	98									7543
(115)	115	116	101									8012
120	120	121	106									8825
(125)	125	126	111									9677
130	130	131	116	14	7	0.5	6	1	7	0.5	7.5	10568
(135)	135	136	121									11499
140	140	141	126									12469
(145)	145	146	131									13478
150	150	151	134	16	8	0.5	7	1	8	0.5	8.5	14103

Table 8.3: Basic dimensions for square threads–coarse series (all dimensions in mm)

Nom. Dia.	*Major Dia* Bolt d	*Major Dia* Nut D	*Minor Dia.* d_1	*Pitch* P	e	r	h_2	b	h_1	a	H	*Area of Core.* mm^2
22	22	22.5	14									164
24	24	24.5	16									201
26	26	25.5	18	8	4	0.25	3.5	0.5	4	0.25	4.25	254
28	28	28.5	20									314
30	30	30.5	20									314
32	32	32.5	22									380
(34)	34	34.5	24	10	5	0.25	4.5	0.5	5	0.25	5.25	452
36	36	36.5	26									531
(38)	38	38.5	28									616
40	40	40.5	28									616
(42)	42	42.5	30									707
44	44	44.5	32									804
(46)	46	46.5	34	12	6	0.25	5.5	0.5	6	0.25	6.25	908
48	48	48.5	36									1018
50	50	50.5	38									1134
52	52	52.5	40									1257
55	55	56	41									1320
(58)	58	59	44									1521
60	60	61	46	14	7	0.5	6	1	7	0.5	7.5	1662
(62)	62	63	48									1810
65	65	66	49	16	8	0.5	7	1	8	0.5	8.5	1886
(68)	68	69	52									2124
70	70	71	54									2290
(72)	72	73	56									2463
75	75	76	59	16	8	0.5	7	1	8	0.5	8.5	2734
(78)	78	79	62									3019
80	80	81	64									3217
(82)	82	83	66									3421
85	85	86	67									3526
(88)	88	89	70									3848
90	90	91	72									4072
(92)	92	93	74	18	9	0.5	8	1	9	0.5	9.5	4301
95	95	96	77									4657
(98)	98	99	80									5027
100	100	101	80									5027
(105)	105	106	85	20	10	0.5	9	1	10	0.5	10.5	5675
110	110	111	90									6362
(115)	115	116	93									6793
120	120	121	98	22	11	0.5	10	1	11	0.5	11.5	7543
(125)	125	126	103									8332
130	130	131	108									9161
(135)	135	136	111									9677
140	140	141	116	24	12	0.5	11	1	12	0.5	12.5	10568
(145)	145	146	121									11499
150	150	151	126									12469

Note: Diameter within brackets are of second preference.

Table 8.4: ISO metric trapezoidal screw threads

Nominal Diameter d			*Pitch*	*Pitch*	*Major*	*Minor Diameter*	
Column 1	*Column 2*	*Column 3*	*p*	*Diameter* $d_2 = D_2$	*Diameter* D_4	d_3	D_1
8			1.5	7.250	8.300	6.200	6.500
	9		1.5	8.250	9.300	7.200	7.500
			2	8.000	9.500	6.500	7.000
10			1.5	9.250	10.300	8.200	8.500
			2	9.000	10.500	7.500	8.000
	11		2	10.000	11.500	8.500	9.000
			3	9.500	11.500	7.500	8.000
12			2	11.000	12.500	9.500	10.000
			3	10.500	12.500	8.500	9.000
	14		2	13.000	14.500	11.500	12.000
			3	12.500	14.500	10.500	11.000
16			2	15.000	16.500	13.500	14.000
			4	14.000	16.500	11.500	12.000
	18		2	17.000	18.500	15.500	16.000
			4	16.000	18.500	13.500	14.000
20			2	19.000	20.500	17.500	18.000
			4	18.000	20.500	15.500	16.000
	22		3	20.500	22.500	18.500	19.000
			5	19.500	22.500	16.500	17.000
			8	18.000	23.000	13.000	14.000
24			3	22.500	24.500	20.500	21.000
			5	21.500	24.500	18.500	19.000
			8	20.000	25.000	15.000	16.000
	26		3	24.500	26.500	22.500	23.000
			5	23.500	26.500	20.5000	21.000
			8	22.000	27.000	17.000	18.000
28			3	26.500	28.500	24.500	25.000
			5	25.500	28.500	22.500	23.000
			8	24.500	30.500	26.500	27.000
	30		3	28.500	30.500	26.500	27.000
			6	27.000	31.000	19.000	20.000
			10	25.000	31.000	19.000	20.000
32			3	30.500	32.500	28.500	29.000
			6	29.000	33.000	25.000	26.000
			10	27.000	33.000	21.000	22.000
	34		3	32.500	34.500	30.500	31.000
			6	31.000	35.000	27.000	28.000
			10	29.000	35.000	23.000	24.000
36			3	34.500	36.500	32.500	33.000
			6	33.000	37.000	29.000	30.000
			10	31.000	37.000	25.000	26.000
	38		3	36.500	38.500	34.500	35.000
			7	34.500	39.000	30.000	31.000
			10	33.000	39.000	27.000	28.000
			3	38.500	40.500	36.500	37.500

40			7	36.500	41.000	32.000	33.000
			10	35.000	41.000	29.000	30.000
			3	40.500	42.500	38.500	39.000
	42		7	38.500	43.000	34.000	35.000
			10	37.000	43.000	31.000	32.000
			3	42.500	44.000	40.000	41.000
44			7	40.000	45.000	36.000	37.000
			12	38.000	45.000	31.000	32.000
			3	44.500	46.500	42.500	43.000
	46		8	42.000	47.000	37.000	38.000
			12	40.000	47.000	33.000	34.000
			3	46.500	48.500	44.500	45.000
48			8	44.000	49.000	39.000	40.000
			12	42.000	49.000	35.000	36.000
			3	48.500	50.000	46.500	47.000
	50		8	46.000	51.000	41.000	42.000
			12	44.000	51.000	37.000	38.000
			3	50.000	52.500	48.500	49.000
52			8	48.000	53.000	43.000	44.000
			12	46.000	53.000	39.000	40.000
			3	53.500	55.500	51.500	52.000
	55		9	50.500	56.000	45.000	46.000
			14	48.000	57.000	39.000	41.000
			3	58.500	60.500	56.500	57.000
60			9	55.500	61.000	50.000	51.000
			14	53.000	62.000	44.000	46.000
			4	63.000	65.000	60.500	61.000
	65		10	60.000	66.000	54.000	55.000
			16	57.000	67.000	47.000	49.000
			4	68.000	70.500	65.500	66.000
70			10	65.000	71.000	59.000	60.000
			16	62.000	72.000	52.000	54.000
			4	73.000	75.500	70.500	71.000
	75		10	70.000	76.000	64.000	65.000
			16	67.000	77.000	57.000	59.000
			4	78.000	80.500	75.500	76.000
80			10	75.000	81.000	69.000	70.000
			16	72.000	82.000	62.000	64.000
			4	83.000	85.500	80.500	81.000
	85		12	79.000	86.000	72.000	73.000
			18	76.000	87.000	65.000	67.000

Table 8.5: Coefficient of friction for power screws

Lubricant	*Coefficient of friction*
Machine oil and graphite	0.17
Lard oil	0.11
Heavy machine oil	0.14

Table 8.6: Coefficient of friction for thrust collars of power screws

Material	*During starting*	*During running*
Soft steel on cast iron	0.170	0.121
Hardened steel on cast iron	0.147	0.092
Soft steel on bronze	0.101	0.084
Hardened steel on bronze	0.081	0.063

Table 8.7: Allowable bearing pressure for nut

Type of machine	*Material of*		*Allowable bearing pressure, p_b MPa*
	Screw	*Nut*	
1. Hand press	Steel	Bronze	17 – 22
2. Screw jack:	Steel	Cast Iron	12 – 17
	Steel	Bronze	11 – 17
3. Hoisting screw	Steel	Cast Iron	4 – 6.8
	Steel	Bronze	5.5 – 9.6
4. Lead screw	Steel	Bronze	1.0 – 1.0

Transmission Shafts

1. Theories of Failure

(*i*) Maximum shear stress theory:

$$\sigma_1 - \sigma_2 = S_{yt}/n \quad ...(9.1)$$

Where σ_1, σ_2 = maximum and minimum principal stresses respectively

S_{yt} = yield strength under simple tension

n = factory of safety

(*ii*) Maximum distortion energy theory:

$$(1/\sqrt{2})\,[(\sigma_1 - \sigma_2)^2 + (\sigma_2 - \sigma_3)^2 + (\sigma_3 - \sigma_1)^2]^{1/2} = S_{yt}/n \quad ...(9.2)$$

For the case of plane stress system, we have

$$[\sigma^2_x + \sigma^2_y - \sigma_x\sigma_y + 3\tau^2_{xy}]^{1/2} = S_{yt}n \quad ...(9.3)$$

2. Combined Bending and Torsion

Let M = bending moment acting on the shaft

T = applied torque to the shaft

d = diameter of shaft

(*a*) Solid shaft

Maximum shear stress in the shaft,

$$\tau_{max} = [16/(\pi d^3)]\ [M^2 + T^2]^{1/2} \quad ...(9.4)$$

According to maximum shear stress theory of failure,

$$[32/(\pi d^3)]\ [M^2 + T^2]^{1/2} = S_{yt}/n \quad ...(9.5)$$

(*b*) Hollow shaft

$$[32/\{\pi d^3_0\,(1 - k^4)\}]\ [M^2 + T^2]^{1/2}] = S_{yt}/n \quad ...(9.6)$$

Where $k = d_i/d_o$

3. A.S.M.E. Code for the Design of Transmission Shafting

$$\tau_{max} = [16/(\pi d^3)]\,[(K_m M)^2 + (K_t T)^2]^{1/2} \quad ...(9.7)$$

The values of K_m and K_t are given in Table 9.3.

4. Combined Axial, Bending and Torsion loading on Hollow Shaft

$$\tau_{max} = [16/\{(\pi d^3)\,(1 - k^4)\}]\ I\{K_m M + \alpha F d_0(1 + k - k^2)/8\}^2 + (K_t T)^2]^{1/2} \quad ...(9.8)$$

Where $k = d_i/d_0$

α = a constant depending upon the magnitude of column action of the shaft length between the bearings.

$= 1/[1 - 0.0044(L/K)]$ for $L/K < 115$

$= S_{yc}\,(L/K)^2/(\pi^2 nE)$ for $L/K > 115$

L = length of the shaft between the supporting bearings

K = radius of gyration of the shaft

$= d^2{}_0\,(1 + k^2)/16$

S_{yc} = yield strength in compression of shaft material

n = 1 for hinged ends = 2.25 for fixed ends

5. Factor of safety

The permissible shear stress τ_d for shafts without keyway is taken as:

$\tau_d = 0.30\ S_{yt}$,

Or $= 0.18\ S_{ut}$, whichever is minimum ...(9.9)

The tensile strength of shafts is taken as:

$\sigma_d = 0.6\ S_{yt}$

Or $= 0.36\ S_{ut}$, whichever is less ...(9.10)

If keyways are present, these values are to be reduced by 25 %.

The static weakening effect of the keyway is given by:

$$k_k = 1.0 - 0.2\,(b/d) - 1.1(t/d) \approx 0.75 \quad ...(9.11)$$

where k_k = Strength of shaft with keyway/strength of shaft without keyway

b, t = width and depth of keyway respectively

d = shaft diameter ...(9.12)

6. Design for Torsional Rigidity

Torsional rigidity, $k_t = T/\theta = GJ/L$ ...(9.12)

Where T = torque applied to the shaft, N.mm

L = length of shaft, mm

G = modulus of rigidity of shaft material, N/mm^2

J = polar moment of inertia of shaft, mm^4

$= \pi d^4/32$

d = shaft diameter, mm

θ = 1 degree per 20 diameters or 3° per metre length of shaft, for line shaft ...(9.13*a*)

= 0.25 per metre length for machine tools ... (9.13*b*)

7. Design for Bending Rigidity

Deflection of a shaft carrying concentrated load W at a distance 'a' from left hand bearing and of length L is given by,

$$\delta = Wa^2b^2/(3EIL) \quad ...(9.14)$$

Where EI = bending stiffness of shaft

b = L – a

Bending rigidity, $k_b = W/\delta = 3EIL/(a^2b^2)$ ...(9.15)

Deflection of a shaft of span L carrying concentrated load W at the middle is given by,

$$\delta = WL^3/(48EI) \quad ...(9.16)$$

Bending rigidity, $k_b = 48EI/L^3$ ...(9.17)

Deflection of a shaft carrying uniformly distributed load of intensity w per unit length over the whole span L, is given by,

$$\delta = 5wL^4/(384EI) \quad ...(9.18)$$

Bending rigidity, $k_b = 384\,EI/(5L^3)$ ...(9.19)

The permissible lateral deflection of various components are as follows:

(*i*) Shafts carrying gears = not more than 3 mm per mm of face width ...(9.20*a*)

or = not more than $0.01 \times m$ ...(9.20*b*)

where m = module of gear teeth

(*ii*) Transmission shaft supported on plain bearings = not more than 0.6 times the distance from the load point to the centre of the bearing in metres. ...(9.21*a*)

Or = (0.001 – 0.003) L ...(9.21*b*)

Where L = span length between the two bearings.

8. Design Against Variable Load

Equivalent mean normal stress,

$$\sigma_{me} = \sigma_m + (S_y/S_e)\,\sigma_a$$

Equivalent alternating normal stress,

$$\sigma_{ae} = (S_e/S_y)\,\sigma_m + \sigma_a$$

The analogous equations for torsional loadings are:

Equivalent mean shear stress,

$$\tau_{me} = \tau_m + (S_{ys}/S_{es})\,\tau_a$$

Equivalent alternating shear stress,

$$\tau_{ae} = (S_{es}/S_{ys})\,\tau_m + \tau_a$$

(a) Combined bending and torsion

(i) **Maximum shear stress theory**

Maximum equivalent mean shear stress,

$$(\tau_{me})_{max} = [(\sigma_{me}/2)^2 + \tau^2_{me}]^{1/2}$$
$$= 0.5S_y/n = S_{ys}/n$$

For a solid circular shaft of diameter d, we have

$$S_{ys}/n = [16/(\pi d^3)]\,[\{M_n + (S_y/S_e)\,M_a\}^2 + \{T_m + (S_{ys}/S_{es})\,T_a\}^2]^{1/2} \quad ...(9.22)$$

(ii) **Distortion energy theory**

$$S_y/n = [\sigma^2_{me} + 3\tau^2_{me}]^{1/2}$$
$$S_y/n = [32/(\pi d^3)]\,[\{M_m + (S_y/S_e)\,M_a\}^2 + 0.75\{T_m + (S_{ys}/S_{es})\,T_a\}^2]^{1/2} \quad ...(9.23)$$

(b) Plane stresses

$$\sigma_{mx} = [(\sigma_x)_{max} + (\sigma_x)_{min}]/2$$
$$\sigma_{ax} = [(\sigma_x)_{max} - (\sigma_x)_{min}]/2$$
$$\sigma_{my} = [(\sigma_y)_{max} + (\sigma_y)_{min}]/2$$
$$\sigma_{ay} = [(\sigma_y)_{max} - (\sigma_y)_{min}]/2$$
$$\tau_{mxy} = [(\tau_{xy})_{max} + (\tau_{xy})_{min}]/2$$
$$\tau_{axy} = [(\tau_{xy})_{max} - (\tau_{xy})_{min}]/2$$

(i) **Maximum shear stress theory**

$$(\tau_m)_{max} = [\{(\sigma_{mx} - \sigma_{my})/2\}^2 + \tau^2_{mxy}]^{1/2}$$
$$(\tau_a)_{max} = [\{(\sigma_{ax} - \sigma_{ay})/2\}^2 + \tau^2_{axy}]^{1/2}$$

Using Soderberg failure criterion, we have

$$S_{ys}/n = (\tau_m)_{max} + (S_{ys}/S_{es})\,(\tau_a)_{max} \quad ...(9.24)$$

(ii) **Distortion energy theory**

$$(\sigma_m)_{max} = [(\sigma^2_{mx} + \sigma^2_{my} - \sigma_{mx}\,\sigma_{my} + 3\tau^2_{mxy}]^{1/2}$$
$$(\sigma_a)_{max} = [(\sigma^2_{ax} + \sigma^2_{ay} - \sigma_{ax}\,\sigma_{ay} + 3\tau^2_{axy}]^{1/2}$$

Using Soderberg failure criterion, we have

$$S_{ys}/n = (\sigma_m)_{max} + (S_y/S_e)\,(\sigma_a)_{max} \quad ...(9.25)$$

9. Critical Speed of Shafts

The critical speed of a shaft of length L carrying a disc at the centre of weight W and supported on anti – friction bearings at the ends is given by,

$$\omega_{cr} = [48EIg/(WL^3)]^{1/2} \quad ...(9.26)$$

Where I = moment of inertia of the shaft about its axis of rotation.

E = Young's modulus of elasticity of shaft material

g = acceleration due to gravity

For a central disc of weight W on a shaft rotating in sleeve (long) bearings,

$$\omega_{cr} = [192EIg/(WL^3)]^{1/2} \quad ...(9.27)$$

When the disc is mounted at a distance 'a' from left bearing and 'b' from right bearing, then

$$\omega_{cr} = [3EIg/(Wa^2b^2)]^{1/2} \text{ for short bearing} \quad ...(9.28)$$

$$= [3EIL^3g/(Wa^2b^2)]^{1/2} \text{ for long bearing} \quad ...(9.29)$$

Critical speed of a shaft of uniformly distributed weight w per unit length supported in short bearing is given by,

$$\omega_s = [3EIg/(5wL^4)]^{1/2} \quad ...(9.30)$$

For a shaft carrying n discs, the critical speed may be found by using Dunkerley's equation:

$$1/\omega^2_{cr} = 1/\omega^2_1 + 1/\omega^2_2 + \ldots\ldots + 1/\omega^2_n + 1/\omega^2_s \quad ...(9.31)$$

Where ω_s = natural frequency of shaft itself.

Table 9.1: Properties of steels used for shafts

Steel Designation	*Tensile strength, S_{ut}, MPa*	*Yield strength, S_{yt}, MPa*
1. 40C8	569 – 667	324
2. 45C8	618 – 696	353
3. 50C8	747 – 765	373
4. 50C12	706 min	392

Table 9.2: Perferred shaft diameters, mm

6, 8, 9, 10, 14, 16, 18, 19, 20, 25, 28, 30, 40, 45, 50, 56, 60, 63
71, 75, 80, 85, 90, 100, 125, 140, 150, 160, 170, 180, 190,
200, 250, 280, 300, 400, 450, 500, 530, 560, 600 and 630 mm

Table 9.3: Values of combined fatigue and shock factors.

Nature of load	K_m	K_t
Gradually applied or steady load	1.5	1.0
Suddenly applied loads with minor shocks only	1.5 – 2.0	1.0 – 1.5
Suddenly applied loads with heavy shocks	2.0 – 3.0	1.5 – 3.0

CHAPTER

10

Keys and Couplings

1. Design of Square and Flat Keys (Fig. 10.1)

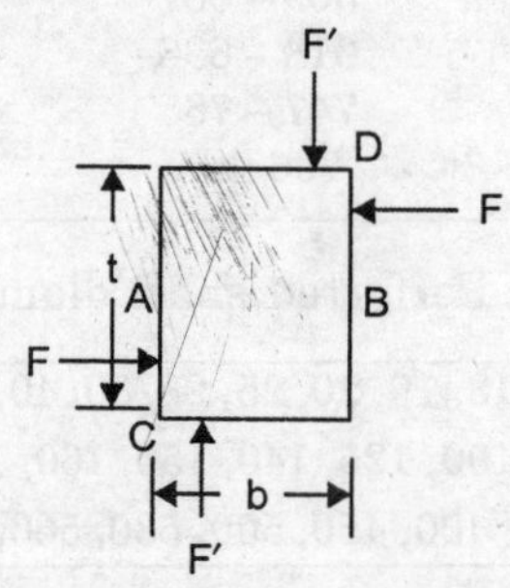

Fig. 10.1 Forces on rectangular key

Shear stress in key, $\tau = 2T/(dbl)$ ...(10.1)

Compressive stress in key,

$$\sigma_c = F/(tl/2) = 2F/(tl) = 4T/(dtl) \qquad ...(10.2)$$

Where b, t, l = width, thickness and length of key respectively

For key is to be equally strong in shear and crushing,

$$2b/t = \sigma_c/\tau \qquad ...(10.3)$$

For $\sigma_c = 2\tau$, b/t = 1, i.e. a square key is the best size.

For key is to be equally strong in shear as the shaft,

$$l = \pi d^2/(8b) \qquad ...(10.4)$$

If $b = d/4$, then $l = \pi d/2 \approx 1.5d$ ...(10.5)

2. Design of Splines (Fig. 10.2)

Let D = major diameter of splines

d = minor diameter of splines

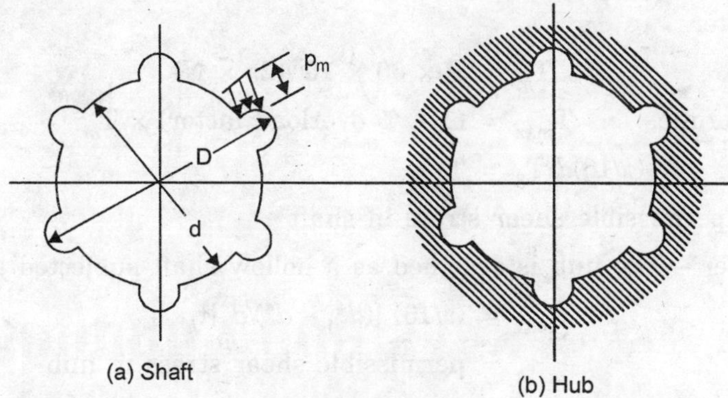

Fig. 10.2 Splines

l = length of hub

n = number of splines

p_m = permissible pressure on splines

$\approx$ 6.5 MPa

A = total area of splines = $(D - d)\, ln/2$

T = torque transmitted by the splines

Mean radius of splines, $R_m = (D + d)/4$

Torque transmitting capacity of splines,

$$T = p_m A R_m = (1/8)\, p_m l n\, (D^2 - d^2) \qquad \text{...(10.6)}$$

3. Shaft Couplings

(*a*) Rigid couplings – (*i*) Sleeve or muff coupling (*ii*) Clamp coupling (*iii*) Flange coupling (*iv*) Solid or forged flange coupling

(*b*) Flexible couplings – (*i*) Oldham's coupling (*ii*) Hooke's coupling (*iii*) Universal flexible coupling (*iv*) Slip coupling

4. Design of Rigid Protected Type Flange Coupling (Fig. 10.3)

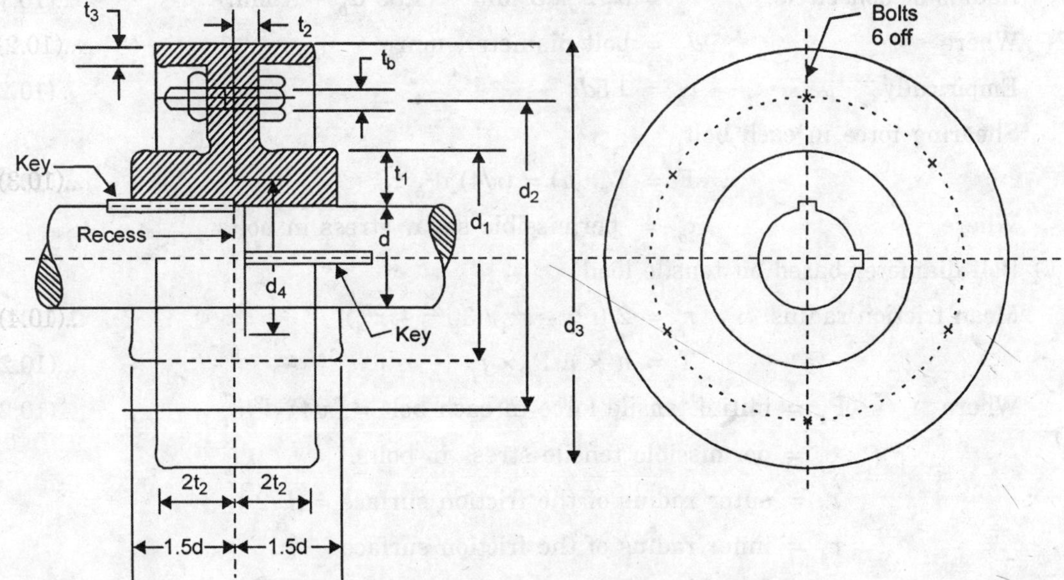

Fig. 10.3 Protected type rigid flange coupling

(*i*) Shaft diameter

Meantorque, $T_m = P \times 60 \times 10^3/(2\pi \times N)$ ...(10.7)

Maximum torque, $T_{max} = (1 + \%\ \text{overload factor}) \times T_m$ ...(10.8)

$(\pi/16)d^3T_8 = T_{max}$ ...(10.9)

Where τ_s = permissible shear stress in shaft

(*ii*) Hub diameter – The hub is designed as a hollow shaft subjected to torsion. Thus

$T_{max} = (\pi/16)\ [(d^4{}_1 - d^4)/d_1]\tau_h$ ...(10.10)

Where τ_h = permissible shear stress in hub

Empirically, $d_1 = 2d$ ...(10.11)

Thickness of hub, $t_1 = (d_1 - d)/2$ ...(10.12)

Diameter of recess in flanges,

$d_4 = d + t_1$ ...(10.13)

Outside diameter of protecting flange,

$d_3 = 4d$ ...(10.14)

(*iii*) Hub length – The hub length is made equal to the key length required.

Width of key, $b = d/4$ ...(10.15)

Lengthofkey, $l = 1.5d$ ...(10.16)

For a rectangular key, $t = d/6$ ...(10.17)

(*iv*) Number of bolts – $n = 4d/150 + 3$ ...(10.18)

Where d = shaft diameter, mm

(*v*) Bolt diameter – The bolts are in single shear and are designed to carry its own share of load.

Radius of bolt circle, $r_2 = d_1/2 + 5\text{ mm} + (1.85\ d_b + 8\text{ mm})$ ...(10.19)

Where d_b = bolt diameter, mm

Empirically, $r_2 = 1.5d$ ...(10.20)

Shearing force in each bolt,

$F = T/(r_2 n) = (\pi/4)\ d^2{}_b\ \tau_b$ ...(10.21)

Where τ_b = permissible shear stress in bolts.

(*vi*) Bolt diameter based on tensile load

Mean friction radius, $r_f = 2\ (r^3{}_3 - r^3{}_4)/[3(r^2{}_3 - r^2{}_4)]$ ...(10.22)

$T = n \times \mu\ P_i \times r_f$ ...(10.23)

Where P_1 = initial tensile force in each bolt = $(\pi/4)\ d^2{}_b\sigma_t$ ...(10.24)

σ_t = permissible tensile stress in bolts.

r_3 = outer radius of the friction surface = $d_3/2$

r_4 = inner radius of the friction surface

μ = coefficient of friction between flanges

(vii) Flange thickness – Considering the bearing failure of flange,

$$F = n \times d_b \times \tau_2 \times \sigma_c \quad ...(10.25)$$

Where σ_c = permissible bearing stress in flange.

Empirically, $$t_2 = 0.5t_1 + 6 \text{ mm} \quad ...(10.26)$$

Where $$t_1 = (d_1 - d)/2 = \text{hub thickness} \quad ...(10.27)$$

Considering shearing of the flange at the junction with the hub,

$$2(\pi d^2{}_1/4) \times \tau \times t_2 = T \quad ...(10.28)$$

(viii) Check for crushing of bolt

Considering crushing of the bolts,

$$nd_b t_2 \, \sigma_{cb} \times d_2/2 = T \quad ...(10.29)$$

Where σ_{cb} = permissible crushing strength of bolts.

(ix) Thickness of protecting flange,

$$t_3 = 0.5t_2 \text{ (min)} \quad \text{or} \quad 0.25d \quad ...(10.30)$$

(x) Hub overlap = 3 mm (min) ...(10.31)

5. Design of Bush Type Flexible Coupling (Fig. 10.4)

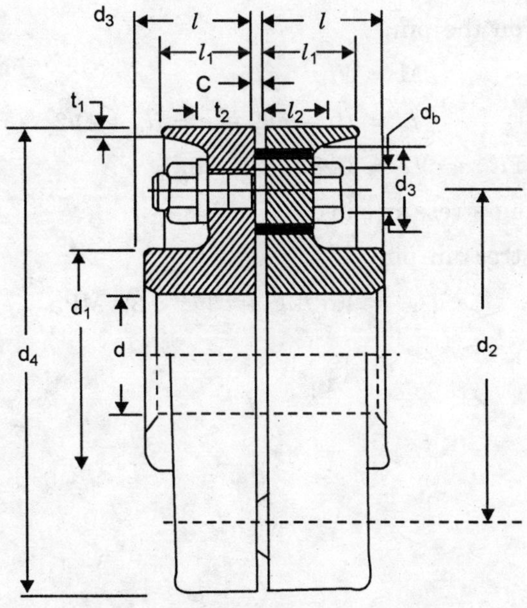

Fig. 10.4 Bush type flexible coupling

(*i*) Number of bolts, $n = 0.04d + 3$ ...(10.32)

where d = diameter of shaft, mm

(*ii*) Outer diameter of hub $d_1 = 2d$ ...(10.33)

(*iii*) Diameter of bolts $d_b = 0.5\ d/\sqrt{n}$ ...(10.34)

Shear force in each bolt, $F = 2T/(nd_2)$

Where d_2 = pitch circle diameter of bolts ...(10.35)

$(\pi/4)\ d^2{}_b\tau = F$ ...(10.36)

Where τ = permissible shear stress in bolts

(*iv*) Outer diameter of rubber bush

d_3 = diameter of pin + 2 × thickness of rubber bush

(*v*) Length of bush

Considering the bearing of bush,

$F = l_2 d_3 p_b$ ...(10.38)

Where p_b = bearing pressure for bush = 2 MPa

(*vi*) Check for pin in bending

Let c = clearance between two flanges. The loaded pin is shown in Fig. 10.5.

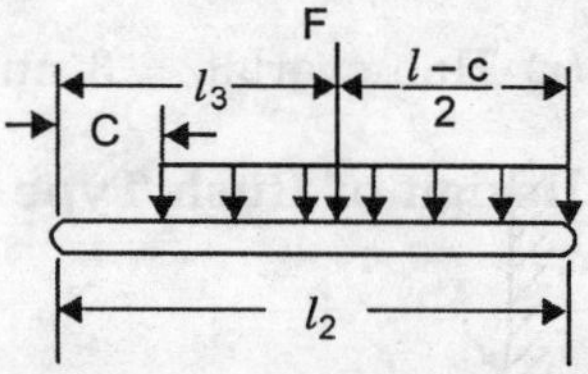

Bending moment on the pin,

$M = Fl_3$

Where $l_3 = (l_2 - c)/2 + c = (l_2 + c)/2$

$F(l_2 + c)/2 = (\pi/32)\ d^3{}_b \times \sigma_b$ ...(10.39)

Where σ_b = bending stress in pin

(*vii*) Maximum shear stress in pin

$\tau_{max} = [(\sigma_b/2)^2 + \tau^2]^{1/2} < 35$ MPa ...(10.40)

Fig. 10.5 Loading on the pin

Table 10.1: Dimensions of Parallel Keys and Keyways

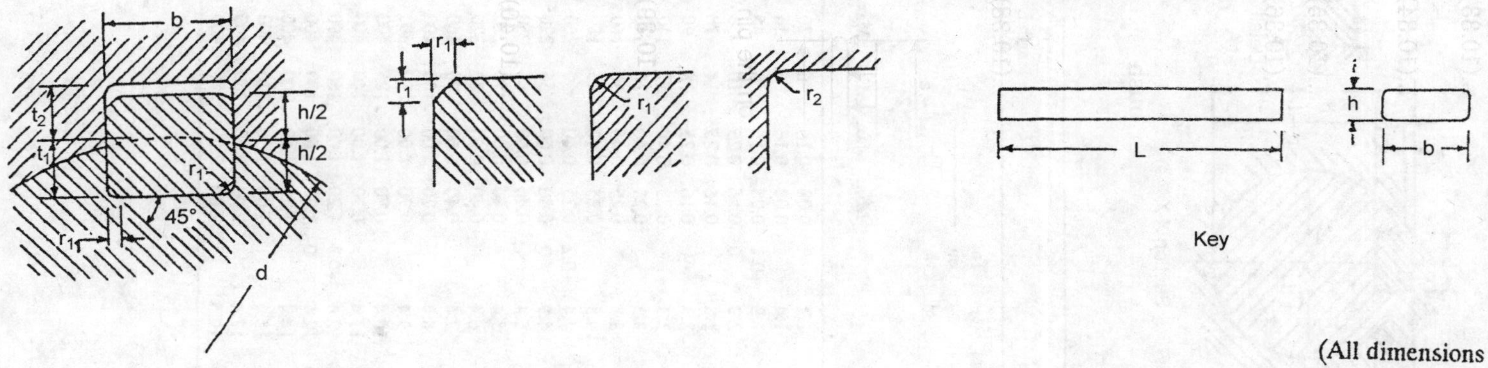

(All dimensions in mm)

For shaft diameters	Above	6	8	10	12	17	22	30	38	44	50	58	65	75	85	95	110	130	150	170	200	230	260	290	330	380	440
	Upto	8	10	12	17	22	30	38	44	50	58	65	75	85	110	150	130	150	170	200	230	260	290	330	380	440	500
Key cross-section	Width, b	2	3	4	5	6	8	10	12	14	16	18	20	22	25	28	32	36	40	45	50	56	63	70	80	90	100
	Height, h	2	3	4	5	6	7	8	8	9	10	11	12	14	14	16	18	20	22	25	28	32	32	36	40	45	50
Keyway depth (Nominal)	in Shaft, t_1	1.2	1.8	2.5	3.0	3.5	4.0	5	5	5.5	6	7	7.5	8.5	9.0	10	11	12	13	15	17	19	20	22	25	28	31
	in Hub, t_2	1	1.4	1.8	2.3	2.8	3.3	3.3	3.3	3.8	4.3	4.4	4.9	5.4	5.9	6.4	7.4	8.4	9.4	10.4	11.4	12.4	13.4	14.4	15.4	17.4	19.5
Tolerance on Keyway depth	t_1	+0.05 −0.00							+0.1 −0.0											+0.15 −0.00							
	t_2	+0.05 −0.00							+0.1 −0.0											+0.15 −0.00							
Chamfer or radius of key	r_1 Max		0.25		0.35				0.55							0.80				1.30			2.00			2.95	
	Min		0.16		0.25				0.40							0.60				1.00			1.60			2.50	
Keyway radius	r_2 Max		0.16		0.25				0.40							0.60				1.00			1.60			2.50	
Length of* key	L Min	6	6	8	10	14	18	22	28	36	45	50	56	63	71	80	90	100	110	125	140	160	180	200	220	250	280
	Max	20	36	45	50	71	90	110	140	160	180	200	220	250	280	320	360	400	400	400	400	400	400	400	400	400	400

IS : 2048—1962

Table 10.2: Dimensions of Parallel Keys and Keyways (All dimensions in mm)

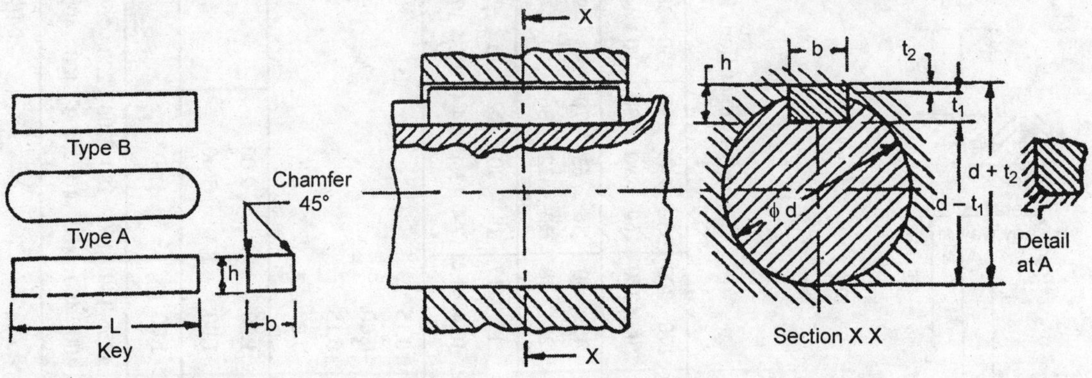

Range of Shaft dia d		Key -		Keyway											Range of Key length L	
Above	Up to	b × h	b	Tol on b					t_1	Tol on t_1	t_2	Tol on t	t			
				Running fit		Light drive fit		Force fit								
				Shaft H9	Hub D10	Shaft N9	Hub Js9	Shaft and hub P9					Min.	Max.	Min.	Max.
6	8	2 × 2	2	+0.025	+0.060	−0.004	+0.012 5	−0.006	1.0		1.2		0.08	0.16	6	20
8	10	3 × 3	3	0	+0.020	−0.029	−0.012 5	−0.031	1.8		1.4		0.08	0.16	6	36
10	12	4 × 4	4						2.5	+0.1	1.8	+0.1	0.08	0.16	8	45
12	17	5 × 5	5	+0.030	+0.078	0	+0.015 0	−0.012	3.0	0	2.3	0	0.16	0.25	10	56
17	22	6 × 6	6	0	+0.030	−0.030	−0.015 0	−0.042	3.5		2.8		0.16	0.25	14	70
22	30	8 × 7	8	+0.036	+0.098	0	+0.018 0	−0.015	4.0		3.3		0.16	0.25	18	90
30	38	10 × 8	10	0	+0.040	−0.036	−0.018 0	−0.051	5.0		3.3		0.25	0.40	22	110
38	44	12 × 8	12						5.0		3.3		0.25	0.40	28	140
44	50	14 × 9	14	+0.043	+0.120	0	+0.021 5	−0.018	5.5		3.8		0.25	0.40	36	160
50	58	16 × 10	16	0	+0.050	−0.043	−0.021 5	−0.061	6.0		4.3		0.25	0.40	45	180
58	65	18 × 11	18						7.0	+0.2	4.4	+0.2	0.25	0.40	50	200
65	75	20 × 12	20						7.5	0	4.9	0	0.40	0.60	56	220
75	85	22 × 14	22	+0.052	+0.149	0	+0.026 0	−0.022	9.0		5.4		0.40	0.60	63	250
85	95	25 × 14	25	0	+0.065	−0.052	−0.026 0	−0.074	9.0		5.4		0.40	0.60	70	280
95	110	28 × 16	28						10.0		6.4		0.40	0.60	80	320
110	130	32 × 18	32						11.0		7.4		0.40	0.60	90	360
130	150	36 × 20	36						12.0		8.4		0.70	1.00	100	400
150	170	40 × 22	40	+0.062	+0.180	0	+0.031 0	−0.026	13.0		9.4		0.70	1.00	110	400
170	200	45 × 25	45	0	+0.080	−0.062	−0.031 0	−0.088	15.0		10.4		0.70	1.00	125	400
200	230	50 × 28	50						17.0		11.4		0.70	1.00	140	400
230	260	56 × 32	56						20.0	+0.3	12.4	+0.3	1.20	1.60	160	400
260	290	63 × 32	63	+0.074	+0.220	0	+0.037 0	−0.032	20.0	0	12.4	0	1.20	1.60	180	400
290	330	70 × 36	70	0	+0.100	−0.074	−0.037 0	−0.106	22.0		14.4		1.20	1.60	200	400
330	380	80 × 40	80						25.0		15.4		2.00	2.50	220	400
380	440	90 × 45	90	+0.087	+0.260	0	+0.043 5	−0.037	28.0		17.4		2.00	2.50	250	400
440	500	100 × 50	100	0	+0.120	−0.087	−0.043 5		31.0		19.5		2.00	2.50	280	400

Table 10.3: Dimensions of Gib-head Keys and Keyways (All dimensions in mm)

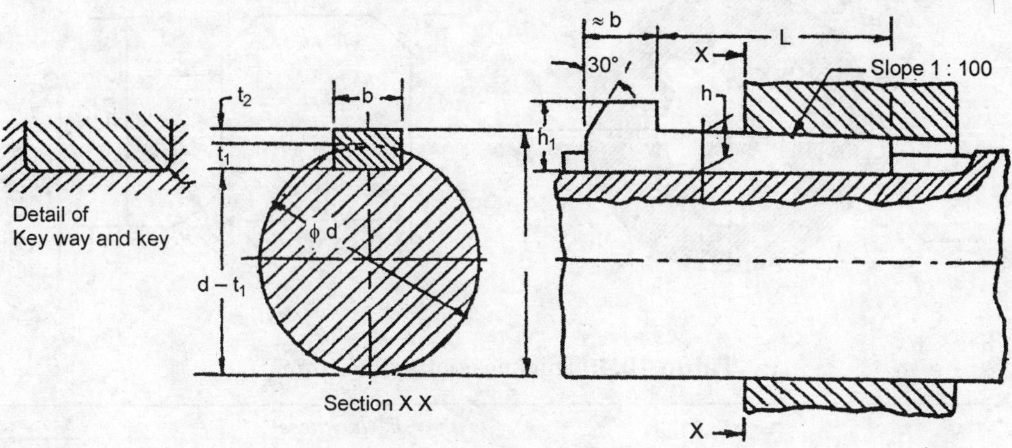

Range of Shaft Dia *d*		Key	Keyway									Range of Key length L		h_1
		b × h	*b*	*Tol on b*	t_1	*Tol on* t_1	t_2	*Tol on* t_2	*r*					
Above	*Up to*			*D10*					*Min.*	*Max.*		*Min.*	*Max.*	
10	12	4 × 4	4		2.5		1.2		0.08	0.16		14	45	7
12	17	5 × 5	5	+0.078	3.0	+0.1	1.7	+0.1	0.16	0.25		14	56	8
17	22	6 × 6	6	+0.030	3.5	0	2.2	0	0.16	0.25		16	70	10
22	30	8 × 7	8		4.0		2.4		0.16	0.25		20	90	11
30	38	10 × 8	10	+0.098	5.0		2.4		0.25	0.40		25	110	12
38	44	12 × 8	12	+0.040	5.0		2.4		0.25	0.40		32	140	12
44	50	14 × 9	14		5.5		2.9		0.25	0.40		40	160	14
50	58	16 × 10	16	+0.120	6.0		3.4		0.25	0.40		45	180	16
58	65	18 × 11	18	+0.050	7.0	+0.2	3.4	+0.2	0.25	0.40		50	200	18
65	75	20 × 12	20		7.5	0	3.9	0	0.40	0.60		56	220	20
75	85	22 × 14	22		9.0		4.4		0.40	0.60		63	250	22
85	95	25 × 14	25	+0.149	9.0		4.4		0.40	0.60		70	280	22
95	110	28 × 16	28	+0.065	10.0		5.4		0.40	0.60		80	320	25
110	130	32 × 18	32		11.0		6.4		0.40	0.60		90	360	28
130	150	36 × 20	36		12.0		7.1		0.70	1.00		100	400	32
150	170	40 × 22	40	+0.180	13.0		8.1		0.70	1.00		110	400	36
170	200	45 × 25	45	+0.080	15.0		9.1		0.70	1.00		125	400	40
200	230	50 × 28	50		17.0		10.1		0.70	1.00		140	400	45
230	260	56 × 32	56		20.0	+0.3	11.1	+0.3	1.20	1.60				50
260	290	63 × 32	63		20.0	0	11.1	0	1.20	1.60				50
290	330	70 × 36	70	+0.220	22.0		13.1		1.20	1.60				56
330	380	80 × 40	80	+0.100	25.0		14.1		2.00	2.50				63
380	440	90 × 45	90	+0.260	28.0		16.1		2.00	2.50				70
440	500	100 × 50	100	+0.120	31.0		18.1		2.00	2.50				80

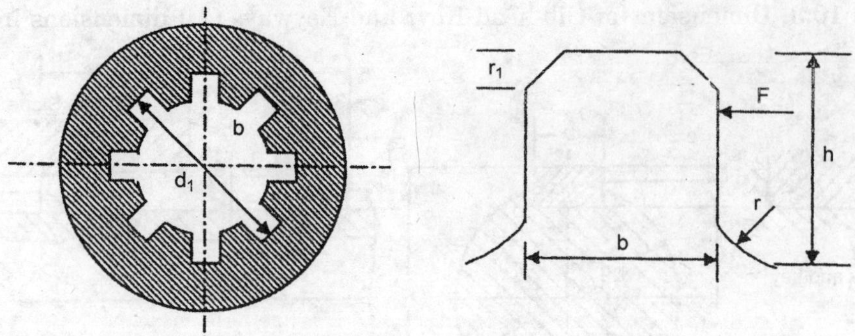

Table 10.4: Thickness of spline, mm

Number of splines	*b*	*h (thickness)*		
		permanent fit	*slides when loaded*	*slids when unloaded*
4	2.241d	0.075d	–	0.125d
6	0.250d	0.050d	0.100d	0.075d
10	0.156d	0.045d	0.095d	0.070d
16	0.098d	0.045d	0.095d	0.070d

Table 10.5: Bearing pressure (steel-steel)

Fit of spline	*Permanent fit*	*Slides when unloaded*	*Slides when loaded*
p	21 MPa	14 MPa	7 MPa

Table 10.6: Dimensions and Tolerances for Straight-sided Spines

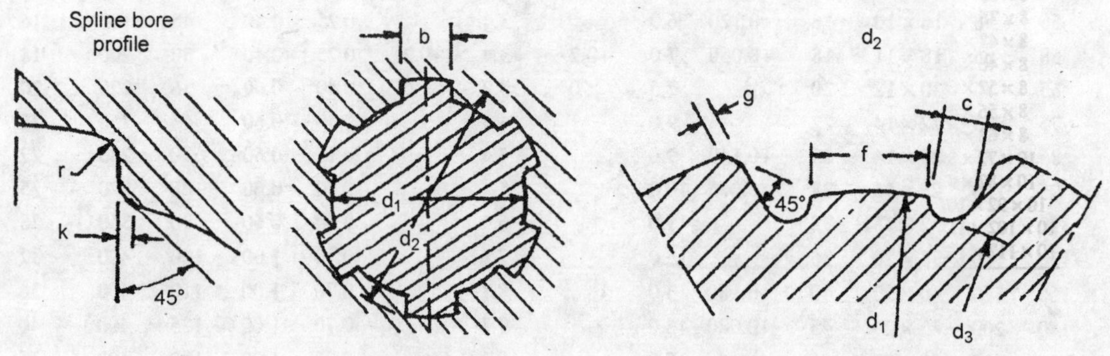

Light Duty Series

Nominal size $n \times d_1 \times d_2$	No. of spline keys n	Centering	d_1	d_2	b	* d_3 Min.	* e Max.	* f	g Max.	k Max.	r Max.
6 × 23 × 26		On	23	26	6	22.1	1.25	3.54	0.3	0.3	0.2
6 × 26 × 30	6	Inside	26	30	6	24.6	1.84	3.85	0.3	0.3	0.2
6 × 28 × 32		diameter	28	32	7	26.7	1.77	4.03	0.3	0.3	0.2
8 × 32 × 36			32	36	6	30.42	1.89	2.71	0.4	0.4	0.3
8 × 36 × 40			36	40	7	34.5	1.78	3.46	0.4	0.4	0.3
8 × 42 × 46			42	46	8	40.4	1.68	5.03	0.4	0.4	0.3
8 × 46 × 50	8	Inside	46	50	9	44.62	1.61	5.75	0.4	0.4	0.3
8 × 52 × 58		diameter	52	58	10	49.7	2.72	4.89	0.5	0.5	0.5
8 × 56 × 62		or Flanks	56	62	10	53.6	2.76	6.38	0.5	0.5	0.5
8 × 62 × 68			62	68	12	59.82	2.48	7.31	0.5	0.5	0.5
10 × 72 × 78			72	78	12	69.6	2.54	5.45	0.5	0.5	0.5
10 × 82 × 88			82	88	12	79.32	2.67	8.62	0.5	0.5	0.5
10 × 92 × 98	10		92	98	14	89.44	2.36	10.08	0.5	0.5	0.5
10 × 102 × 108			102	108	16	99.9	2.23	11.49	0.5	0.5	0.5
10 × 112 × 120			112	120	18	108.8	3.23	10.72	0.5	.5	0.5

Medium Duty Series:

Nominal size $n \times d_1 \times d_2$	No. of spline keys n	Centering	d_1	d_2	b	* d_3 Min.	* e Max.	* f	g Max.	k Max.	r Max.
6 × 11 × 14			11	14	3	9.9	1.55	–	0.3	0.3	0.2
6 × 13 × 16			13	16	3.5	12.0	1.5	0.32	0.3	0.3	0.2
6 × 16 × 20			16	20	4	14.54	2.1	0.16	0.3	0.3	0.2
6 × 18 × 22		On	18	22	5	16.7	1.95	0.45	0.3	0.3	0.2
6 × 21 × 25	6	Inside	21	25	5	19.5	1.98	1.95	0.3	0.3	0.2
6 × 23 × 28		Diameter	23	28	6	21.3	2.3	1.34	0.3	0.3	0.2
6 × 26 × 32			26	32	6	23.4	2.94	1.65	0.4	0.4	0.3
6 × 28 × 34			28	34	7	25.9	2.94	1.70	0.4	0.4	0.3
8 × 32 × 38			32	38	6	29.4	3.3	0.15	0.4	0.4	0.3
8 × 36 × 42			36	42	7	33.5	3.01	1.02	0.4	0.4	0.3
8 × 42 × 48			42	48	8	39.5	2.91	2.57	0.4	0.4	0.3
8 × 46 × 54	8	On	46	54	9	42.7	4.1	0.86	0.5	0.5	0.5
8 × 52 × 60		Inside	52	60	10	48.7	4.0	2.44	0.5	0.5	0.5
8 × 56 × 65		Diameter	56	65	10	52.2	4.74	2.5	0.5	0.5	0.5
8 × 62 × 72		or	62	72	12	57.8	5.0	2.4	0.5	0.5	0.5
10 × 72 × 82		Flanks	72	82	12	67.4	5.43	–	0.5	0.5	0.5
10 × 82 × 92			82	92	12	77.1	5.4	3.0	0.5	0.5	0.5
10 × 92 × 102	10		92	102	14	87.3	5.2	4.5	0.5	0.5	0.5
10 × 102 × 112			102	112	16	97.7	4.9	6.3	0.5	0.5	0.5
10 × 112 × 125			112	125	18	106.3	6.4	4.4	0.5	0.5	0.5

Heavy Duty Series:

Nominal size $n \times d_1 \times d_2$	No. of spline keys n	Centering	d_1	d_2	b	d_i Min.	g Max.	k Max.	r Max.
10 × 16 × 20			16	20	2.5	14	0.3	0.3	0.15
10 × 18 × 23			18	23	3	15.6	0.3	0.3	0.15
10 × 21 × 26		On	21	26	3	18.44	0.3	0.3	0.15
10 × 23 × 29	10	Inside	23	29	4	20.5	0.3	0.3	0.15
10 × 26 × 32		Diameter	26	32	4	23	0.4	0.4	0.15
10 × 28 × 35		or Flanks	28	35	4	24.4	0.4	0.4	0.25
10 × 32 × 40			32	40	5	28	0.4	0.4	0.25
10 × 36 × 45			36	45	5	31.3	0.4	0.4	0.25
10 × 42 × 52			42	52	6	36.9	0.5	0.4	0.4
10 × 46 × 56			46	56	7	40.9	0.5	0.5	0.4
10 × 52 × 60			52	60	5	47	0.5	0.5	0.4
16 × 56 × 65	16		56	65	5	50.6	0.5	0.5	0.4
16 × 62 × 72			62	72	6	56.1	0.5	0.5	0.4
16 × 72 × 82		On	72	82	7	65.9	0.5	0.5	0.4
20 × 82 × 92		Flanks	82	92	6	75.6	0.5	0.5	0.4
20 × 92 × 102			92	102	7	85.5	0.5	0.5	0.4
20 × 102 × 115	20		102	115	8	93.7	0.5	0.5	0.4
20 × 112 × 125			112	125	9	103.7	0.5	0.5	0.4

These values have been calculated on the basis that the spline shaft profiles are produced by the hobbing process.

Cotter Joints

(a) Spigot and Socket Joint (Fig. 11.1)

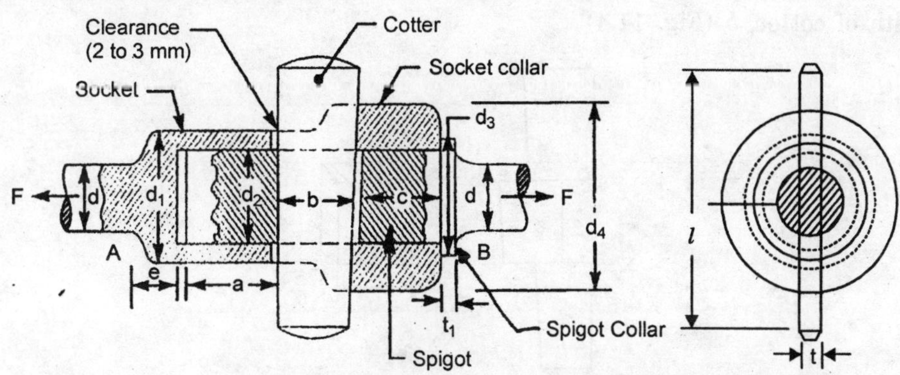

Fig. 11.1 Socket and spigot cotter joint.

1. Diameter of rods, d.

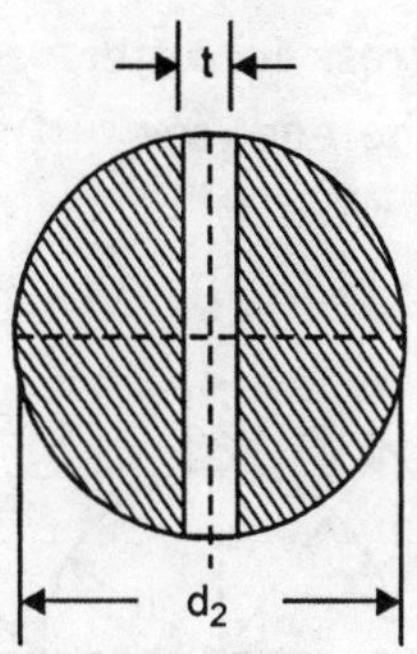

Fig. 11.2

$$d = [4F/(\pi\sigma_t)]^{1/2} \qquad \text{...(11.1)}$$

2. Diameter of spigot, d_2. (Fig. 11.2)

$$[(\pi/4)d^2{}_2 - d_2t]\ \sigma_t = F \qquad \text{...(11.2)}$$

Where t = thickness of cotter = $d_2/4$ ...(11.3)

3. Allowable compressive stress in rod or cotter.

$$\sigma_c = F/(d_2 t) \quad \text{...(11.4)}$$

4. Outside diameter of socket, d_1. (Fig. 11.3)

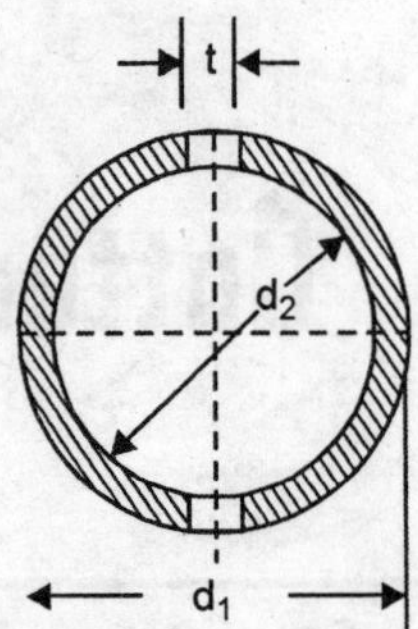

Fig. 11.3

$$[(\pi/4)(d^2_1 - d^2_2) - (d_1 - d_2)t]\ \sigma_t = F \quad \text{...(11.5)}$$

5. Width of cotter, b (Fig. 11.4)

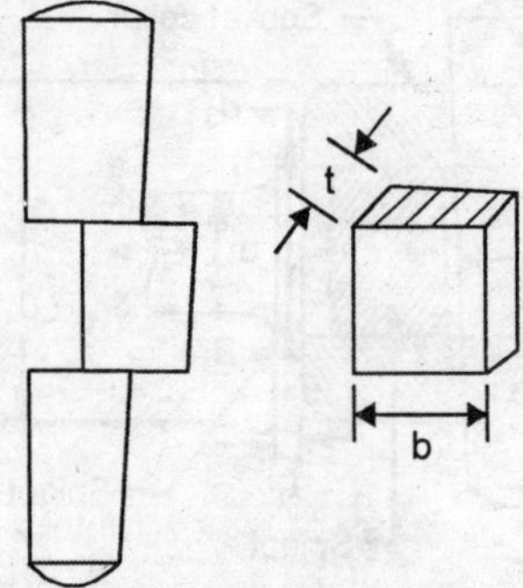

Fig. 11.4

$$b = F/(2t\tau) \quad \text{...(11.6)}$$

where τ = permissible shear stress in cotter

6. Length of spigot from slot to the end, a

$$a = F/(2d_2\tau) \quad \text{...(11.7)}$$

7. Socket collar diameter, d_4 (Fig. 11.5)

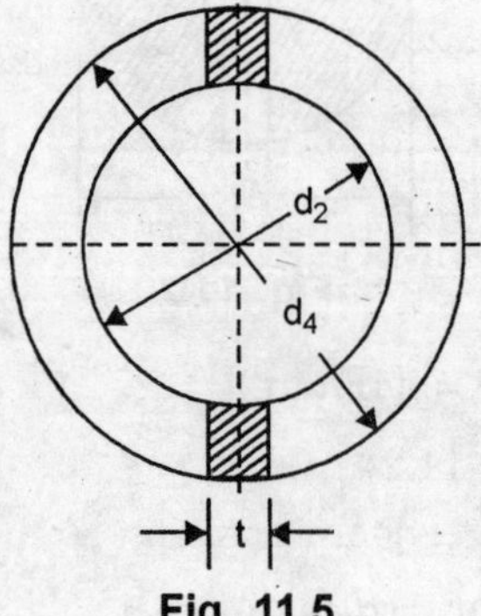

Fig. 11.5

$$d_4 = [F/(t\sigma_c)] + d_2 \quad \text{...(11.8)}$$

8. Socket collar thickness, c

$$c = F/[2(d_4 - d_2)\tau] \qquad ...(11.9)$$

9. Diameter of spigot collar, d_3 (Fig. 11.6)

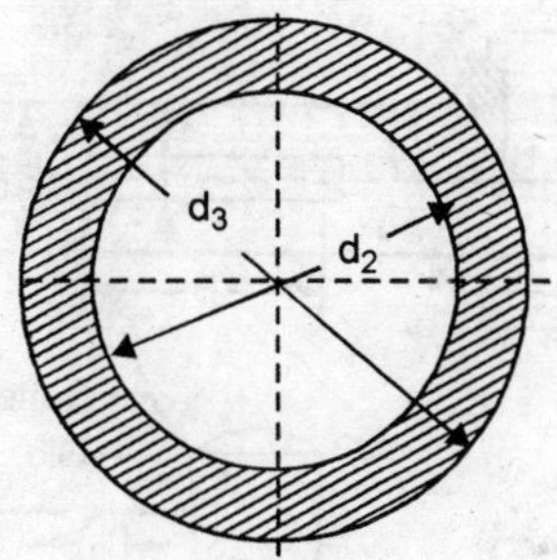

Fig. 11.6

$$[(\pi/4)\ (d^2{}_3 - d^2{}_3)\ \sigma_c = F \qquad ...(11.10)$$

10. Thickness of spigot collar, t_1 (Fig. 11.7)

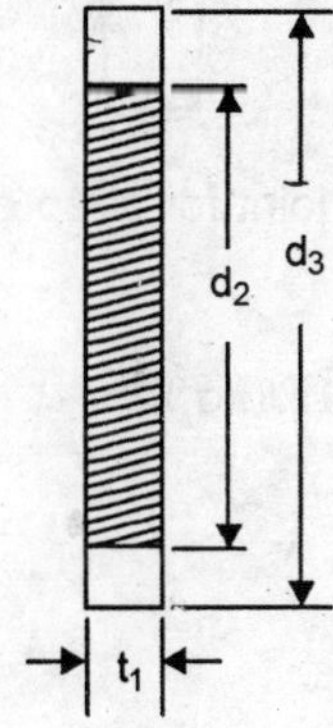

Fig. 11.7

$$t_1 = F/(\pi d_2 \tau) \qquad ...(11.11)$$

11. Bending stress in cotter. (Fig. 11.8)

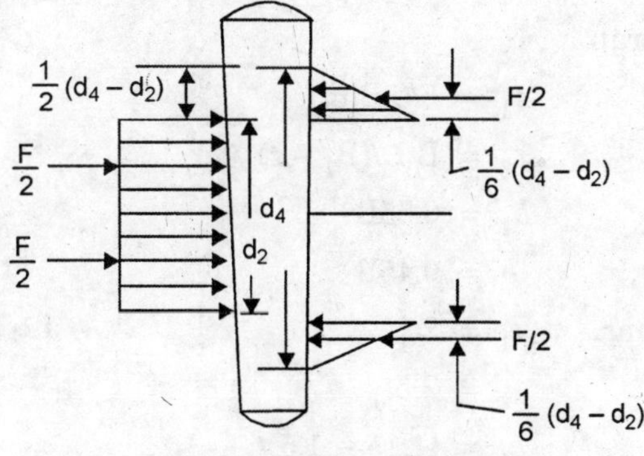

Fig. 11.8 Load distribution on the cotter.

Bending stress, $$\sigma_b = [F/(4tb^2)]\ (2d_4 + d_2) \qquad ...(11.12)$$

(b) Gib and Cotter Joint

(i) Strap End of Connecting Rod (Fig. 11.9)

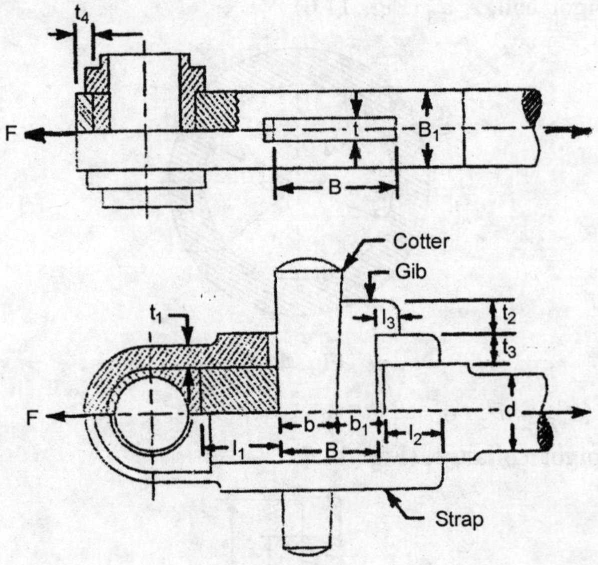

Fig. 11.9 Gib and cotter joint for strap end of a connecting rod.

1. Diameter of rod, d

$$d = [4F/(\pi\sigma_t)]^{1/2} \quad ...(11.13)$$

2. Width of strap, B_1

$$B_1 = d \quad ...(11.14)$$

3. Thickness of cotter, t

$$t = B_1/4 \quad ...(11.15)$$

4. Thickness of gib = t ...(11.16)

5. Height of gib head, $t_2 = t$ ...(11.17)

   Length of gib head, $l_3 = t$ ...(11.18)

6. Thickness of strap

$$t_1 = F/(2B_1\sigma t) \quad ...(11.19)$$

$$t_3 = B_1 t_1/(B_1 - t) \quad ...(11.20)$$

7. Width of gib, $b_1 = 0.55B$ ...(11.21)

   Width of cotter, $b = 0.45B$ ...(11.22)

8. Other dimensions. $l_1 = 2t_1$ ...(11.23)

$$l_2 = 2.5t_1 \quad ...(11.24)$$

$$t_4 = (1.15 - 1.5)t_1 \quad ...(11.25)$$

(ii) For Square Rods (Fig. 11.10)

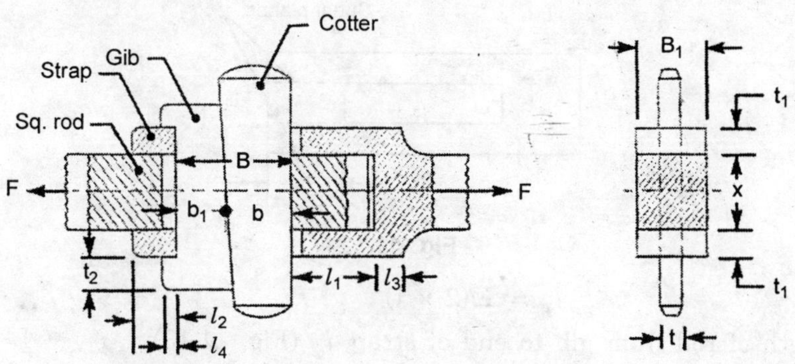

Fig. 11.10 Gib and cotter joint for square rods.

1. Size of rod, x

$$x = (F/\sigma_t)^{1/2} \qquad ...(11.26)$$

2. Width of strap, $B_1 = x$ (Fig. 11.11) ...(11.27)

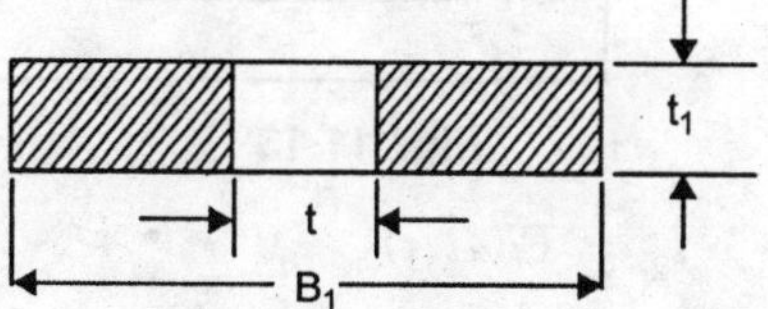

Fig. 11.11

Thickness of cotter, $t = B_1/4$ ...(11.28)

Thickness of gib $= t$ ...(11.29)

Height of gib, $t_2 = t$ ...(11.30)

Length of gib head, $l_4 = t$ ...(11.31)

3. Width of gib and cotter

$$B = F/(2t\tau) \qquad ...(11.32)$$

Width of gib, $b_1 = 0.55B$ for a single gib ...(11.33*a*)

$= 0.30B$ for a double gib ...(11.33*b*)

Width of cotter, $b = 0.45B$ for a single gib ...(11.34*a*)

$= 0.40B$ for a double gib ...(11.34*b*)

4. Thickness of strap, t_1

$$t_1 = F/[2(x - t)\sigma_t] \qquad ...(11.35)$$

5. Check for strap or gib in crushing

$$\sigma_c = F/(2tt_1) \qquad ...(11.36)$$

6. Length of rod from cotter to end, l_1 (Fig. 11.12)

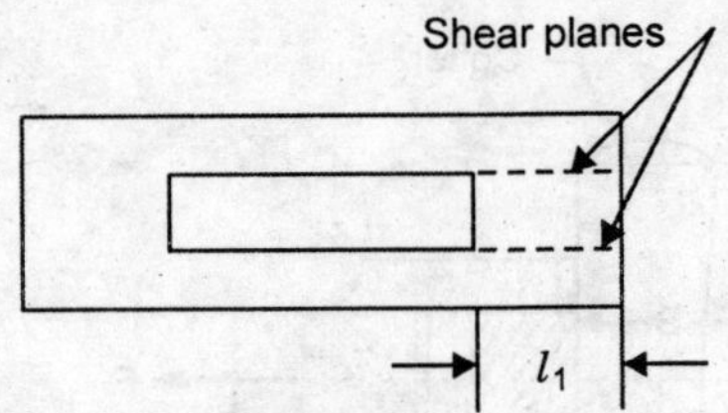

Fig. 11.12

$$l_1 = F/(2 \times \tau) \qquad ...(11.37)$$

7. Length of rod from gib to end of strap, l_2 (Fig. 11.13)

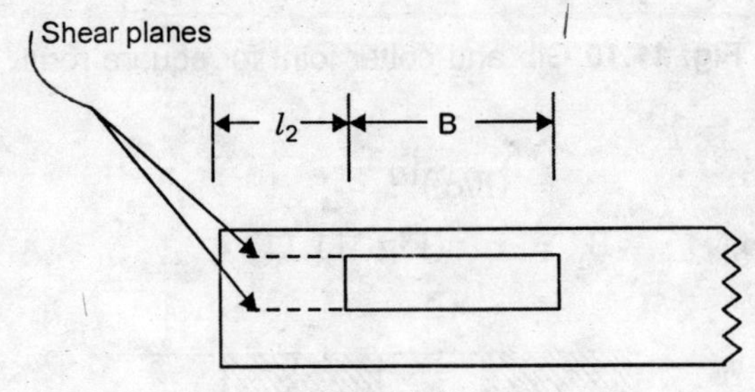

Fig. 11.13

$$l_2 = F/(4t_1\tau) \qquad ...(11.38)$$

8. Other dimensions.

$$l_3 = 2x/3 \qquad ...(11.39)$$

CHAPTER

12

Kunckle Joint

Knuckle Joint (Fig. 12.1)

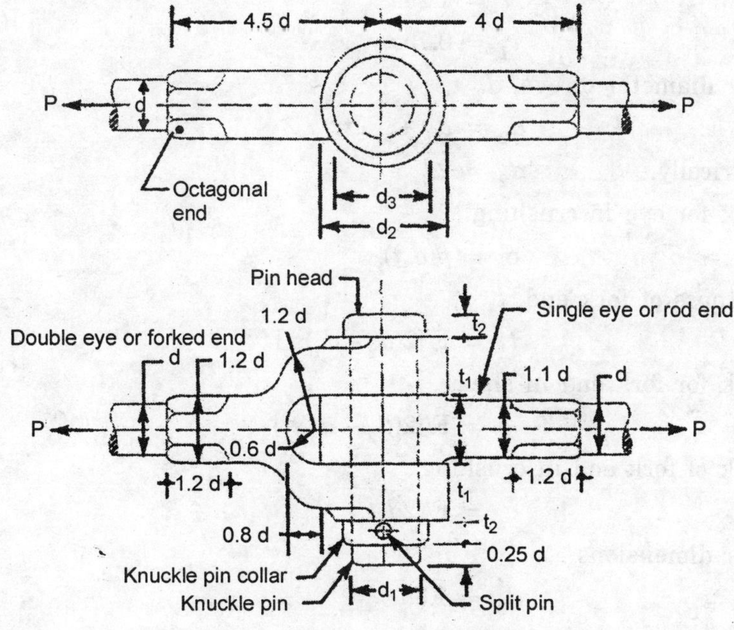

Fig. 12.1 Kunckle joint.

1. Diameter of rod, d

$$d = [4F/(\pi\sigma_t)]^{1/2} \qquad ...(12.1)$$

2. Diameter of knuckle pin, d_1

$$d_1 = [2F/(\pi\tau)]^{1/2} \qquad ...(12.2a)$$

Empirically, $$d_1 = d \qquad ...(12.2b)$$

3. Bending stress in pin (Fig. 12.2)

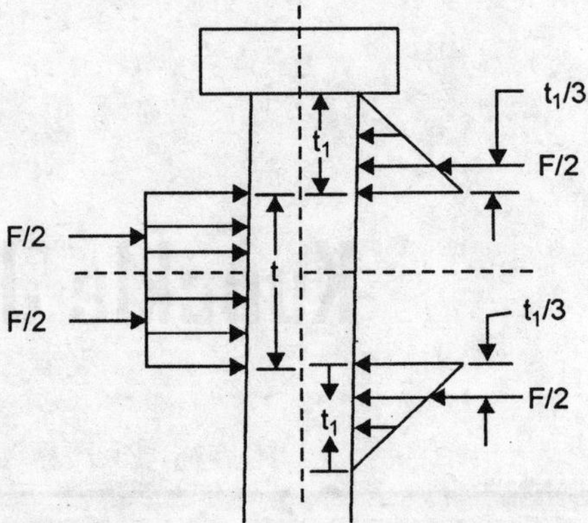

Fig. 12.2 Load distribution on the pin.

$$\sigma_b = 16F(t_1/3 + t/4)/(\pi d^3{}_1) \quad ...(12.3)$$

$$t = 1.25d \quad ...(12.4)$$

$$t_1 = 0.75d \quad ...(12.5)$$

4. Outer diameter of eye, d_2

$$d_2 = d_1 + F/(t\tau) \quad ...(12.6a)$$

Empirically, $$d_2 = 2d \quad ...(12.6b)$$

5. Check for eye in crushing.

$$\sigma_c = f/(d_1 t) \quad ...(12.7)$$

6. Thickness of fork end, t_1

$$t_1 = F/[2(d_2 - d_1)\sigma_t] \quad ...(12.8)$$

7. Check for fork end in shear

$$\tau = F/[2(d_2 - d_1)t_1] \quad ...(12.9)$$

8. Check of fork end in crushing

$$\sigma_c = F/(2d_1 t_1) \quad ...(12.10)$$

9. Other dimensions

$$d_3 = 1.5d \quad ...(12.11)$$

$$t_2 = 0.5d \quad ...(12.12)$$

CHAPTER

13

Pipe Joints

1. Stresses in Pipes

Hoop stress in a thin pipe of diameter d and thickness t under internal pressure p,

$$\sigma_\theta = pd/(2t) \quad ...(13.1)$$

Maximum hoop stress in a thick pipe,

$$(\sigma_\theta)_{max} = p[(k^2 + 1)/(k^2 - 1)] \quad ...(13.2)$$

Where k = outside radius/inside radius = r_o/r_i

2. Thickness of Pipe

$$t = pd/(2\sigma_\theta) + c, \text{ for a thin pipe} \quad ...(13.3)$$

where c = corrosion allowance

$$t = r_i[\{(\sigma_\theta + p)/(\sigma_\theta - p)\}^{1/2} - 1], \text{ for a thick pipe} \quad ...(13.4)$$

3. Diameter of Pipe

$$d = [4Q/(\pi V)]^{1/2} \quad ...(13.5)$$

Where Q = discharge through the pipe, and V = velocity of fluid.

4. Design of Pipe Joints

(*i*) Circular Flanged Joint (Fig. 13.1)

1. Pipe thickness

$$t = p_i d_i/(2\sigma_t) + c \quad ...(13.6)$$

where $c \approx 9$ mm

2. Outside diameter of pipe

$$d_o = d_i + 2t \quad ...(13.7)$$

$$D_1 = d_o + \text{clearance between bolt hole edge and pipe} \quad ...(13.8)$$

3. Core diameter of bolts,

$$d_c = D_1[p_i/(n \times \sigma_t)]^{1/2} \quad ...(13.9)$$

Where D_1 = diameter of circle touching the bolt holes

n = number of bolts

p_i = inside fluid pressure

σ_t = permissible tensile stress in bolts

4. Pitch circle diameter of bolts

$D = d_o + d_b + 2 \times$ clearance between bolt head and pipe ($\approx$ 20 mm) ...(13.10*a*)

Exact value of

$$D_1 = D - d_b \quad ...(13.10b)$$

Tensile force in each bolt,

$$F = (1/n)\,[(\pi/4)\,D^2{}_1\,p_1] \quad ...(13.11)$$

Angle between two bolts,

$$\theta = 360°/n \quad ...(13.12)$$

Pitch of bolts,

$$p = \pi D/n \quad ...(13.13a)$$

$$= (20 \text{ to } 30)\sqrt{d_b} \text{ for a leak proof joint} \quad ...(13.13b)$$

5. Outside diameter of flange

$D_2 = D + 2d_b + 2 \times$ clearance between bolt head and flange edge ...(13.14)

6. Thickness of flange

Distance from outside of pipe to bolt centre,

$$a = (D - d_o)/2 \quad ...(13.15)$$

$$M_{max} = Fa \quad ...(13.16)$$

Depth of flange,

$$b = d_o \tan(\theta/2) \quad ...(13.17)$$

Section modulus of flange,

$$z = bt^2{}_f/6 \quad ...(13.18)$$

$$M_{max} = \sigma_b z \quad ...(13.19)$$

7. Width of flange

$$B = (D_2 - d_o)/2 \quad ...(13.20)$$

(*ii*) Oval Flanged Joint (Fig. 13.3)

1. Core diameter of bolts

$$d_c = D_1\,[p_i/(2\sigma_t)]^{1/2} \quad ...(13.21)$$

$$d_b \geq 12 \text{ mm}$$

2. Thickness of flange

$$t_f = [3Fa/(\sigma_t \times b)]^{1/2} \quad ...(13.22)$$

$$a = (D - d_o)/2 \quad ...(13.23)$$

b = width of flange at section tangential to pipe

3. Empirical dimensions

$$d_b = 0.75\ t + 10 \text{ mm} \quad ...(13.24)$$

$$t_f = 1.5\ t + 3 \text{ mm} \quad ...(13.25)$$

$$D_2 = d_i + 2t + 4.6\ d_b \quad ...(13.26)$$

$$D = D_2 - (3t + 20 \text{ mm}) \quad ...(13.27)$$

***(iii)* Square Flanged Pipe** (Fig. 13.5)

1. Core diameter of bolts

$$d_c = D_1 [p_i/(4\sigma_t)]^{1/2} \quad ...(13.28)$$

2. Thickness of pipe

$$t = (d_i/2)\ [\{(\sigma_t + p_i)/(\sigma_t - p_i)\}^{1/2} - 1] \quad ...(13.29)$$

3. Outside diameter of pipe

$$d_o = d_i + 2t \quad ...(13.30)$$

4. Minimum diagonal length of bolt centres

$l = d_o + 2d_b + 2 \times$ clearance between pipe and bolt head ...(13.31)

5. Distance between bolt centres,

$$l_1 = l/\sqrt{2} \quad ...(13.32)$$

6. Width of flange

$l_2 = l_1 + 2d_b + 2 \times$ clearance between nuts and flange endge ...(13.33)

7. Thickness of flange

Bending moment due to force in two bolts about section through pipe centre.

$$M_1 = FI_1$$

$$F = (\pi/16)\ D^2{}_1 p_i$$

Mean diameter of pipe threads,

$$d_m = (d_1 + d_o)/2$$

Centroid of radius of arc from pipe centre at which load due to fluid pressure is concentrated.

$$r = 0.6366 \times d_m/2$$

Bending moment due to fluid pressure

$$M_2 = 2Fr$$

Resultant bending moment,

$$M = M_1 - M_2$$

Width of flange at section A-A,

$$b = I_2 - d_o$$

$$t_f = [6M/(d\sigma_b)]^{1/2} \quad ...(13.34)$$

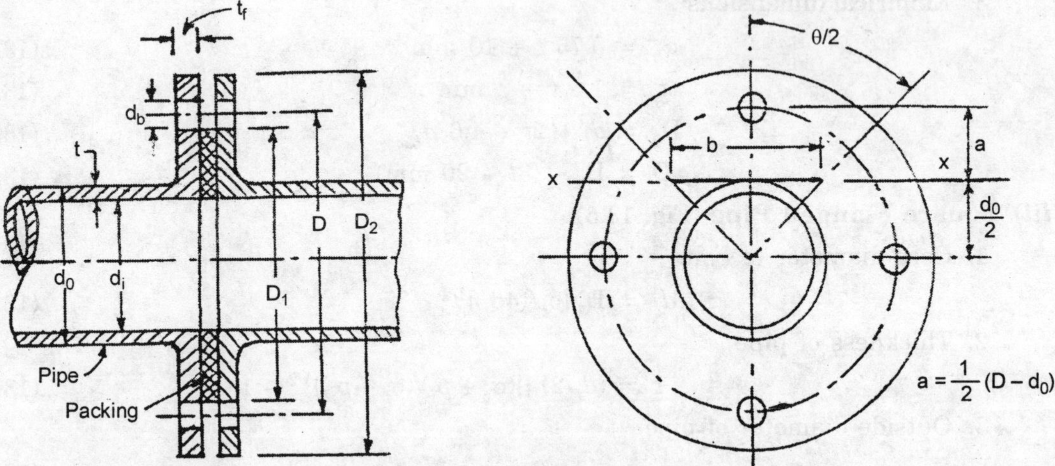

Fig. 13.1. Circular flanged joint

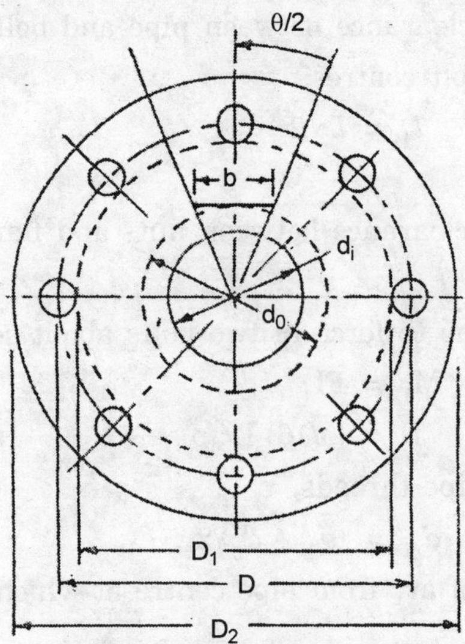

Fig. 13.2

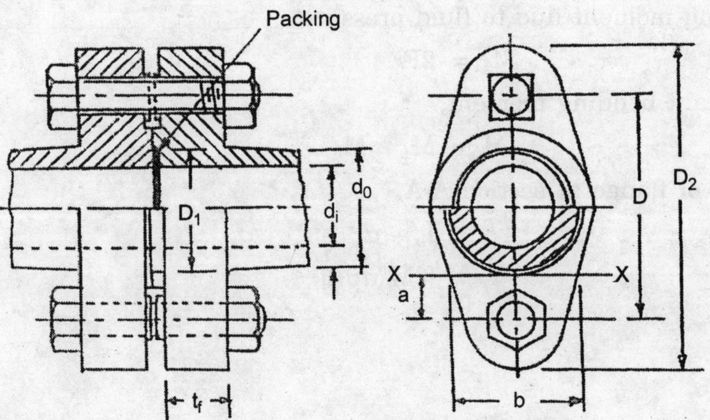

Fig. 13.3. Oval flanged pipe joint

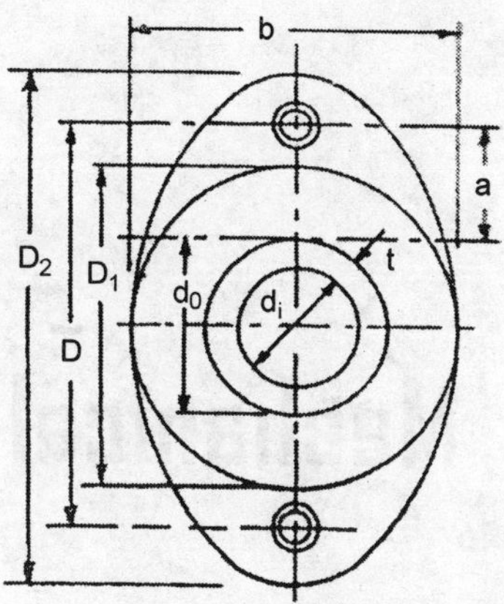

Fig. 13.4. Elliptical flange profile

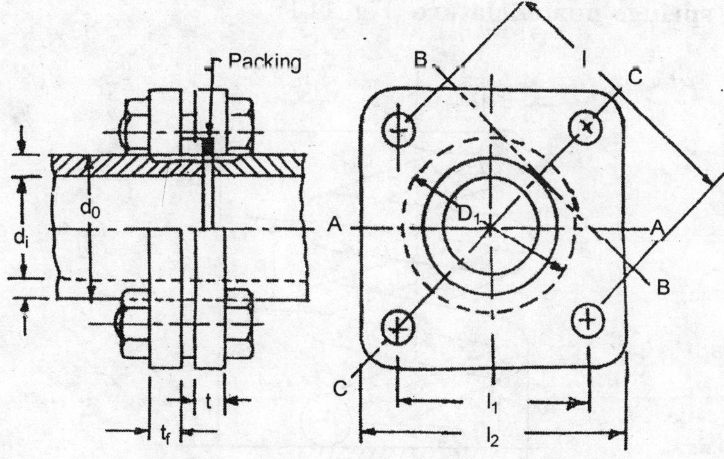

Fig. 13.5. Square flanged pipe joint

CHAPTER

14

Mechanical Springs

1. Compression springs nomenclature (Fig. 14.1)

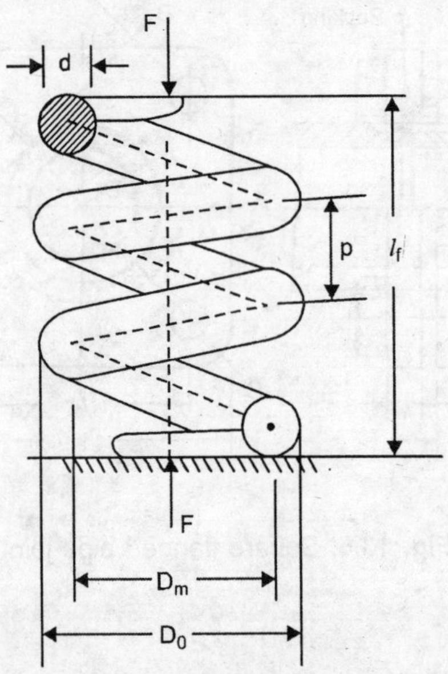

Fig. 14.1 Helical compression spring.

$$\text{Solid length, } l_s = id \quad \text{...(14.1)}$$

Where i = total number of coils

d = wire diameter.

Free length,

$$l_f = id + 1.15\ \delta_{max} \quad \text{...(14.2)}$$

Spring index

$$C = D_{mean}/d \qquad ...(14.3)$$

Where D_m = mean coil diameter

Spring stiffness (or rate),

$$k = F/\delta \qquad ...(14.4)$$

Coil pitch,

$$p = l_f/(i - 1) \qquad ...(14.5a)$$

$$= (l_f - l_s)/i + d \qquad ...(14.5b)$$

2. Design of Helical Compression Springs Against Static Load

(*i*) Stresses in the spring. (Fig. 14.2)

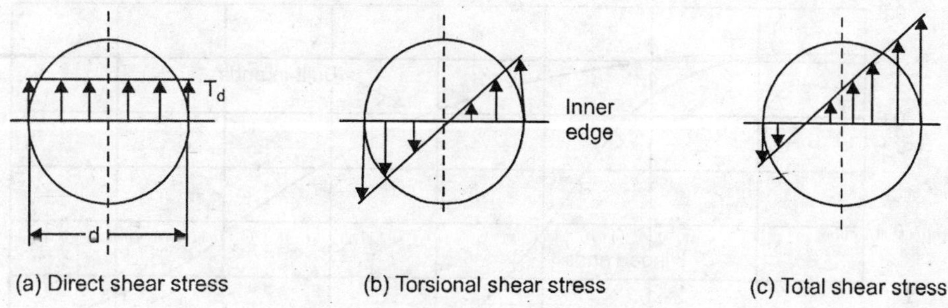

Fig. 14.2 Stresses in spring wire.

Direct shear stress in spring wire,

$$\tau_d = 4F/(\pi d^2) \qquad ...(14.6)$$

Torsional shear stress in the spring wire,

$$\tau_s = 8FD_m/(\pi d^3) \qquad ...(14.7a)$$

$$= 8FC/(\pi d^2) \qquad ...(14.7b)$$

Maximum shear stress in spring wire,

$$\tau_{max} = [8FD_m/(\pi d^3)]\,[1 + 0.5/C] \qquad ...(14.8a)$$

$$= K_s \times [8FD_m/(\pi d^3)] \qquad ...(14.8b)$$

$$= K_s\,[8FC/(\pi d^2)] \qquad ...(14.8c)$$

Shear stress factor,

$$K_s = 1 + 0.5/C \quad \text{(Table 14.1)} \qquad ...(14.9)$$

Wahl factor,

$$K_w = (4C - 1)/(4C - 4) + 0.615/C \quad \text{(Table 14.1)} \qquad ...(14.10)$$

$$K_w = K_s K_c \qquad ...(14.11)$$

Curvature correction factor,

$$K_c = K_w/K_s \quad \text{(Table 14.1)} \qquad ...(14.12)$$

Shear stress considering direct shear stress and curvature of wire,

$$\tau_{max} = K_w\,[8FC/(\pi d^2)] \qquad ...(14.13)$$

(*ii*) Deflection of spring,

$$\delta = 8FD^3_m\, i/(Gd^4) \quad ...(14.14a)$$

$$= 8FC^3 i/(Gd) \quad ...(14.14b)$$

(*iii*) Stiffness of spring,

$$k = Gd^4/(8D^3_m i) \quad ...(14.15a)$$

$$= Gd/(8C^3 i) \quad ...(14.15b)$$

(*iv*) Critical buckling load,

$$F_{cr} = kk_l l_f \quad ....(14.16)$$

Where k = spring stiffness

k_l = buckling factor depending upon the ratio l_f/D_m (Fig. 14.3)

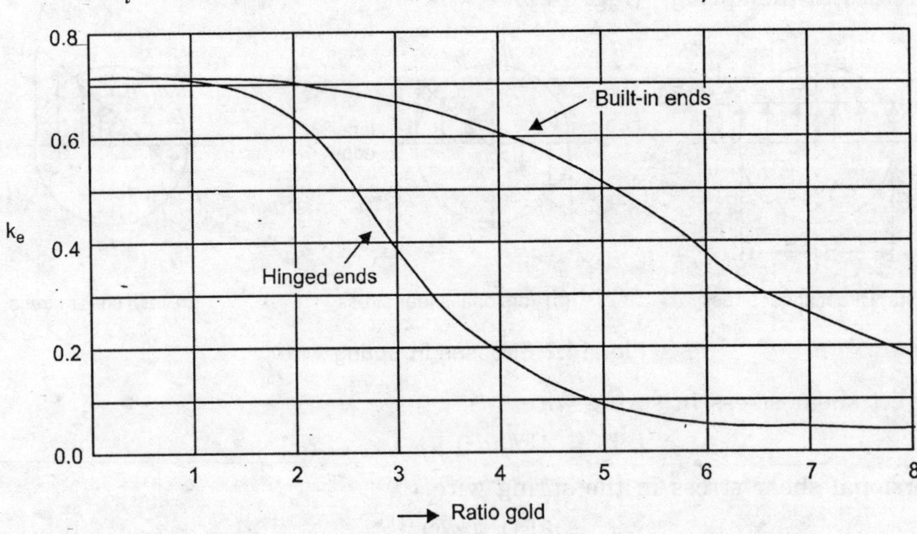

Fig. 14.3 Curve for buckling load correction factor.

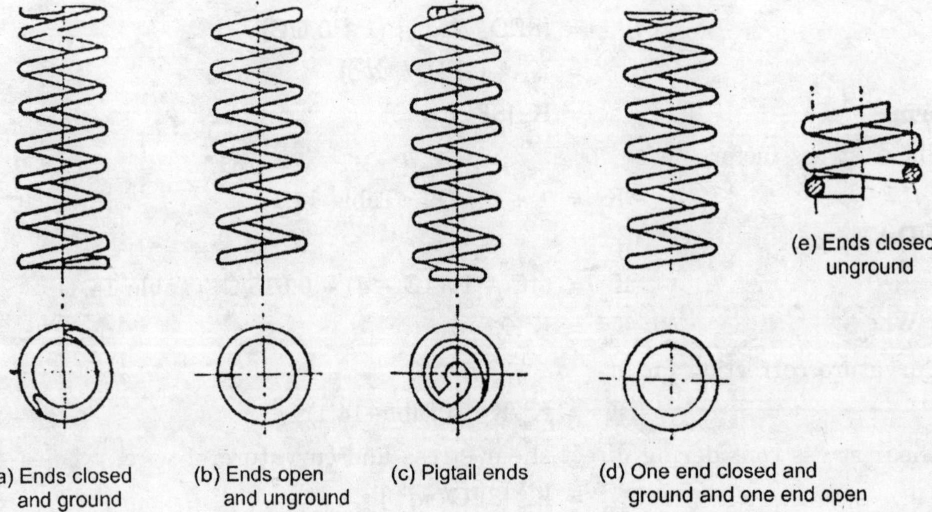

Fig. 14.4 Spring ends.

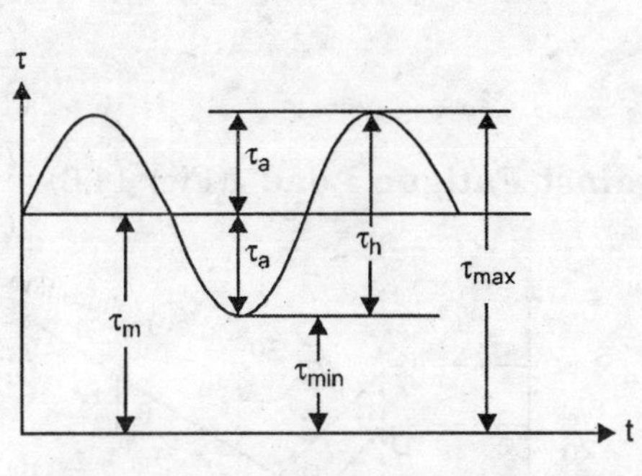

(a) Two-way shear loading

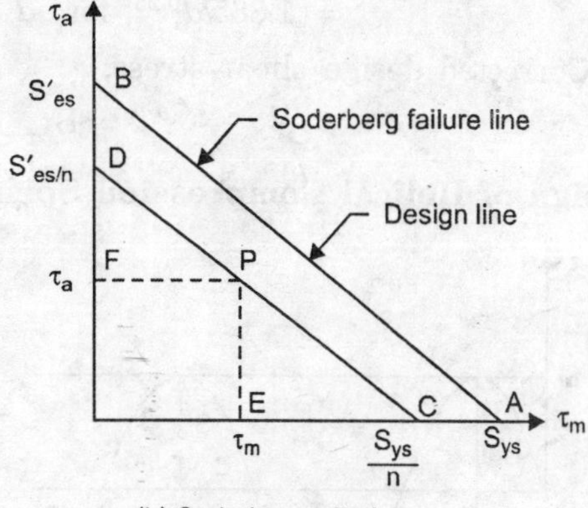

(b) Soderberg criterion

Fig. 14.5 Two-way shear design of spring wire.

l_f = free length of spring

For safe in buckling,

$$F_{cr} > F_{max} \quad \text{...(14.17)}$$

If $l_f/D_m \leq 2.6$, guide is not necessary ...(14.18*a*)

> 2.6, guide is necessary ...(14.18*b*)

(*v*) Surge in springs:

(*a*) Spring guided at both ends

$$f_n = [d/(\pi i D^2_m)] [G/(8\rho)]^{1/2} \text{ Hz} \quad \text{...(14.19a)}$$

(*b*) Spring has one end fixed and the other free

$$f_n = [d/(\pi i D^2_m)] [G/(32\rho)]^{1/2} \text{ Hz} \quad \text{...(14.19b)}$$

Where ρ = density of spring material, kg/m³.

For safe design,

$$f_n > 20 \times f \quad \text{...(14.20)}$$

Where f = forcing frequency

3. Permissible stresses for plain carbon steels spring wires,

$$S_{yt} = 0.75 S_{ut} \quad \text{...(14.21a)}$$

$$S_{ys} = 0.577 S_{yt} \quad \text{...(14.21b)}$$

Design shear stress,

$$\tau_d = S_{ys}/n \quad \text{...(14.22)}$$

When n = factor of safety

$\tau_s = 0.55 S_{ut}$ for alloy steels ...(14.23*a*)

$= 0.8 S_{ys}$ for light service $< 10^4$ cycles ...(14.23*b*)

$= 0.6 S_{ys}$ for average service $\geq 10^4$ and $< 10^5$ cycles ...(14.23*c*)

$= 0.4 S_{ys}$ for $\geq 10^6$ cycles ...(14.23*d*)

Size factor, $B = 0.86 + 1.75/d$ for $d < 12.5$ mm ...(14.24*a*)

$$= 1.885/d^{0.25} \quad \text{for} \quad d \geq 12.5 \text{ mm} \qquad ...(14.24b)$$

Corrected design shear stress,

$$\tau'_d = B\tau_d \qquad ...(14.25)$$

4. Design of Helical Compression Springs Against Fatigue Load (Fig. 14.6)

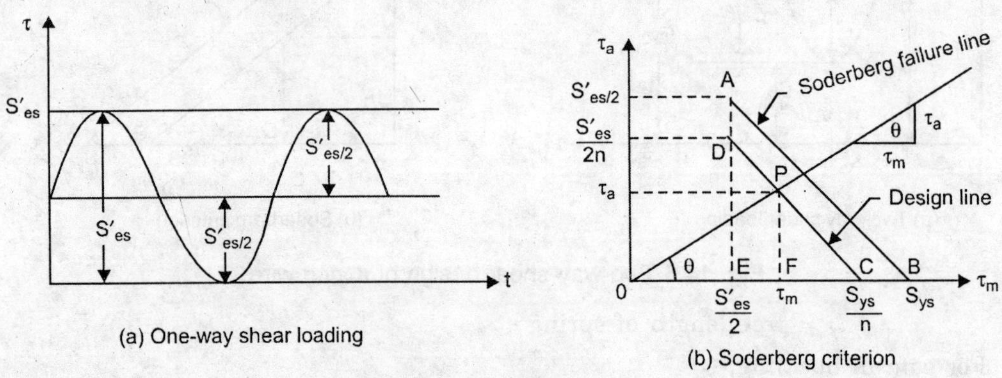

Fig. 14.6 One-way shear design of spring wire.

For one – way shear, Soderberg failure line is,

$$(\tau_m - \tau_a)/S_{ys} + 2\tau_a/S'_{es} = 1/n \qquad ...(14.26)$$

Where $\tau_m = K_s\,[8F_m D_m/(\pi d^3)]$

$\tau_a = K_w\,[8F_a D_m/(\pi d^3)]$

$F_m = (F_{max} + F_{min})/2$, and $F_a = (F_{max} - F_{min})/2$

The wire diameter,

$$d^3 = K_w[8F_m\,D_m/(\pi\tau_m)] \qquad ...(14.27)$$

The endurance limit of steel spring wires is given by,

$$S'_{es} = 0.21\,S_{ut} \text{ for cold drawn steel wires} \qquad ...(14.28a)$$

$$= 0.22\,S_{ut} \text{ for oil-hardened and tempered steel wires} \qquad ...(14.28b)$$

The yield shear strength of steel spring wires is given by,

$$S_{ys} = 0.42\,S_{ut} \text{ for cold drawn steel wires} \qquad ...(14.29a)$$

$$= 0.45\,S_{ut} \text{ for oil-hardened and tempered steel wires} \qquad ...(14.29b)$$

5. Design of Helical Extension

(i) Static Load

Initial tension,

$$F_i = F - \delta k \qquad ...(14.30a)$$

$$= F - Gd^4\delta/(8D_m^3 i_f) \qquad ...(14.30b)$$

The total number of coils in an extension spring with initial tension,

$$i_t = L_k/d - 1 \qquad ...(14.31)$$

where L_k = body length of coils when not loaded but subject to initial tension.

$$L_k = (i_t + 1)\,d_{max} \qquad ...(14.32)$$

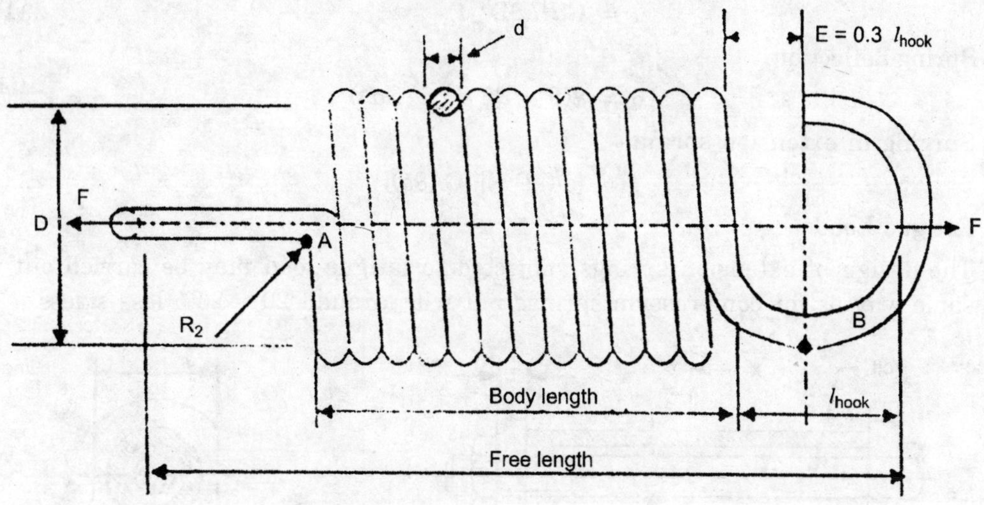

Fig. 14.7 Helical extension spring.

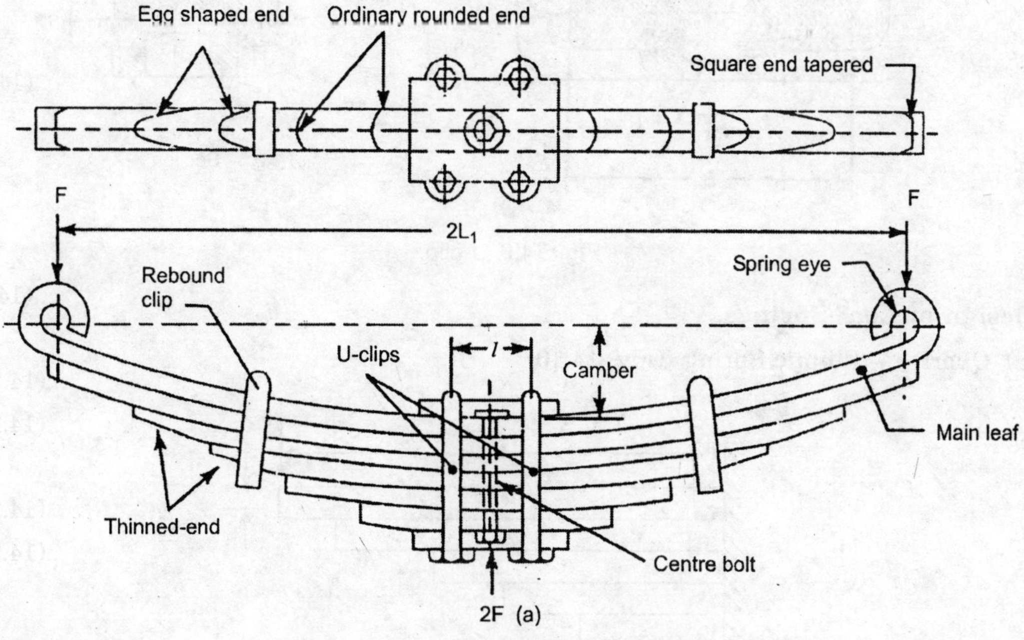

Fig. 14.8 Semi-elliptical ieaf spring.

The unloaded length of the spring L_i for equal loops measured between the inside edge of the hooks is given by,

$$L_i = L_k + 2L_h \quad ...(14.33)$$

Where L_h = distace of inside edge of hook from body of spring $\approx D_i$ ...(14.34)

Spring stiffness,

$$k = (F - F_i)/\delta \quad ...(14.35a)$$

$$= Gd^4/(8D^3_m i) \quad ...(14.35b)$$

Spring deflection,

$$\delta = 8(F - F_i)\, d^3_m i/(Gd^4) \quad ...(14.36)$$

Surging in extension spring,

$$f = [d/(\pi D^2 i)]\, [G/(8\rho)]^{1/2} \quad ...(14.37)$$

(*ii*) Fatigue Load

The design of extension springs subjected to fatigue load may be carried out in the same way as for compression springs but with around 20 – 25% less stress.

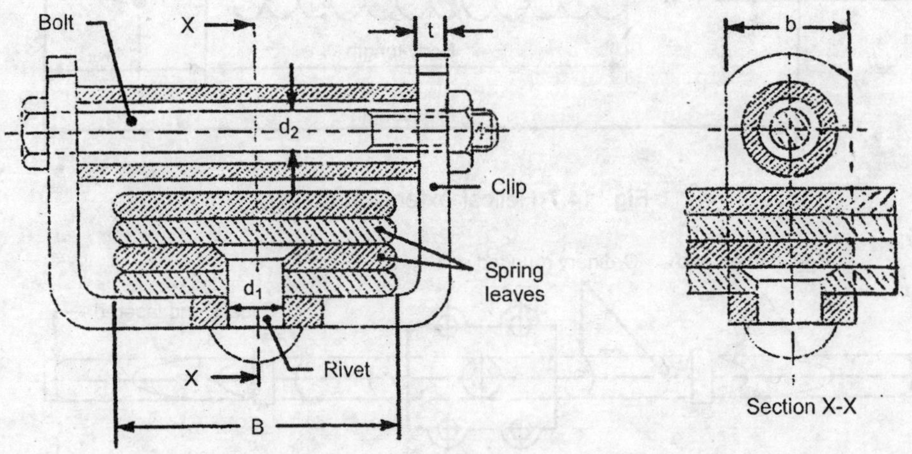

Fig. 14.9 U-clip.

6. Design of Leaf Springs

(*a*) Quarter – Elliptic Spring (Fig. 14.10)

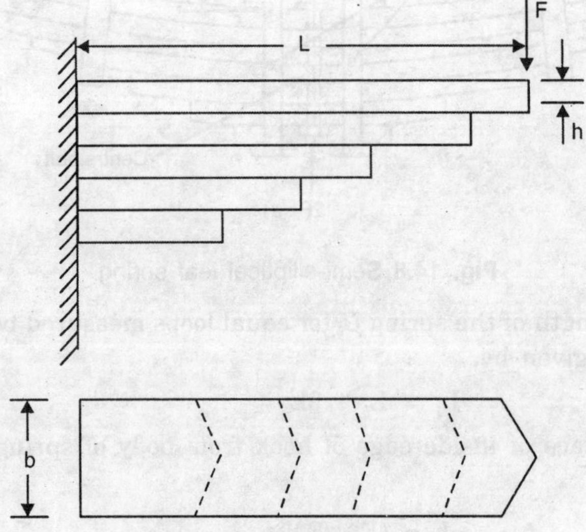

Fig. 14.10 Quarter-elliptic leaf spring.

Bending stress,

$$\sigma = 6\ FL/(nbh^2) \quad ...(14.38)$$

Deflection,

$$\delta = 6FL^3/(nEbh^3) \quad ...(14.39)$$

(*b*) Semi-Elliptic Leaf Spring (Fig. 14.8)

Bending stresses,

$$\sigma = 3FL/(2nbh^2) \quad ...(14.40)$$

Where F = load on the spring

L = length of the longest leaf

b = width of the leaf

h = thickness of leaf

n = nuber of graduated leaves

Deflection,

$$\delta = 3FL^3/(8\ nEbh^3) \quad ...(14.41)$$

Where E = modulus of elasticity of spring material.

(*c*) Graduated and Full Length Leaves (Fig. 14.11)

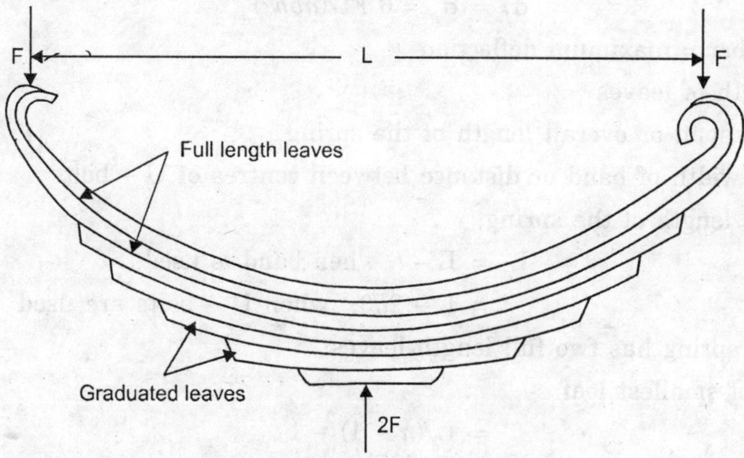

Fig. 14.11 Semi-elliptical spring having full length and graduated leaves.

Let n_f = number of full length leaves

n_g = number of graduated leaves

(*i*) Load shared by the full length leaves,

$$F_f = 3n_f\ F/(2n_g + 3n_f) \quad ...(14.42)$$

Load shared by the graduated leaves,

$$F_g = 2n_g F/(2n_g + 3n_f)] \quad ...(14.43)$$

(*ii*) Stresses in the leaves.

$$\sigma_g = 12\ FL/[(2n_g + 3n_f)\ bh^2] \quad ...(14.44)$$

$$\sigma_f = 18\ FL/[2n_g + 3n_f)\ bh^2] = 1.5\ \sigma_g \quad ...(14.45)$$

Deflection,

$$\delta_g = \delta_r = 12\ FL^3/[(2n_g + 3n_f)\ Ebh^3] \quad ...(14.46)$$

(*iii*) Initial gap or nip in the leaves, (Fig. 14.12)

$$c = 2FL^3/(nEbh^3) \quad ...(14.47)$$

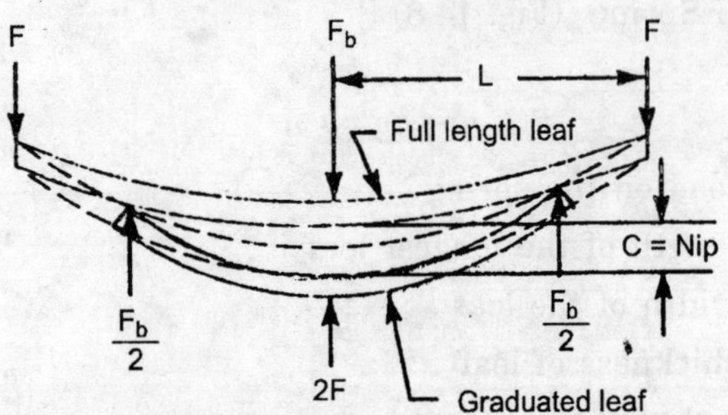

Fig. 14.12 Nip in a leaf spring.

(*iv*) Bolt load required to close the gap,

$$F_b = 2n_g n_f F/[n(2n_g + 3n_f)] \quad ...(14.48)$$

(*v*) Maximum stress in spring,

$$\sigma_f = \sigma_g = 6\ FL/(nbh^2) \quad ...(14.49)$$

(*vi*) Camber = maximum deflection

(*vii*) Length of leaves

Let L = span or overall length of the spring

l = width of band or distance between centres of U – bolts

Effective length of the spring,

$$L_e = L - l, \text{ when band is used} \quad ...(14.50a)$$

$$= L - 2l/3, \text{ when U – bolts are used} \quad ...(14.50b)$$

If a leaf spring has two full length leaves:

Length of smallest leaf

$$= L_e/(n - 1) + 1 \quad ...(14.51a)$$

Where

$$n = n_f + n_g$$

Length of next leaf

$$= 2L_e/(n - 1) + 1 \quad ...(14.51b)$$

Length of $(n - 1)$th leaf

$$= (n - 1)\ L_e/(n - 1) + 1 \quad ...(14.51c)$$

Length of master leaf

$$= L + 2\pi\ (d + h) \quad ...(14.51d)$$

Where d = inside diameter of eye

h = thickness of master leaf

Table 14.1: Spring stress factors

Spring index	*Wahl factor*	*Direct shear stress factor* K_s	*Coil curvature factor* K_c
5.0	1.311	1.100	1.192
5.5	1.278	1.091	1.171
6.0	1.253	1.083	1.157
6.5	1.231	1.077	1.143
14.0	1.213	1.071	1.133
14.5	1.197	1.067	1.122
8.0	1.184	1.063	1.114
8.5	1.172	1.059	1.107
9.0	1.162	1.056	1.100
10.0	1.145	1.050	1.090

Table 14.2: Applications of steel wires for helical springs

<table>
<tr><th>Type of wire</th><th colspan="4">Applications</th></tr>
<tr><td rowspan="2">(i) Patented and cold drawn steel</td><td>Grade 1</td><td>Grade 2</td><td>Grade 3</td><td>Grade 4</td></tr>
<tr><td>Springs subjected to static or low – load cycles</td><td>Springs subjected to moderate – load cycles</td><td>Highly stressed static springs or subjected to moderate dynamic loads</td><td>Springs subjected to severe stresses</td></tr>
<tr><td rowspan="2">(ii) Unalloyed, oil – hardened and tempered spring steel,</td><td colspan="2">Grade SW</td><td colspan="2">Grade VW</td></tr>
<tr><td colspan="2">Springs subjected to moderate fluctuating stresses</td><td colspan="2">Springs subjected to high amplitude of fluctuating stresses</td></tr>
<tr><td rowspan="2">(iii) Alloyed, oil – hardened and tempered steel</td><td colspan="2">1 S and 2 S</td><td colspan="2">1 D and 2 D</td></tr>
<tr><td colspan="2">For static working springs</td><td colspan="2">For dynamic working springs</td></tr>
<tr><td>(iv) Stainless steel for normal corrosion resistance</td><td colspan="4">For corrosion resistance environment applications</td></tr>
</table>

Table 14.3: Modulus of rigidity G for spring wires

Material	*G N/mm²*
Patent and cold drawn steel wires—unalloyed, Grades 1, 2, 3 and 4 as per IS : 4454 (Part I)—1981.	81,370
Oil hardened and tempered spring steel wire and valve spring wire-unalloyed, Grades SW and VW as per IS : 4454 (Part II)—1975.	81,370
Alloyed, oil hardened tempered valve spring wire and spring steel wire for use under moderately elevated temperatures, Grades IS, 2S, 1D and 2D as per IS : 4454 (Part III)—1975.	81,370
Hot rolled steel for hot formed springs, Grades 1, 2 as per IS : 3431—1965.	80,000
Hot rolled steel for hot formed springs, Grades 98C6, 113 C6, 55 Si 7, 60 Si 7, 50 Cr 4V2 and 60 Cr 4V2 as per IS : 3195—1975.	80,000
Stainless spring steel wire, as per IS : 4454 (Part IV)—1975.	
Grade 1 : Tempered	73,530
Untempered	69,610
Grade 2 : Tempered	78,430
Untempered	73,530
Hard drawn brass wires as per IS : 4076—1976	35,000
Phosphor bronze wires as per IS : 1385—1968	42,000

Table 14.4: Physical properties of patented and cold drawn, steel wires unalloyed for cold formed springs

Uncoated wire dia d (nominal), mm	*Tensile, strength, N/mm², Min.* Grade 1	Grade 2	Grade 3	Grade 4
0.07	—	—	2550	—
0.08	—	—	2450	—
0.09	—	—	2530	—
0.10	—	—	2530	—
0.11	—	—	2520	—
0.12	—	—	2520	—
0.14	—	—	2510	—
0.16	—	—	2500	—
0.18	—	—	2500	—
0.20	—	—	2490	2700
0.22	—	—	2480	2680
0.25	—	—	2470	2670
0.28	—	—	2460	2660

0.30	1720	2060	2460	2660
0.32	1710	2050	2450	2650
0.34	1710	2050	2440	2640
0.36	1700	2040	2430	2630
0.38	1700	2040	2430	2620
0.40	1700	2040	2430	2620
0.43	1690	2030	2420	2610
0.45	1680	2020	2410	2600
0.48	1680	2020	2400	2590
0.50	1670	2010	2390	2580
0.53	1660	2000	2380	2570
0.56	1660	2000	2370	2560
0.60	1650	1990	2360	2550
0.63	1640	1980	2340	2540
0.65	1640	1980	2330	2540
0.70	1630	1970	2320	2530
0.75	1620	1960	2300	2500
0.80	1610	1950	2280	2480
0.85	1600	1930	2260	2460
0.90	1590	1920	2250	2440
0.95	1580	1910	2250	2420
1.00	1570	1900	2240	2400
1.05	1560	1890	2210	2380
1.10	1550	1880	2190	2370
1.20	1540	1860	2170	2340
1.25	1530	1850	2140	2320
1.30	1520	1840	2130	2300
1.40	1500	1800	2110	2290
1.50	1490	1800	2100	2260
1.60	1470	1780	2080	2250
1.70	1460	1760	2050	2220
1.80	1440	1750	2030	2190
1.90	1430	1730	2010	2180
2.00	1420	1720	1990	2160
2.10	1410	1700	1960	2130
2.25	1400	1680	1940	2100
2.40	1380	1660	1910	2070
2.50	1370	1640	1890	2050
2.60	1360	1620	1860	2030

2.80	1340	1600	1840	2000
3.00	1320	1570	1830	1980
3.20	1310	1550	1790	1960
3.40	1290	1530	1760	1920
3.60	1270	1510	1750	1920
3.80	1260	1490	1720	1860
4.00	1250	1480	1700	1840
4.25	1250	1460	1680	1820
4.50	1230	1440	1660	1800
4.75	1210	1420	1620	1770
5.00	1190	1390	1600	1750
5.30	1170	1370	1570	1720
5.60	1150	1350	1550	1690
6.00	1130	1320	1530	1670
6.20	1120	1310	1500	1640
6.50	1110	1290	1480	1620
7.00	1090	1260	1460	1610
7.50	1070	1250	1430	1570
8.00	1050	1220	1400	1540
8.50	1050	1200	1370	1500
9.00	1000	1180	1350	1480
9.50	990	1150	1310	—
10.0	980	1130	1290	—
10.5	—	1100	—	—
11.0	—	1080	—	—
12.0	—	1040	—	—
12.5	—	1030	—	—
13.0	—	1020	—	—
14.0	—	990	—	—
15.0	—	970	—	—
16.0	—	960	—	—
17.0	—	950	—	—

Table 14.5: Physical properties of oil hardened and tempered spring steel wire and valve spring wire unalloyed

Wire Dia (Nominal)	*Tolerance*	*Tensile* strength N/mm², min For grades*	
mm	*mm*	*SW***	*VW†*
1.00			
1.05		1760–1960	1670–1810
1.10			
1.20			
1.25	± 0.015	1720–1960	1620–1760
1.30			
1.40			
1.50			
1.60			
1.70	± 0.020	1670–1860	1570–1690
1.80			
1.90			
2.00			
2.10		1620–1760	1520–1620
2.25			
2.40			
2.50			
2.50	± 0.030	1570–1720	1470–620
2.80			
3.00		1520–1670	1430–1530
3.20			
3.40			
3.60		1480–1630	1400–1500
3.80			
4.00			
4.25			
4.50	± 0.040	1440–1590	1370–1470
4.75			
5.00			
5.30			
5.60			

6.00		1400–1500	1340–1440
6.30			
6.50			
7.00	± 0.06	1360–1510	1300–1400
7.50			
8.00			
8.50			
9.00		1290–1440	—
9.50	±0.08		
10.0			
10.5		1250–1400	—
11.0			
12.0	±0.12		
12.5			
13.0			
14.0			

∗ Yield strength = 0.9 × Ultimate tensile strength.

∗∗ For moderately stressed springs.

† For higher dynami stressed and moderate impact loaded springs.

Table 14.6: Physical properties of oil hardened and tempered steel wire-alloyed

Wire Dia (Nominal) mm	*Tolerance mm*	*Tensile* strength N/mm², min For grades*				*Reduction in area*
		*IS**	*ID†*	*2S**	*2D†*	*percent, Min*
1.00		—	—	2060	2060	
1.05						
1.10	± 0.015	1860	1760	1960	1960	
1.20						
1.25						
1.30						
1.40						
1.50						
1.60						
1.70	± 0.020					50
1.80						
1.90						

2.00						
2.10						
2.25						
2.40	± 0.030	1760	1670	1910	1910	
2.50						
2.60						
2.80						
3.00						
3.20						
3.40						
3.60						
3.80		1670	1570	1860	1860	
4.00						
4.25	± 0.040					45
4.50						
4.75						
5.00						
5.30						
5.60		1520	1470	1760	1760	
6.00						
6.30						
6.50	± 0.060	1470	1420	1720	1720	42
7.00						
7.50		—	—	1670	1670	
8.00						

* For static working springs.

† For dynamic working springs.

Table 14.7: Specification for steel wires for cold formed stainless spring steel wire for normal corrosion resistance

Wire dia (Normal) mm	*Tolerance mm*	*Tensile Strength N/mm², Min*
0.10		2060
0.11	± 0.005	
0.125		2010
0.14		
0.16		
0.18		
0.20		
0.22		
0.25		1960
0.28		
0.30		
0.32		
0.34		
0.36		
0.38		
0.40		
0.43	± 0.010	1910
0.45		
0.48		
0.50		
0.53		
0.56		
0.60		
0.63		1860
0.65		
0.70		
0.75		
0.80		
0.85		
0.90		

Wire dia (Normal) mm	*Tolerance mm*	*Tensile Strength N/mm², Min*
1.50		
1.60		
1.70	± 0.020	1670
1.80		
1.90		
2.00		
2.10		
2.25		
2.40		
2.50	± 0.030	1579
2.60		
2.80		
3.00		
3.15		
3.20		
3.40		
3.60		
3.80		
4.00		
4.25	± 0.040	1470
4.50		
4.75		
5.00		
5.30		
5.60		
6.00		
6.30		

0.95			6.50		
1.00			7.00	± 0.060	1320
1.05			8.00		
1.10	± 0.015	1760	8.50		
1.15			9.00		
1.20			9.50		
1.25			10.00	± 0.080	1270
1.30					
1.40					

Table 14.8: Total number of coils for helical compression springs.

Type of ends	*Total number of coils, i_t*
1. Hot coiled springs	
Closed (squared) and ground	$i_f + 1.50$
Open and unground	$i_t + 1.00$
Pigtail	$i_f + 1.50$
One end closed and ground and one end open	$i_f + 1.67$
2. Cold coiled springs	
Squared	$i_f + 2$
Squared and ground	$i_f + 2$
Plain	i_f
Plain and ground	i_f

Table 14.9: Permissible stresses in extension springs

Type of steel wire	S_{yt}/S_{ut}	S_{ys}/S_{ut}
1. Cold drawn plain carbon steel	0.75	0.45
2. Hardened and tempered steel	0.75	0.50
3. Stainless steel	0.55	0.35

Table 14.10: Size of flats for laminated springs for automotive suspension

Width, mm = 40, 45, 50, 55, 60, 65, 70, 75, 80, 90, 100, 120 and 150
Thickness, mm = 4, 5, 6, 7, 8, 10, 11, 12, 13, 14 and 16

Table 14.11: Standard sections of steel flats for the manufacture of laminated springs for railways

All dimensions in mm

Width	*Tickness*	*Width*	*Thickness*	*Width*	*Thickness*
50	10 13	90	6 8 10 11 13 16 19	115	10 11 13 16 19
63	3 8 10 11 13			120	16 19
75	6 8 10 11 13 16	100	8 10 11 13 16 19	130	10 13 16
				140	11 13
				150	11 13 16

IS : 3885 (Part 1)—1977

Table 14.12: Steel for manufacture of laminated springs (Railway rolling stock) - Rib and groove sections

Thickness, mm = 6, 8, 10, 11, 13, 14, 16 and 19
Width, mm = 50, 63, 75, 90, 100, 115, 120, 125, 140 and 150

Grade	Designation	Tensile strength, MPa
1	55C6	1180 — 1420 H & T
2	75C6	1180 — 1570 H & T
4	55Si7	1275 — 1471 H & T

Table 14.13: Factor and safety for laminated springs

Type of load / service	*Factor of safety*
Static	1.5
Railroad service	2 – 2.5
Automobile suspension:	
Rear	2.25 – 3.5
Front	3.25 – 3.5
Dynamic load	1.25 – 1.5

Table 14.14: Physical properties of cold-rolled steel strips for springs

Material grades	*Designation*	*Annealed*			*Hardened and Tempered*		
		Vickers Hardness HV max	Yield stress N/mm² min	Tensile strength N/mm² max	Vickers Hardness HV	Yield stress N/mm² min	Tensile strength N/mm²
1	45C8	185					
2	55C6	195					
3	65C6	205	270	590	350–425	1030–1420	1180–1420
4	70C6	215				1180	
5	75C6	220				1470	1180–1570
6	80C6	220	290	640	350–475	1470	1570–1760
7	85C6	220				1570	1570–1960
8	98C6	225	310	690	475–540	1760	1670–2960
9	55Si7	240	340	780	475–570		1880–2960
10	50Cr4	240	340	780	505–650		
11	50Cr4V2	240	340	780	555–670		
Standard thickness (mm): 0.10, 0.15, 0.20, 0.25, 0.30, 0.40, 0.50, 0.60, 0.80, 1.00, 1.25, 1.50, 1.80 2.00, 2.50, 3.00, 3.55, 4.00 and 5.00							

IS : 2507—1975

Table 14.15: Dimensions for centre bolts for leaf springs (All dimensions in mm)

Width of leaves	≤ 60	≤ 80	≤ 80	≥ 80 ≤ 100	≥ 90	≥ 90	≥ 90	≥ 90	≥ 90
Diameter of bolt	M8	M10	M12	M14 × 1.5	M16	M16×1.5	M18×1.5	M20×2	M24×2
Diameter of bolt head	12	15	17	20	20	22	25	28	34
Length of bolt head	10	10	10	11	11	11	12	14	18

Table 14.16: Recommended bore diameter of eye, mm

Eye bore diameter, mm = 18, 20, 22, 23, 25, 27, 28, 30, 32, 35, 38, 50, 55, 60

Table 14.17: Dimensions for U – clips, rivets and bolts (All dimensions in mm)

Spring width, B	*Clip section, b × t*	*Rivet diameter, d_1*	*Bolt diameter, d_2*
Upto 50	20 × 4	6	6
50, 55 and 60	25 × 5	8	8
76, 70, 75 and 80	25 × 6	10	8
90, 100 and 125	32 × 4	12	10

Flat and V-Belts

A. Flat Belts

1. Clasification of flat belts

Light: $v \leq 12$ m/s and upto 7.5 kW ...(15.1*a*)

Medium: $v = 12 - 24$ m/s and 7.5 – 15 kW ...(15.1*b*)

Heavy: $v > 24$ m/s and above 15 kW ...(15.1*c*)

2. Speed ratio (Fig. 15.1)

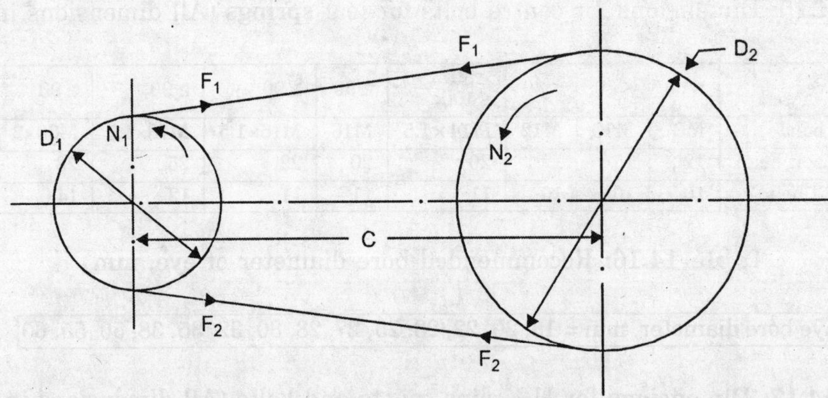

Fig. 15.1 Open flat belt drive.

Neglecting slip and thickness of belt, the speed ratio,

$$i = N_2/N_1 = D_1/D_2 \quad \text{...(15.2a)}$$

where N_1, N_2 = rpm of smaller and bigger pulley respectively

D_1, D_2 = diameter of smaller and bigger pulley respectively, mm.

Considering slip and thickness of belt, speed ratio,

$$i = (1 - s/100)\,[(D_1 + t)/(D_2 + t)] \quad \text{...(15.2b)}$$

where t = thickness of belt.

s = total percentage slip on the pulleys

$$= s_1 + s_2 - 0.01\, s_1 s_2 \qquad \text{...(15.3)}$$

s_1, s_2 = % slip on smaller and bigger pulley respectively

3. Angle of arc of contact

For open belt drive, anlge of arc of contact,

$$\theta = \pi \pm 2 \sin^{-1} [(D_2 - D_1)/(2C)] \text{ rad} \qquad \text{...(15.4}a\text{)}$$

Where C = centre distance

Use +ve sign for the bigger pulley and –ve sign for the smaller pulley.

For crossed belt drive,

$$\theta = \pi \pm 2 \sin^{-1} [(D_1 + D_2)/(2C)] \text{ rad} \qquad \text{...(15.4}b\text{)}$$

4. Belt tensions and power transmitted neglecting centrifugal tension

At low speeds (v < 12 m/s), the effect of centrifugal tension may be neglected. The ratio of belt tensions, is,

$$F_1/F_2 = \exp(\mu\theta) \qquad \text{....(15.5)}$$

Where F_1, F_2 = belt tension on the tight and slack side respectively

μ = coefficient of friction between belt and pulley (Table 15.1)

= 0.54 – 0.712/(v + 2.542) for tanned leather belts on C.I. pulleys, v in m/s ...(15.6)

Peripheral speed of belt,

$$v = \pi D_1 N_1/(60 \times 10^3) \text{ m/s} \qquad \text{...(15.7)}$$

Corrected power,

$$P_c = P \times K_s/K_j \qquad \text{...(15.8)}$$

Where K_s = service correction factor (Table 15.2)

K_j = correction factor for belt joint (Table 15.3)

Power transmitted,

$$P_c = (F_1 - F_2)\, v/10^3 \text{ kW} \qquad \text{...(15.9)}$$

Power transmitted per square mm of belt cross-section,

$$P_c = [\sigma_1 v/10^3]\, [1 - \exp(-\mu\theta)] \qquad \text{...(15.10)}$$

Where $\sigma_1 = F_1/(bt)$ is the allowable stress in the belt material

b = belt width, mm

5. Belt tensions and power transmitted considering centrifugal tension (v > 12 m/s)

$$(F_1 - F_c)/(F_2 - F_c) = \exp(\mu\theta) \qquad \text{...(15.11)}$$

Where F_c = centrifugal tension per unit belt length

$$= \rho b t v^2 \qquad \text{...(15.12)}$$

ρ = density of belt material, kg/m^3 (Table 15.4)

Power transmitted,

$$P_c = [(F_1 - F_c)\, v/10^3]\, [1 - \exp(-\mu\theta)] \text{ kW} \quad \text{...(15.13)}$$

6. Condition for maximum power transmission

$$v = [\sigma_d/(3\rho)]^{1/2} \quad \text{...(15.14)}$$

where σ_d = design stress for the belt material per unit width of belt

7. Initial tension

$$F_i = (F_1 + F_2)/2 \quad \text{...(15.15)}$$

8. Width of belt

Neglecting centrifugal tension,

Width of belt,

$$b = F_1/\sigma_d \quad \text{...(15.16)}$$

Considering centrifugal tension,

$$bt = [(10^3\, P_c)/\{v(\sigma_d - \rho v^2)\}]\ [1/\{1 - \exp(-\mu\theta)\}] \quad \text{...(15.17)}$$

Refer to Tables 15.5, 15.7, 15.8 and 15.9 for standard width of belt.

9. Maximum stress in the belt

σ_{max} = stress due to maximum belt tension + centrifugal stress + bending stress while turning over the pulleys

$$= F_1/(bt) + \rho v^2 + Et/D_1 \quad \text{...(15.18)}$$

10. Length of belt

The length of open belt is,

$$L_o = (\pi/2)(D_1 + D_2) + 2C + (D_2 - D_1)^1/(4C) \quad \text{...(15.19a)}$$

The length of crossed – belt (Fig. 15.2),

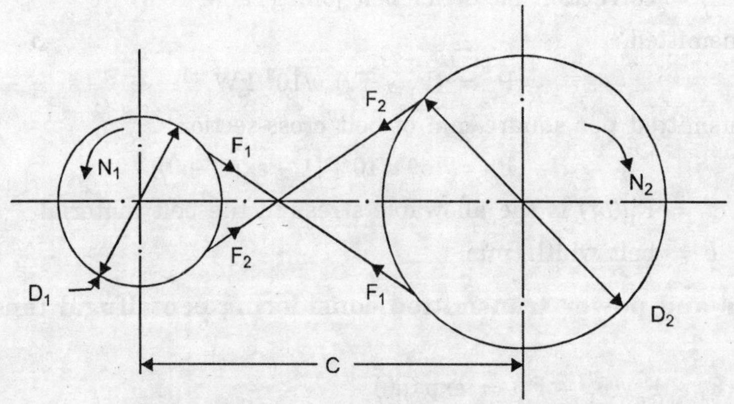

Fig. 15.2 Crossed flat belt drive.

$$L_c = (\pi/2)(D_1 + D_2) + 2C + (D_2 + D_1)^2/(4C) \quad \text{...(15.19b)}$$

11. Centre distance

$$C = L_o/4 - 0.3925\,(D_1 + D_2) + [\{L_o/4 - 0.3925\,(D_2 - D_1)\}^2 - (D_2 - D_1)^2/8]^{1/2} \quad ...(15.20)$$

12. Pulleys for Flat Belts

(*a*) Cast Iron Pulley

A cast iron pulley is shown in Fig. 15.3

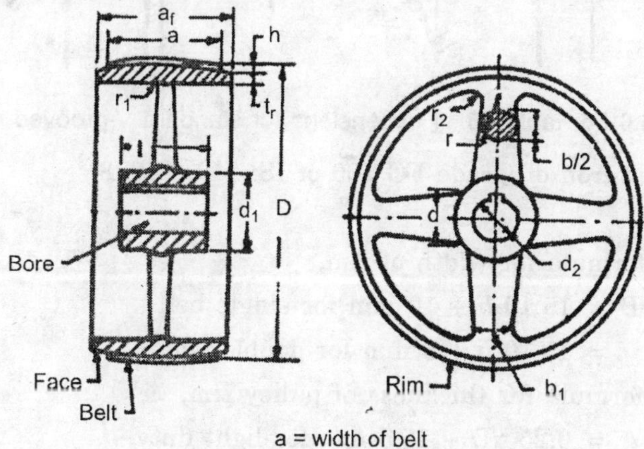

Fig. 15.3 Cast iron flat belt pulley.

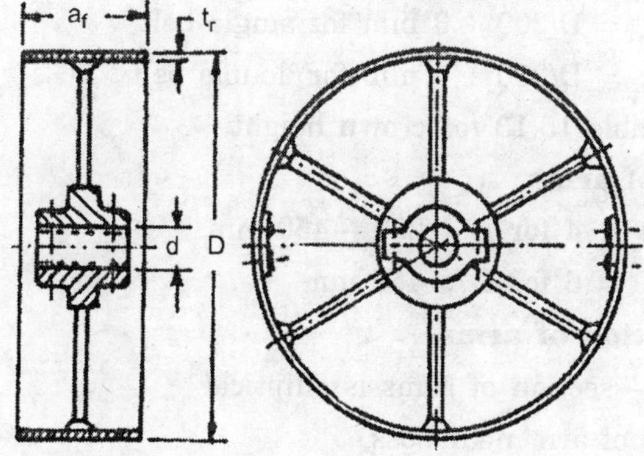

Fig. 15.4 Mild steel flat belt pulley.

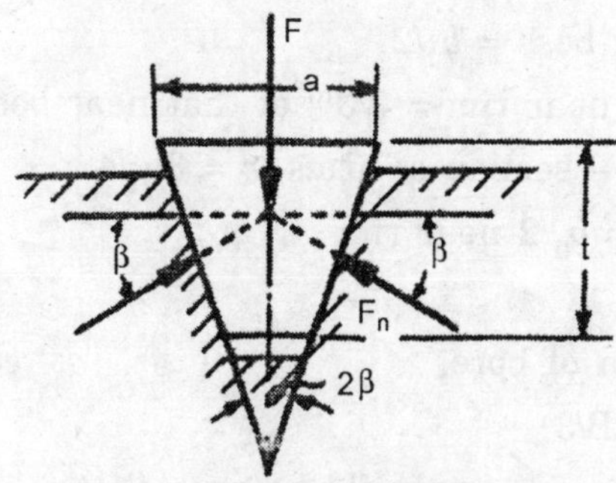

Fig. 1.5.5 V-belt in pulley groove

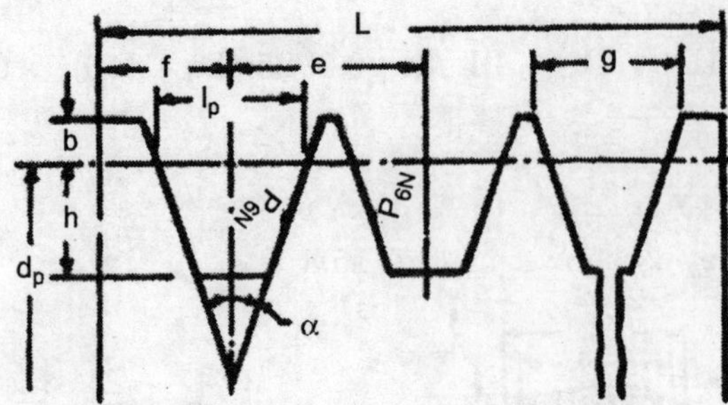

Fig. 15.6 For table 15.31 dimensions for standard V-grooved pulley.

Material: Cast iron of Grade FG 200 of IS: 210 – 1978

***(i)* Rim size**

Bartho's formula for width of rim,

$$B = 15.19\, b + 10 \text{ mm for single belt} \quad ...(15.21a)$$

$$= 15.10\, b + 5 \text{ mm for double belt} \quad ...(15.21b)$$

Bartho's formula for thickness of pulley rim,

$$t_r = 0.25 \sqrt{D} + 15.5 \text{ mm for light duty} \quad ...(15.22a)$$

$$= 0.375 \sqrt{D} + 3.2 \text{ mm for heavy duty} \quad ...(15.22b)$$

For cast iron flat pulleys, rim thickness,

$$t_r = D/200 + 3 \text{ mm for single belt} \quad ...(15.23a)$$

$$= D/200 + 6 \text{ mm for double belt} \quad ...(15.23b)$$

Refer to Table 15.13 for crown height.

***(ii)* Number of arms**

$$n = 4 \text{ for } D = 200 - 450 \text{ mm} \quad ...(15.24a)$$

$$= 6 \text{ for } D > 450 \text{ mm} \quad ...(15.24b)$$

***(iii)* Cross-section of arms**

The cross – section of arms is elliptical.

Major axis of arm near boss,

$$b_1 = 2.94\, [BD/(4n)]^{1/3} \text{ for single belt} \quad ...(15.25a)$$

$$= 2.94\, [BD/(2n)]^{1/3} \text{ for double belt} \quad ...(15.25b)$$

Minor axis near boss = $b_1/2$ ...(15.26)

Cross – section near rim = 2/3rd of that near boss ...(15.27)

Radius of cross – section of arms, $r = 3b_1/4$ ...(15.28)

Radius $r_1 = r_2 = b_1/2$ near rim ...(15.29)

***(iv)* Size of boss**

Minimum length of bore,

$$l = 2B/3 \quad ...(15.30)$$

The diameter d_1 of boss should be such that,

$$(d_1 - d_2)/2 = 0.412\ (BD)^{1/3} + 6 \text{ mm for single belt} \quad ...(15.31a)$$

$$= 0.529\ (BD)^{1/3} + 6 \text{ mm for double belt} \quad ...(15.31b)$$

(*b*) Steel Pulley

A steel pulley is shown in Fig. 15.3

Material: Mild steel conforming to IS: 226 – 1975

Thickness of rim,

$$t_f = 5 \text{ mm} \quad ...(15.34)$$

Length of boss

$$= ½ \times \text{width of face} \quad ...(15.35)$$

= 100 mm minimum for pulleys with 19 mm diameter spokes

= 138 mm minimum for pulleys with 22 mm diameter spokes

Cross section of spokes = circular

Rows of arms = single for pulleys upto 300 mm width ...(15.36*a*)

= two for width more than 300 mm ...(15.36*b*)

13. Shaft and Key size

Resultant pull on pulley due to belt tensions,

$$P = [F/(K - 1)]\ [K^2 + 1 + 2K \cos (180° - \theta)]^{1/2} \quad ...(15.37)$$

Where $K = (F_1 - F_c)/(F_2 - F_c)$, $F = F_1 - F_2$

Let W = weight of the pulley, and α = angle between P and W

Resultant force on the pulley,

$$S = [P^2 + W^2 + 2PW \cos \alpha]^{1/2} \quad ...(15.38)$$

Let x = overhang of pulley

$$M = Sx, \text{ bending moment due to S} \quad ...(15.39)$$

Torque in the shaft,

$$T = (F_1 - F_2)\ D/2 \quad ...(15.40)$$

Equivalent torque,

$$T_e = [(K_m M)^2 + (K_t T)^2]^{1/2} \quad ...(15.41)$$

Where K_m and K_t are shock and fatigue factors for bending and torsion respectively.

Let d_2 = diameter of shaft

$$T_e = (\pi/16)\ d^3_2\ \tau \quad ...(15.42)$$

Where τ = permissible shear stress for the shaft material.

A suitable key for the shaft and pulley assembly is a square key with a gib head.

14. Diameter of motor pulley

Diameter of motor pulley,

$$D_1 = (12000 \text{ to } 1400)\ [P/N_1]^{1/3} \quad ...(15.43)$$

15. Selection of Flat – Belts from Manufacturer's Catalogue

(*i*) Calculate the maximum power transmitted by the belt from the rated power.

$$(\text{kW})_{max} = F_a\ (\text{kW}) \quad ...(15.44)$$

Where F_a = load correction factor

(*ii*)
$$(\text{kW})_{corrected} = F_a F_d\ (\text{kW}) \quad ...(15.45)$$

Where F_d = arc of contact factor

(*iii*) Calculate the corrected power rating for the belt.

Corrected kW rating

$$= 0.0118\ v/5.08 \quad \text{for HI-SPEED belt} \quad ...(15.46a)$$

$$= 0.0147\ v/5.08 \quad \text{for FORT belt} \quad ...(15.46b)$$

Where v = belt velocity, m/s

(*iv*) Calculate the product of (width × number of piles),

With × number of plies = corrected power/corrected kW rating ...(15.47)

(*v*) Select the standard belth width and number of plies.

16. Creep in Belts

$$\text{Creep} = (F_1 - F_2)/(btE) \quad ...(15.48)$$

Where F_1, F_2 = tension on the tight and slack sides respectively

b, t = width and thickness of belt respectively

E = Young's modulus of elasticity of belt material

B. V - Belts

16. Equivalent coefficient of friction (Fig. 15.5)

$$\mu_e = \mu/\sin\beta \quad ...(15.49)$$

where 2β = pulley groove angle = 40°

17. Number of belts

$$n(F_1 - F_c)\,[1 - 1/\exp(\mu_e\,\theta)] = 10^3 \times P \times K_s/v \quad ...(15.50)$$

where $v = \pi D_1 N_1/(60 \times 10^3)$ m/s

D_1 = diameter of smaller pulley, mm

N_1 = rpm of smaller pulley

K_s = correction factor according to service

$$\theta = 2\cos^{-1}[(D_2 - D_1)/(2C)] \quad ...(15.51)$$

According to IS; 2494 – 1974, the number of belts are given by,

$$n = P \times K_a/(\text{kW} \times K_c \times K_d) \quad ...(15.52)$$

where P = drive power, kW

kW = rating of V – belts at 180° arc of contact (Table 15.25)

K_a = correction factor for industrial service (Table 15.26)

K_c = correction factor for belt length (Table 15.28)

K_d = correction factor for arc of contact (Table 15.29)

Equivalent pitch diameter,

$$d_e = d_p \times K_b$$

K_d = small diameter factor (Table 15.27)

18. Power rating of V – belts

$$kW = (nav/10^3)\,[\sigma_d - \rho v^2/10^6]\,[1 - 1/\exp(\mu_e\theta)] \quad ...(15.53)$$

where σ_d = design stress = 2.26 N/mm^2

ρ = mass density of belt material, kg/m^3

a = area of cross-section of belt, mm^2

19. Initial tension

$$F_i = (F_1 + F_2)/2 \quad ...(15.54)$$

20. Optimum velocity of belt for maximum power transmission

$$v = (F_i/3m)^{1/2} \quad ...(15.55)$$

where m = mass per meter length of belt

21. Ratio of belt tensions

$$(F_1 - mv^2)/(F_2 - mv^2) = \exp(\mu_e\theta) \quad ...(15.56)$$

22. Pitch length of belt

$$L_p = 2C + 15.57\,(D_1 + D_2) + (D_2 - D_1)^2/(4C) \quad ...(15.57)$$

23. Centre distance

$$C = A + (A^2 - B)^{1/2} \quad ...(15.58)$$

where $A = L_p/4 - \pi\,(D_1 + D_2)/8$

$B = (D_2 - D_1)^2/8$

$$C_{max} = 2(D_1 + D_2) \quad ...(15.59)$$

$$C_{min} = 0.55\,(D_1 + D_2) + t \quad ...(15.60)$$

Refer to table 15.24 for selection of centre distance.

24. Pulley size

Refer to Tables 15.23 and 15.31.

Table 15.1: Coefficient of friction for belts

Belt material	*Pulley material*						
	Cast iron, steel			*Wood*	*Compressed paper*	*Leather face*	*Rubber face*
	Dry	*Wet*	*Greasy*				
Leather, Oak tanned	0.25	0.20	0.15	0.30	0.33	0.38	0.40
Leather, chrome tanned	0.35	0.32	0.22	0.40	0.45	0.48	0.50
Canvas, stitched	0.20	0.15	0.12	0.23	0.25	0.27	0.30
Cotton, woven	0.22	0.15	0.12	0.25	0.28	0.27	0.30
Camel hair, woven	0.35	0.25	0.20	0.40	0.45	0.45	0.45
Rubber	0.30	0.18	—	0.32	0.35	0.40	0.42
Balats	0.32	0.20	—	0.35	0.38	0.40	0.42

Table 15.2: Service correction factor, K_s

Service condition	K_s
(*a*) Atmosphere condition:	
Clean, scheduled maintenance on large drives	1.2
Normal	1.0
Oily, wet or dusty	0.7
(*b*) Angle of centre line:	
Horizontal to 60° from horizontal	1.0
60° to 75° from horizontal	0.9
75° to 90° from horizontal	0.8
(*c*) Pulley material:	
Fibre on motor and small pulleys	1.2
Cast iron on steel	1.0
(*d*) Service:	
Temporary or infrequent	1.2
Normal	1.0
Important or continuous	0.8
(*e*) Peak loads:	
Steady belt loads as obtained with steam engines, turbine, diesel and multi-cylinder gas engines, fans, centrifugal pumps, and steady line shaft loads.	1.0
Jerky belt loads, as obtained with large induction motors; compensator-started, shunt wound, direct current motors; single cylinder gas engines; reciprocating machines; and machines developing series of peak loads, such as compressors, rock crushers and punch presses etc.	0.8
Shock and reversing belt loads on all motors 10 hp. and under, all cross-the-line start motors, wound-rotor (slip-ring) motors, synchronous motors, and reversing loads such as printing presses, elevator service and laundary washers etc.	0.6

Table 15.3: Correction factors K_j for belt joints

Type of Joint	K_j
Cemented by belt maker	1.00
Cemented	0.98
Wire laced by machine	0.90
Wire laced by hand	0.82
Rawhide faced	0.60
Metal belt hoops	0.35

Table 15.4: Properties of belt materials

Material	*Design strength* N/mm^2	*Endurance strength* N/mm^2	*Elastic modulus* N/mm^2	*Density* kg/m^3	*Maximum velocity*
Rubber	1.6	6.0	10.0	$1.1 - 1.2 \times 10^{-3}$	30
Leather	2.0	6.0	30.0	$1.0 - 1.1 \times 10^{-3}$	30 – 50
Fabric	1.5	3.0	15.0	$0.9 - 1.0 \times 10^{-3}$	20 – 25
Plastic	4.0	6.0	60.0	$1.0 - 1.1 \times 10^{-3}$	60

Table 15.5: Widths of friction surface rubber transmission belting

Nominal belt width, mm	*Tolerance, mm*
25, 32, 40, 50, 63	± 2.0
75, 80, 90, 100, 112, 125	± 3.0
140, 160, 180, 200, 224, 250	± 4.0
280, 315, 355, 400, 450, 500	± 5.0

Table 15.6: Tensile strength of solid woven impregnated cotton* and hair** belting

Type of belt	*Minimum tensile strength,* N/mm^2
Cotton	51.68
Hair:	
Grade 1	37.04
Grade 2	28.86

* IS : 529–1976

** IS : 530–1976

Table 15.7: Dimensions for solid woven impregnated cotton* and hair** belting for power transmission

Pulley width mm	*Nominal belt width mm*	*Tolerance on width mm*	*Nominal belt thickness, mm*				
			5 + 0.5 – 0.0	*6.5 + 0.5 – 0.3*	*8 + 0.6 – 0.3*	*10 + 0.7 – 0.5*	*12 + 0.8 – 0.6*
32	25		*x*	—	—	—	—
40	32		*y*	—	—	—	—
50	40		*x*	—	—	—	—
63	50	± 1.6	*x*	*x*	—	—	—
71	63		*x*	*x*	—	—	—
80	71		—	*y*	—	—	—
90	80		—	*y*	—	—	—
100	90		—	*y*	*y*	—	—
112	100		—	*x*	*x*	—	—
125	112		—	*x*	*x*	—	—
140	125		—	*x*	*x*	*x*	—
160	140		—	*y*	*y*	*y*	—
180	160		—	—	*y*	*y*	—
200	180	± 2.4	—	—	*y*	*y*	—
224	200		—	—	*x*	*x*	—
250	224		—	—	*x*	*x*	—
280	250		—	—	—	*x*	*y*
315	280		—	—	—	*y*	*y*
355	315	± 3.2	—	—	—	*y*	*y*
400	355		—	—	—	*x*	*y*
450	400		—	—	—	*x*	*x*
500	450		—	—	—	—	*x*
560	500		—	—	—	—	*x*
675	600		—	—	—	—	*z*

x = preferred size, *y* = permissible size, *z* = discouraged size.

*IS : 529—1976, **IS : 530—1976

Table 15.8: Specification for vegetable tanned leather belting

Type of belt	*Width of belt mm*	*Length of splice mm*	*Tensile strength N/mm²*
Single	Upto and including 125	110	
	Over 25 upto and including 40	115	
	Over 40 upto and including 50	125	
	Over 50 upto and including 71	140	24.05
	Over 71 upto and including 80	150	
	Over 90 upto and including 140	180	
	160 and above	200	
Double	Upto and including 125	100	
	Over 125	125	24.05
Splices (single and double)			20.20

IS : 2240—1976

Table 15.9: Leather belt data

Grade of belt	*Thickness in mm*				*Width mm*	*Increment mm*
	Single ply pulley dia upto 200 mm	*Double ply pulley dia upto 300 mm*	*Triple ply pulley dia upto 600 mm*	*Quadruple pulley dia above 600 mm*		
Light	3	6	–	–	12 to 24 24 to 102 102 to 198	3 6 12
Medium	4	8	12.5	17.5	200 to 800 8000 to 1400	25 50
Heavy	5	10	15	20	1500 to 2100	100

Table 15.10: Weight and tensile strength of rubber transmission belting

Type of duck	*Nominal mass g/m²*	*Tensile strength, N/mm of width*	
		Warp	*Weft*
Soft 28 32	810 915	60.11 67.37	28.83 31:77
Hard 31 34	910 970	60.11 67.37	33.73 43.25

IS : 1370—1976

Table 15.11: Rating in Kilowatts per mm width of rubber transmission belting at 180° are of contact

Type of duck	*Mass of fabric g/m²*	*Pulley dia mm*	*Belt speed m/s*	*Rating per mm width per ply in kW*
Soft 28 32	810 915	250 250	5 5	0.012 0.015
Hard 31 34	910 970	250 250	5 5	0.012 0.015

IS : 1370—1976

Table 15.12: Power rating of HI – SPEED and FORT flat belts

Type of belt	*Power rating in kW per mm per ply*	*Corrected kW rating*	*Optimum belt speed, m/s*
HI – SPEED FORT	0.0118 0.0147	0.0118 × V/5.08 0.0147 × V/5.08	17.8 – 22.0 17.8 – 22.0

Table 15.13: Thickness of friction surface: Rubber transmission belting

Ply construction	*Nominal belt thickness, mm*	*Tolerance, mm*
3 ply	3.9	± 0.5
4 ply	5.1	± 0.7
5 ply	6.4	± 0.8
6 ply	7.7	± 0.9
7 ply	9.1	± 1.0
8 ply	10.4	± 1.1
9 ply	11.7	± 1.2
10 ply	13.0	± 1.3

IS : 1370—1976

Table 15.14: Crown of cast iron and mild steel flat pulleys of diameters from 400 mm to 2000 m

Nominal D mm	*Crown h of pulley of width*						
	125 and smaller	*140 and 160*	*180 and 200*	*224 and 250*	*280 and 315*	*355*	*400 and larger*
400	1	1.2	1.2	1.2	1.2	1.2	1.2
450	1	1.2	1.2	1.2	1.2	1.2	1.2
500	1	1.5	1.5	1.5	1.5	1.5	1.5
560	1	1.5	1.5	1.5	1.5	1.5	1.5
630	1	1.5	2	2	2	2	2
710	1	1.5	2	2	2	2	2
800	1	1.5	2	2.5	2.5	2.5	2.5
900	1	1.5	2	2.5	2.5	2.5	2.5
1000	1	1.5	2	2.5	3	3	3
1120	1.2	1.5	2	2.5	3	3	3.5
1250	1.2	1.5	2	2.5	3	3.5	4
1400	1.5	2	2.5	3	3.5	4	4
1600	1.5	2	2.5	3	3.5	4	5
1800	2	2.5	3	3.5	4	4.5	5
2000	2	2.5	3	3.5	4	4.5	6

All dimensions in mm

IS : 1691—1980

Table 15.15: Width and diameters of flat cast iron and mild steel pulleys

Width, mm	*Tolerance, mm*	*Nominal dia, mm*	*Tolerance, mm*
20, 25, 32,	± 1	40	± 0.5
40, 50, 63, 71		45, 50	± 0.6
80, 90, 100	± 1.5	56, 63	± 0.8
112, 121, 140		71, 80	± 1.0
160, 180, 200	± 2	90, 100, 112	± 1.2
224, 250, 280, 315		125, 140	± 1.9
355, 490, 450	± 3	160, 180, 200	± 2.0
500, 560, 630		224, 250	± 2.5
Material of C.I. Pulley		280, 315, 355	± 3.0
Grade FG 200 of IS : 210—1978		400, 450, 500	± 4.0
		560, 630, 710	± 5.0
Material for M.S. pulley:		800, 900, 1000	± 6.3
IS : 226 – 1975		1120, 1250, 1400	± 8.0
		1600, 1800, 2000	± 10.0

IS : 1691—1980

Table 15.16: Width of flat pulleys

Belt width, mm	*Pulley to be wider than the belt width by, mm*
Upto and including 125	13
From 125 upto and including 250	25
From 250 upto and including 375	38
From 375 upto and including 500	50

IS : 2122—1973

Table 15.17: Number of arms in mild steel pulley

Diameter of pulley, mm	*No. of arms*	*Dia of arms, mm*
280 to 500	6	19
560 to 710	8	19
800 to 1000	10	22
1120	12	22
1250	14	22
1400	16	22
1600	18	22
1800	18	22
2000	22	22

IS : 1691—1980

Table 15.18: Minimum pulley diameters for given belt speeds and belt plies in mm

No. of piles	*Maximum belt speed, m / s*				
	10	*15*	*20*	*25*	*30*
3	90	100	112	140	180
4	140	160	180	200	250
5	200	224	250	315	355
6	250	315	355	400	405
7	355	400	450	500	560
8	450	500	560	630	710
9	560	630	710	800	900
10	630	710	800	900	1000

IS : 1370—1976

Table 15.19: Load correction factor, F_a

	Type of load	F_a
Normal load	Exact peak load is known	1.0
Steady load	Screens, centrifugal pumps and fans, agitators, belt conveyor, evaporators, light machine tools, printing and textile machinery	1.2
Intermittent load	Heavy duty fans and blowers, reciprocating pumps and compressors, saw mill and paper mill machinery, brick work machinery, heavy machine tools, elevators and line shafts	1.3
Shock load	Vacuum pumps, rolling mills, tube and ball mills, grinders, hammers, automatic machinery	1.5

Table 15.20: Arc of contact correction factor, F_d

Arc of contact	F_d	*Arc of contact*	F_d
90°	1.68	190°	0.97
120°	1.33	200°	0.94
130°	1.26	210°	0.91
140°	1.19	220°	0.88
150°	1.13	230°	0.84
160°	1.08	240°	0.82
170°	1.04	250°	
180°	1.00		

Table 15.21: Standard width of dumlop HIGH SPEED and dunlop FORD flat belt

3 ply mm	*4 ply mm*	*5 ply mm*	*6 ply mm*	*8 ply mm*
25	25	-	-	-
32	32	-	-	-
40	40	-	-	-
44	44	-	-	-
50	50	-	-	-
63	63	-	-	-
76	76	76	-	-
90	90	90	-	-
100	100	100	100	-
-	112	112	112	-
-	125	125	125	-
-	140	-	-	-
-	152	152	152	-
-	-	180	180	-
-	200	200	200	200
-	-	224	-	-
-	-	250	250	250
-	-	-	-	305
-	-	-	-	355
-	-	-	-	400

Table 15.22: Classification of V-belts

Type of belt	*Is Code No.*	*Cross-section symbol*	*Nominal top width, a, mm*	*Nominal thickness, t, mm*	*Angle of V-Belt, 2β degree*
Industrial purposes	IS : 2494—1974	A	13	8	
		B	17	11	40
		C	22	14	
		D	32	19	
		E	38	23	
Automotive fan	IS : 5635—1970	FB 10	10	8	
		FB 13	13	11	40
Agricultural Machinery	IS : 8777—1978	HI	25.4	12.7	
		HJ	31.8	15.1	
		HK	38.1	17.5	40
		HL	44.5	19.8	
		HM	50.8	22.2	

Table 15.23: Properties of V-belts for industrial purposes

Cross-section of belt symbol	*Area of cross-section* mm^2	*Max. working load* F_1, *N*	*Power* range kW*	*Min. groove pulley dia, mm*	*Weight per m. N*	*Density of belt material* kg/m^3	*Max. number of pulley grooves*
A	80.7	200	Upto 3.42	75	0.882	1115	6
B	143.0	300	0.5-6.61	137	1.628	1160.8	9
C	236.7	750	0.80-13	229	2.962	1275.8	14
D	476.6	1400	1.65-32.2	330	5.923	1267.3	14
E	681.4	1800	3.48-51.8	533	8.885	1329.6	20

IS : 9515—1980

Table 15.24: Dimensions for standard V-grooved pulleys

Groove section	*Pitch width* l_p *mm*	*Minimum height of groove above pitch line, b, mm*	*Minimum depth of groove below pitch line, h, mm*	*Centre to centre distance of grooves, e, mm*	*Edge of pulley to first groove centre, f mm*
A	11	3.3	8.7	15 ± 0.3	$10 {+2 \atop -1}$
B	14	4.2	10.1	19 ± 0.4	$12.5 {+2 \atop -1}$
C	19	5.7	14.3	25.5 ± 0.5	$17 {+2 \atop -1}$
D	27	8.1	19.9	37 ± 0.6	$24 {+3 \atop -1}$
E	32	9.6	23.4	44.5 ± 0.7	29 + 4

IS : 3142—1965

Table 15.25: Selection of centre distance

Speed ratio (D/d)	*Recommended C/D ratio*
1	1.50
2	1.20
3	1.00
4	0.95
5	0.85
6	0.85
7	0.85
8	0.85
9	0.85

Table 15.26: Power rating of V-belts in kW

Belt cross-section symbol	*Formula*	*Maximum value of d_e in formula, mm*
A	$P(kW) = \left(0.45v^{-0.09} - \frac{19.62}{d_e} - 0.765\times10^{-4}v^2\right)v$	125
B	$P(kW) = \left(0.79v^{-0.09} - \frac{50.8}{d_e} - 1.32\times10^{-4}v^2\right)v$	175
C	$P(kW) = \left(1.47v^{-0.09} - \frac{142.7}{d_e} - 2.34\times10^{-4}v^2\right)v$	300
D	$P(kW) = \left(3.22v^{-0.09} - \frac{506.7}{d_e} - 4.78\times10^{-4}v^2\right)v$	425
E	$P(kW) = \left(4.58v^{-0.09} - \frac{952}{d_e} - 7.05\times10^{-4}v^2\right)v$	700

Where
P = power at 180° arc of contact for a belt of average length.
v = belt speed in m/s.
d_e = equivalent pitch diameter
= $d_p \times Kb$
d_p = pitch diameter of the smaller pulley
Kb = small diameter factor (Table 15.27)

IS : 2494—1974

Table 15.27: Correction factors for industrial service, K_a

Severity of service	*Type of driven machines*	*A.C. motors; normal torque, squirrel cage, synchronous and split phase. D.C. motors, shunt wound, multiple cylinder internal combustion engines over 600 rpm.*			*Type of driving unit A.C. motors, high torque, high slip repulsion, induction, single phase, series wound and sliprings. D.C. motors, series wound and compound wound, single and multi-cylinder I.C. engines under 600 rpm, line shaft, clutches, brakes.*		
		Operational hours per day			*Operational hours per day*		
		Upto 10 hrs.	*Over 10 to 16 hrs.*	*Over 16 hrs. and continuous service*	*Upto 10 hrs.*	*Over 10 to 16 hrs.*	*Over 16 hrs. and continuous service*
Light duty	Agitators for liquids, blowers and exhausters, centrifugal pumps and compressors, fans upto 7.5 kW and light duty conveyors.	1.0	1.1	1.2	1.1	1.2	1.3
Medium duty	Belt conveyors for sand grains etc., dough mixers, fans over 7.5 kW, generators, line shafts, laundry machinery, machine tools, punches, presses and shears, printing machinery, positive displacement rotary pumps, and revolving and vibrating screens.	1.1	1.2	1.3	1.1	1.2	1.3
Heavy duty	Brick machinery, bucket elevators, conveyors (dragpan-screw), hammer mills, paper mill beaters, piston pumps, positive displacement blowers, pulverizers, saw mill and textile machinery	1.2	1.3	1.4	1.4	1.5	1.6
Extra heavy duty	Crushers (gyratory jaw-roll), mills (ball-rod-tube), hoists	1.3	1.4	1.5	1.5	1.6	1.8

Table 15.28: Small diameter factor, K_b

Speed ratio range	K_b
1.000 to 1.019	1.00
1.020 to 1.032	1.01
1.033 to 1.055	1.02
1.056 to 1.081	1.03
1.082 to 1.109	1.04
1.110 to 1.142	1.05
1.143 to 1.178	1.06
1.179 to 1.222	1.07
1.223 to 1.274	1.08
1.275 to 1.340	1.09
1.341 to 1.429	1.10
1.430 to 1.562	1.11
1.563 to 1.814	1.12
1.815 to 2.948	1.13
2.949 and over	1.14

IS : 2494—1974

Table 15.29: Correction factors for belt length, K_c

Nominal inside length mm	*Belt cross-section symbol*				
	A	*B*	*C*	*D*	*E*
(1)	*(2)*	*(3)*	*(4)*	*(5)*	*(6)*
610	0.80				
660	0.81				
711	0.82				
787	0.84				
813	0.85				
889	0.87	0.81			
914	0.87				
965	0.88	0.83			
991	0.88				
1016	0.89	0.84			
1067	0.90	0.85			
1092	0.90				
1168	0.92	0.87			
1219	0.93	0.88			
1295	0.94	0.89	0.80		

1372		0.90			
1397	0.96	0.90			
1422	0.96	0.90			
1473	0.97				
1524	0.98	0.92	0.82		
1600	0.99				
1651	1.00	0.94			
1727	1.00	0.95	0.85		
1778	1.01	0.95			
1905	1.02	0.97	0.87		
1981	1.03	0.98			
2032	1.04				
2057	1.04	0.98	0.89		
2159	1.05	0.99	0.90		
2286	1.06	1.00	0.91		
2438	1.08		0.92		
2464		1.02			
2540		1.03			
2667	1.10	1.04	0.94		
2845	1.11	1.05	0.95		
3048	1.13	1.07	0.97	0.86	
3150			0.97		
3251	1.14	1.08	0.98	0.87	
3404			0.99		
3658		1.11	1.00	0.90	
4013		1.13	1.02	0.92	
4115		1.14	1.03	0.92	
4394		1.15	1.04	0.93	
4572		1.16	1.05	0.94	
4953		1.18	1.07	0.96	
5334		1.19	1.08	0.96	0.94
6045			1.11	1.00	0.96
6807			1.14	1.03	1.99
7569			1.16	1.05	1.01
8331			1.19	1.07	1.03
9093			1.21	1.09	1.05
9855			1.23	1.11	1.07
10617			1.24	1.12	1.09
12141				1.16	1.12
13665				1.18	1.14
15189				1.20	1.17
16713				1.23	1.19

IS : 2494—1974

Table 15.30: Correction factors for arc of contact factor, K_b

Arc of contact on small pulley, degree	*Correction factor, K_b*	
	VV	*V-flat*
180	1.00	0.75
177	0.99	0.76
174	0.99	0.76
171	0.98	0.77
169	0.97	0.78
166	0.97	0.79
163	0.96	0.79
160	0.95	0.80
157	0.94	0.80
154	0.93	0.81
151	0.93	0.81
148	0.92	0.83
145	0.91	0.83
142	0.90	0.4
139	0.89	0.85
136	0.88	0.85
133	0.87	0.86
130	0.86	0.86
127	0.85	0.85
123	0.83	0.83
120	0.82	0.82
117	0.81	0.81
113	0.80	080
110	0.78	0.78
106	0.77	0.77
103	0.75	0.75
99	0.73	0.73
95	0.72	0.72
91	0.70	0.70
87	0.68	0.68
83	0.65	0.65

IS : 2494—1974

Table 15.31: Nominal inside length, nominal pitch length and permissible length variations for V-blets

Nominal inside length, mm	*Nominal pitch length, mm, cross-section symbol*					*Pitch length*	*Variation*
	A	*B*	*C*	*D*	*E*	*PLL**	*MVL***
(1)	*(2)*	*(3)*	*(4)*	*(5)*	*(6)*	*(7)*	*(8)*
610		645					
660		696				+ 11.4	
711		747				– 6.4	
787		823				+ 12.5	
813		848				– 7.5	
889		925	932				
914		950					2.5
965		1001	1008				
991		1026					
1016		1051	1059			+ 14.0	
1067		1102	1110			– 8.9	
1092		1128					
1168		1204	1212				
1219		1255	1262				
1295		1331	1339	1531		+ 16.0	
1372			1415			– 9.0	
1397		1433	1440				
1422		1451	1466				
1473		1509					
1524		1560	1567	1580			5.0
1600		1636				+ 17.8	
1626		1661				– 12.5	
1651		1687	1694				
1727		1763	1770	1783			
1778		1814	1821				
1905		1941	1948	1991			
1981		2017	2024				
2032		2068					
2057		2093	2101	2113		+ 30	

2159		2195	2202	2215		− 16	7.5
2286		2322	2329	2342			
2438		2474		2494			
2464	—	2507	—	—	—	+ 34	
2540	—	2583	—	—	—	− 18	
2667	2703	2710	2723	—	—		
2845	2880	2888	2901	—	—		
3048	3084	3091	3104	3127	—		
3150				3205			
3251		3287	3294	3307	3330	+ 38	
3404				3459		− 21	12.5
3658			3693	3701	3736		
4013		4056	4069	4092		+ 43	
4115		4158	4171	4194		− 24	
4394		4437	4450	4473			
4572		4615	4628	4651			12.5
4953		4996	5009	5032	—	+ 49	
5334		5377	5090	5413	5426	− 28	
6045			6101	6124	6137		
6807			6863	6886	6889	+ 56	
7569			7625	7648	7661	− 32	
8331			8387	8410	8423	+ 65	
9093			9149	9172	9185		15
9855				9934	9947	− 37	
10617				10696	10709	+ 76	
12141				12220	12233	− 43	17.5
13665				13744	13757	+ 89	
15189				15268	15281	− 50	
16713				16792	16805	+ 105	
						− 59	

IS : 2494—1974

PLL* Pitch length limits, MVL* maximum variation in length within a method set.

Table 15.32 : Recommended standard pulley pitch diameters

Series of pitch diameters			*Degree of preference* for pitch diameters, according to groove section*				
Nominal value mm	*Pitch dia limits, mm*						
	Max.	*Min.*	*A*	*B*	*C*	*D*	*E*
(1)	*(2)*	*(3)*	*(4)*	*(5)*	*(6)*	*(7)*	*(8)*
75	75	76.5	3				
80	80	81.3	3				
85	85	86.4	3				
90	90	91.4	1				
95	95	96.5	2				
100	100	101.6	1				
106	106	107.7	2				
112	112	113.8	1				
118	118	119.9	2				
125	125	127.0	1	2			
132	132	134.1	2	2			
140	140	142.2	1	1			
150	150	152.4	2	2			
160	160	162.6	1	1			
170	170	172.7	3	2			
180	180	182.9	1	1			
190	190	193.0	3	3			
200	200	203.2	1	1	1		
212	212	215.4			2		
224	224	227.6	2	2	1		
236	236	239.8			2		
250	250	254.0	1	1	1		
265	265	269.2			2		
280	280	284.5	2	2	1		
300	300	304.8	2	2	2		
315	315	320.0	1	1	1		
355	355	360.7	2	2	2	1	
375	375	381.0		2	2	2	
400	400	406.4	1	1	1	1	
425	425	431.8			2		

450	450	457.2	2	2	2	1	
475	475	482.6			2		
500	500	508.8	1	1	1	1	1
530	530	538.5		3	3	3	2
560	560	569.0	2	2	2	2	1
600	600	609.6		2	2	2	2
630	630	640.0	1	1	1	1	1
670	670	680.7					2
710	710	721.4	2	2	2	2	1
750	750	762.0		2	2	2	3
800	800	812.8	3	1	1	1	1
900	900	914.4		2	2	2	2
1000	1000	1016.0		1	1	1	1
1060	1060	1077.0				2	
1120	1120	1137.9		3	2	2	2
1250	1250	1270.0			1	1	1
1400	1400	1422.4			2	2	2
1500	1500	1524.0				2	2
1600	1600	1625.6			1	1	1
1800	1800	1828.4				2	2
1900	1900	1930.4					2
2000	2000	2032.0				1	1
2240	2240	2275.8					2
2500	2500	2540.0					1

IS : 3142—1965

*1 = Recommended, 2 = Second preference, 3 = Not recommended.

CHAPTER 16

Chains and Ropes

A. Chains

1. Transmission ratio

$$i = n_1/n_2 = d_2/d_1 = z_2/z_1 \quad \text{...(16.1)}$$

where n = rpm, d = pitch diameter of sprocket, z = number of teeth on sprocket

Refer to Table 16.1 for preferred transmission ratios.

2. Average chain speed

$$v = pz_1n_1/(60 \times 10^3) \text{ m/s} \quad \text{...(16.2)}$$

where p = pitch of chain, mm

Refer to Table 16.2 for maximum speed of chains.

3. Minimum number of teeth on sprocket

$$Z_{min} = 17, Z_{max} = 114$$

$$Z_{min} = 4d_r/p + 5 \text{ for pitches of 16.54 mm} \quad \text{...(16.3a)}$$

$$= 4d_r + 4 \text{ for pitches of 3.175 to 5.785 mm} \quad \text{...(16.3b)}$$

Where d_r = roller pin diameter, mm

For recommended z_1, refer to Tables 16.3 and 16.4.

4. Maximum sprocket speed

(*a*) Based on allowable amount of impact between a roller and a sprocket, the speed of sprocket, according to Bartlett formula, is given by:

$$n \leq 1170\,[A/(W_f p)\,]^{1/2} \text{ rpm} \quad \text{...(16.4a)}$$

where W_f = weight of chain, N/m

$A = b_1 \times d_r$ = projected area of roller, m^2

b_1 = length of roller, m

p = pitch, m

$v_{max} \leq 19.48 z_1 [A/(W_f p)]^{1/2}$ m/s ...(16.4*b*)

(*b*) Based on energy of impact per tooth per minute,

$n \leq (1437/p)\ [A/(W_f)]^{1/2}$ rpm ...(16.5*a*)

$v_{max} \leq 24 z_1 [A/(W_f)]^{1/2}$ m/s ...(16.5*b*)

(*c*) Based on the effect of centrifugal force,

$n < (36350/p)\ (W_f z_1)\ ]^{1/2}$ rpm ...(16.6*a*)

$v_{max} \leq 600\ [A z_1/(W_f)]^{1/2}$ m/s ...(16.6*b*)

Refer to Table 16.2 for maximum speed of sprocket pinion.

5. Chain pitch

$p \leq 250\ (90/n_1)^{2/3}$ mm ...(16.7)

Refer to Table 16.5 for standard pitches.

6. Sprocket pitch diameter

$d = p/\sin(180°/z)$ ...(16.8*a*)

Root diameter, $d_f = d - d_r$ ...(16.8*b*)

7. Centre distance

$C_{min} = K_{min}\ C'$

Where $C' = (d_{a1} + d_{a2})/2$

d_a = tip diameter of sprocket

$= p/\tan(180°/z) + 0.6\ p$

K_{min} = a constant

$C_{max} = 80\ p$

Cetre distance in pitches, C_p = (20 – 30) p for highly pulsating loads ...(16.10)

= (30 – 50) p for uniform loads

Refer to Table 16.6 for centre distance.

8. Chain length

Chain length in pitches,

$L_p = 2C_p \cos\alpha + (z_1 - z_2)/2 + \alpha(z_1 - z_2)/180°$ ...(16.12*a*)

$= 2C_p + (z_1 + z_2)/2 + k(z_1 - z_2)^2/C_p$ ...(16.12*b*)

Where $\alpha = \sin^{-1}[(d_2 - d_1)/C_p]$

k = a variable dependent on $(Z_1 - z_2)/C_p$

Chain tength, $L = 2C\cos\alpha + z_1 p[(180° + 2\alpha)/360°] + z_2 p\ [(180° - 2\alpha)/360°]$ ...(16.13)

$L = mp$ ...(16.14)

Where m = number of chain links

$= 2C/p + (z_1 + z_2)/2 + p(z_1 - z_1)^2/(4\pi^2 C)$

Exact centre distance,

$$C = (p/4)\ [m - (z_1 + z_2)/2\ [\{m - (z_1 + z_2)/2\}^2 - 8\{(z_2 - z_1)/(2\pi)\}^2]^{1/2} \quad \text{...(16.15)}$$

9. Allowable chain pull

$$F_a = F_u/K_f \quad \text{...(16.16)}$$

Where F_u = ultimate strength of chain, N

K_f = working factor of safety

= 5 for z upto 40 and v upto 0.5 m/s

= 18 for z upto 10 and v upto 6 m/s

10. Power transmitted

$$P = F_v/(10^3 \times K_s)\ \text{kW} \quad \text{...(16.17)}$$

Where K_s = service factor (Table 16.9)

11. Design power

Design power = rated power × K_S

Rated power per strand, kW = $p^2\ [0.98v - (0.1988v)\ \{1 + 50 \sin^2 (90°/z_1)\}]$ ...(16.18)

12. Actual factor of safety

$$K_{fa} = F_u/(F + F_c + F_s)$$

Where $F = 10^3 \times P \times K_s/v$, is the tangential force in N

$F_c = W_f\ v^2/g$, is the centrifugal force in N

$F_s = K_{sg}\ W_f C$, is the sag force in N

W_f = weight of chain per meter, N/m

K_{sg} = coefficient of sag (Tabte 16.10)

For a safe design, $K_{fa} > K_f$

13. Number of strands.

If $F > F_a$, then number of strands,

$$i = F/F_a \quad \text{...(16.20)}$$

14. Bearing pressure on roller

$$p_b = F/A \quad \text{...(16.21)}$$

where $A = b_1 d_r$ = area of roller.

Power transmitted on the basis of allowable stress,

$$P = \sigma_b A v/K_s \quad \text{...(16.22)}$$

Where σ_b = allowable bearing stress, N/mm²

15. Load on shaft.

$$F_{shaft} = K_L F \quad \text{...(16.23)}$$

Where K_L = load factor (Table 16.12)

16. Sprocket proportions.

Refer to Tables 16.8, 16.9, 16.14 and 16.15.

B. ROPES

17. Selecting Fibre Ropes

Factor of safety $\approx$ 35

Speed of rope = 15 – 30 m/s

Diameter of rope sheave = (36 – 40) d

Where d = rope diameter

Number of grooves on sheave < 24

Length of rope,

$$L = \pi(D_2 + D_2)/2 + (D_2 - D_1)\sin^{-1}[D_2 - D_1]/(2C)] + 2C\cos\beta \quad ...(16.24)$$

Where D_1, D_2 = diameters of smaller and bigger pulley respectively

C = centre distance

β = semi – groove angle of pulley

Minimum centre distance,

$$C_{min} = (D_2 - D_1) + 1.5D_2 \quad ...(16.25)$$

Considering centrifugal tension,

$$(F_1 - F_c)/(F_2 - F_c) = \exp(\mu_e\,\theta) \quad ...(16.26)$$

Where μ_e = n/sin p

Total power transmitted, $P = n\,(F_1 - F_2)\,v/10^3$ kW ...(16.27)

Where n = number of ropes

v = rope velocity, m/s

Fibre ropes are joined together either by splicing or with the help of eyelet thimbles.

18. Steel Wire Ropes

(*i*) **Number of bends** – Refer to Table 16.19.

(*ii*) **Diameter of wire rope,** $d = 1.5\,d_w\sqrt{i}$ ...(16.28)

where i = number of wires in the rope

d_w = wire diameter

(*iii*) **Stress in the wire rope,**

$$\sigma = \sigma_{br}/K_f = W/A + d_w\,E_r/D_{min} \quad ...(16.29)$$

Where σ_{br} = ultimate breaking strength of the rope wire material, N/mm^2

K_f = rope factor of safety (Table 16.20)

A = area of useful cross – section of the rope, mm^2 (Table 16.21)

W = tension in the rope, N

E_r = 3E/8, corrected elastic modulus of rope, N/mm^2 (Table 16.22)

(iv) Maximum allowable tension in the rope:

$$W_a = F_{br}/K_f \text{ N} \quad ...(16.30)$$

Where F_{br} = breaking strength of the rope, N (Table 16.23 and 16.24)

(v) Steel rope weight and breaking load

The approximate steel rope weight is given by,

$$W_r = Kd^2 \text{ N/m length} \quad ...(16.31)$$

Where K = empirical factor for weight of steel cored ropes (Table 16.22)

d = diameter of rope, mm

Minimum breaking load,

$$F_{br} = K'\, d^2\, R_0 / 10^3 \text{ N} \quad ...(16.32)$$

where R_0 = tensile grade of wire, N/mm^2

K = empirical factor for the minimum breaking load for a given rope construction (Table 16.22)

(vi) Stresses in Hoisting Ropes

(a) Direct tensile stress

Effective load, $W_e = W + W_r$

$$\sigma_d = (W + W_r)/A \quad ...(16.33)$$

Where W = load lifted

W_r = self weight of the rope

A = area of cross – section of the rope

$= i\pi d^2_w/4$

(b) Bending stress, $\sigma_b = d_w E_r/D \quad ...(16.34)$

Where D = sheave diameter

Bending load, $W_b = \sigma_b \times A$

Effective load, $W_e = W + W_r + W_b$

(c) Starting stresses

1. Rope starts without slack

Acceleration, $a = v/1$

Acceleration load, $W_a = (W + W_r)\, a/g$

Effective load, $W_e = W + W_r + W_b + W_a \quad ...(16.35)$

2. Rope starts with slack 'h

$$W_{st} = (W + W_r)\, \{1 + v^2_{st}\, E_r/(\sigma L g)\}^{1/2}] \quad ...(16.36)$$

Where v_{st} = velocity of rope at the instant when the rope is taut

$= (2ah)^{1/2}$

Static stress in rope, $\sigma = (W + W_r)/A$

L = length of the rope

Effective load during starting,

$$W_e = W_{st} + W_b \quad ...(16.37)$$

(d) Stresses due to change in speed or sudden stopping of hoisting drum

1. Let the speed of rope change from v_1 to v_2 in time t. Then acceleration of rope,

$$a = (v_2 - v_1)/t$$

Acceleration load, $W_a = (W + W_r)\ a/g$

Effective load, $W_e = W + W_r + W_a + W_b$ ...(16.38)

2. Let x = distance in which the load being lifted is stopped.

L = length of the rope

Then $x_s = (W + W_f)L/(AE_r)$

$$x/x_s = \sigma_{sud}/\sigma \quad ...(16.39)$$

where σ_{sud} = sudden stress in the rope

Sudden load, $W_{sud} = \sigma_{sud} \times A$

Effective load, $W_e = W + W_r + W_b + W_{sud}$ ...(16.40)

(vii) Factor of safety

Factor of safety = Breaking load/Maximum effective load on the rope ...(16.41)

19. Wire Rope Sheares (Table 16.26)

Unit bearing pressure, $p = \dfrac{W}{l_a d_a}$

What l_a = length of axle bushing.

d_a = diameter of sheave axle

$$\frac{l_a}{d_a} = 1.5 \text{ to } 1.8 \quad ...(16.43)$$

20. Steel wire Rope Drums (Fig. 16.14, Table 16.29 and Fig. 16.15)

Number of turns on drum for one rope,

$$z = \frac{H^2}{\pi D} + 2 \quad ...(16.44)$$

where i = ratio of pully system

D = drum diameter

H = height to which load is raised

Length of drum, $L = \left(\dfrac{H^2}{\pi D} + 7\right) s$ for one rope ...(16.45a)

$= \left(\dfrac{2H^2}{\pi D} + 12\right) s + l_1$ for two ropes ...(16.45) where s = pitch of helix

l_1 = space in the middle

wall thickness of cost iron drums,

$$w = 0.02D + 6 \text{ to } 10 \text{ mm} \qquad ...(16.46)$$

Outside diameter of drum,

$$D_o = D + 6d \qquad ...(16.47)$$

21. Stresses in Rope Drms

Compressive stress in drum, $$\sigma_C = \frac{W}{w \times s} \qquad ...(16.48)$$

Maximum bending stress, $$\sigma_b = \frac{8\ WLD}{\pi(D^4 - D_i^4)} \qquad ...(16.49)$$

Maximum torque on drum, $$T_{max} = \frac{W(D + d)^2}{2} \qquad ...(16.50)$$

Maximum shear stress, $$\tau_{max} = \frac{16T_{max} \times D}{\pi(D^4 - D_i^4)} \qquad ...(16.51)$$

Maximum stress in drum, $$\sigma_{max} = \sqrt{\sigma_b^2 + 4\tau_{max}^2 + \sigma_c^2} \qquad ...(16.52)$$

22. Holding Capacity of Wire Rope Reel (Table 16.31)

$$L = \frac{\pi(A + D_r) \times A \times C}{1000\ d} \text{ m}$$

where $A = \frac{1}{2}(D_d - D_r)$, D_d = drum dia, mm, D_r = reel dia, mm, C = width between reel flanges, d = rope dia.

Table 16.l: Preferred transmission ratios for chain drive, ₹ 20 series

i = 1, 1.12, 1.25, 1.4, 1.6, 1.8. 2, 2.25, 3.15, 4, 4.5, 5. 5.6, 6.3 and 7.1.

Table 16.2: Maximum velocity for various types of chains, rpm

Type of chain	*Number of sprocket teeth*	*Chain pitch p, nun*				
	z	*12*	*15*	*20*	*25*	*30*
Roller chain	15	2300	1900	1350	1150	1000
	19	2400	2000	1450	1200	1050
	23	2500	2100	1500	1250	1100
	27	2550	2150	1550	1300	1100
	30	2600	2200	1550	1300	1100
Silent chain	17 to 35	12.7	15.87	19.05	25.4	31.75
		3300	2650	2200	1650	1300

Table 16.3: Preferred range of teeth on roller chain sprockets

sprockets for steet roller chains
z = 17, 19, 21, 23, 25, 38, 57, 76, 95 and 114
Sprockets for extended pitch precision chains
z = 7.9, 10, 11, 13, 19, 27, 38 and 57

Table 16.4: Recommended number of teeth on smaller sprocket

Type of chain	*Number of teeth at speed ratio*				
	1—2	*2—3*	*3—4*	*4—5*	*5—7*
Roller	31—27	27—25	25—23	23—21	21—17
Silent	40—35	35—31	31—27	27—23	23—19

Table 16.5: Standard pitches for transmission steel roller chains

Preferred pitches (mm)— 8.0, 9.325, 12.70, 15.875, 19.05, 25.40, 31.75, 38.10, 44.45. 50.80, 63.50, 76.20, 88.90, 101.60 and 114.30

IS: 2403—1975

Table 16.6: Centre distance in chain drive

Transmission ratio	*K* *min*	*Minimum centre distance, C_{min}* *mm*
3	1 + (30 to 50/C*)	$(d_{a1} + d_{a2})$ + (30 to 50mm)
3 to 4	1.2	$0.3\,(d_{a1} + d_{a2})$
4 to 5	1.3	$0.65\,(d_{a1}, + d_{a2})$
5 to 6	1.4	$0.7\,(d_{a1} + d_{a2})$
6 to 7	1.5	$0.75\,(d_{a1} + d_{a2})$

c* = ½ $(d_{a1} + d_{a2})$, d_a = tip dia. of sprocket.

Table 16.7: Condensed values of k in chain length, pitches

$\frac{z_1 - z_2}{C_p}$	*0.1*	*1.0*	*2.0*	*3.0*	*4.0*	*5.0*	*6.0*
k	0.02533	0.02538	0.02525	0.02584	0.02631	0.02704	0.02828

Table16.8: Factor of safety K, for chain drives

Type of chain	*RPM of smaller sprocket*								
	50	*200*	*400*	*600*	*800*	*1000*	*1200*	*1600*	*2000*
Buch roller chain									
p = 12 to 15 mm	7	7.8	8.55	9.35	10.2	11.0	11.7	13.2	14.8
= 20 to 25 mm	7	8.2	9.35	10.3	11.7	12.9	14.0	16.3	–
= 30 to 35 mm	7	8.55	10.2	13.2	14.8	16.3	19.5	—	–
Silent chain									
p = 12.7 to 15.87	20	22.2	24.2	28.7	29.0	31.0	33.4	37.8	42.0
= 19.05 to 25.4 mm	20	23.4	26.7	30.0	33.4	36.8	40.0	46.5	53.5

Table 16.9: Service Factor K,

Service factor, K_s	=	$K_1 K_2 K_3 K_4 K_5\ K_6$
Load factor K_1	=	1 .0 for constant load
	=	1 .25 for variable load with mild shock
	=	1 .50 for heavy shock loads.
Factor for distance regulation K_2	=	1 .0 for adjustable supports
	=	1.1 for chains with idlers
	=	1 .25 for fixed supports
Factor for centre distance K_3	=	1.25 for $c_p < 25$
	=	1.0 for $c_p = 30$ to 50
	=	0.8 for $c_p = 60$ to 80
Factor for position of shaft K_4	=	1 .0 for inclination upto 60° to horizontal.
	=	1 .25 for inclination over 60°
Lubrication factor K_5	=	0.8 for continuous lubrication
	=	1.0 for drop lubrication
	=	1.5 for periodic lubrication.
Rating factor K_6	=	1.0 for 8 hours/day
	=	1.25 for 16 hours/day
	=	1.50 for continuous service
Design power = Rated power × K_s		

Table 16.10: Coefficient for sag

Correction Factor	*Position of chain drive*			
	Horizontal	*Upto 40°*	*More than 40°*	*Vertical*
K_{sg}	6	4	2	1

Table 16.11: Allowable bearing stress p_b on rollers, N/nun^2 (For $K_s = 1$ and $z_1 = 15$ to 30)

Pitch mm	*Speed of small sprocket, rpm*										
	<50	*200*	*400*	*600*	*800*	*1000*	*1200*	*1600*	*2000*	*2400*	*2800*
9.525 12.70 15.875 19.05 34.32	34.32	30.9	28.14	25.7	23.73	21.97	20.6	18.14	16.18	14.7	13.43

Table 16.12: Load factor, K_L

Load factor	*Position of chain drive*			
	Horizontal or upto 40°		*Vertical or more than 40°*	
	Constant load	*Shoch load*	*Constant load*	*Shock load*
K_L	1.15	1.3	1.05	1.15

Table 16.13: Tooth width of simple chain wheels ana multiplex chain wheels.

Dimensions	*P ≤ 12.7 mm*	*Tol*	*P > 12.7 mm*	*Tol*
bf_1 for simple chain wheels	0.93 b_1		0.95 b_1	
bf_1 for duplex and triplex chain wheels	0.91 b_1	$h14$	0.93 b_1	$h14$
bf_1 for quadruplex chain wheels and above	0.88 b_1		0.95 b_1	

Table 16.14: Chain Dimensions, Measuring Loads and Breaking Loads of Base Chains (Transmission steel roller chains) (Refer Fig. 16.8)

All dimensions in millimetres

ISO Chain Number	*Pitch*	*Roller Dia*	*Width between Inner Plates*	*Bearing Pin Body Diameter*	*Chain Path Depth*	*Inner Plate mediate*	*Outer Inter Plate Depth*	*Cranked Link Dimensions**			*Transverse Pitch*	*Width Over Inner Link*	*Width between Outer Plates*
	p	d_r *Max*	b_1 *Min*	d_p *Max*	h_1 *Min*	h_2 *Max*	h_3 *Max*	l_1 *Min*	l_2 *Min*	c	p_t	b_2 *Max*	b_3 *Min*
05B	8.00	5.00	3.00	231	7.37	7.11	7.11	3.71	3.71	0.08	5.64	4.77	4.90
06B	9.525	9.53	5.72	3.28	8.52	8.26	8.26	437	432	0.08	10.24	853	8.66
08B	12.70	8.51	7.75	4.45	12.07	11.81	10.92	5.66	6.12	0.08	13.92	11.30	11.43
10B	15.875	10.16	9.65	5.08	14.99	14.73	13.72	7.11	7.62	0.10	16.59	13.28	13.41
12B	10.05	12.07	11.63	5.72	16.39	16.13	16.13	833	833	0.10	19.46	15.62	15.75
16B	25.40	15.88	17.02	8.28	21.34	21.08	21.08	11.15	11.15	0.13	31.88	25.45	2558
20B	31.75	19.05	1956	10.19	26.28	26.42	26.42	13.89	13.89	0.15	36.45	29.01	29.14
24B	38.10	25.40	25.40	14.63	33.73	33.40	33.40	17.25	17.55	0.18	4836	37.92	38.05
28B	44.45	27.94	30.99	15.90	37.46	37.08	37.08	1951	1951	0.20	5956	46.58	46.71
32B	50.80	29.21	30.98	17.81	42.72	42.29	42.29	22.20	22.50	0.20	58.55	45.57	45.70
40B	63.50	39.76	38.10	22.89	54.49	52.96	52.96	27.76	27.76	0.20	72.29	55.75	55.88
48B	76.20	48.26	45.72	29.24	6452	63.88	63.88	33.45	33.45	0.20	91.21	7056	70.69
56B	88.90	53.98	53.34	34.32	78.64	77.85	77.85	40.61	40.61	0.20	106.60	71.33	81.46
64B	101.60	63.50	60.96	39.40	91.08	90.17	90.17	47.07	47.07	0.20	119.89	92.02	92.15
72B	114.30	72.39	68.51	44.48	104.67	102.63	103.63	53.37	5337	0.20	136.27	103.81	103.94

Table 16.14: continued.

ISO Chain Number	Width Over Bearing Pins			Additional Width for Joint Fastener+	Measuring Load, N			Breaking Load, kN			Average weight W_p N/m		
	Simple	Duplex	Triplex		Simple	Duplex	Triplex	Simple	Duplex	Triplex	Simple	Duplex	Triplex
	b_4 Max	b_5 Max	b_6 Max	b_7 Max				Max	Min	Min			
05B	8.6	14.3	19.9	3.1	49	98	147	4.51	7.84	11.18	1.76	3.24	5.10
06B	13.5	23.8	34.0	33	78.5	14.7	21.6	8.92	16.96	24.96	4.02	7.55	10.69
O8B	17.0	31.0	44.9	3.9	127.5	255	382	17.85	31.18	44.52	6.86	12.94	19.12
10B	19.6	36.2	52.8	4.1	196.1	392	588	22.26	4452	66.78	8.92	17.85	26.77
12B	22.7	42.2	61.7	4.6	284.4	559	843	28.93	57.86	88.79	11.47	23.14	34.71
16B	36.1	68.0	99.9	5.4	500.1	1002	1491	42.27	84.53	126.80	26.48	51.97	76.98
20B	43.2	79.7	116.1	6.1	774.7	1559	2334	64.53	129.05	193.58	35.79	70.91	116.70
24B	53.4	101.8	150.2	6.6	1108	2226	3334	97.87	195.74	293.62	67.17	142.20	196.13
28B	65.1	124.7	184.3	7.4	1510	3020	4440	129.05	258.11	387.17	—	—	—
32B	67.4	126.0	184.5	7.9	2000	4001	6001	160.07	338.13	507.20	—	—	—
40B	82.6	154.9	227.2	10.2	3118	6227	9346	262.52	525.05	78757	—	—	—
48B	99.1	190.4	281.6	105	4452	8895	13347	400.40	800.71	1201.02	—	—	—
56B	114.6	2212	—	11.7	6090	12,189	—	542.70	1085.40	—	—	—	—
64B	130.9	250.8	—	13.9	7963	15,926	—	711.77	1423.43	—	—	—	—
72B	147.4	283.7	—	14.3	10,101	20,192	—	89858	1797.71	—	—	—	—

IS-24Q3—1975

Table 16.15: Chain Dimensions, Measuring Loads and Breaking Loads for extended pitch transmission precision roller chains

All dimensions in millimetres

Chain Number	Pitch	Roller dia	Width between Inner Plates	Bearing pin body dia	Bush bore	Chain path depth	Plate depth	Cranked Link Dimension	Width over Inner Link plates	Width between outer pin	Width Over Bearing	Addl. Width for Joint* Fastner	Measuring load	Breaking load	Weight per m
	p	d_r Max	b_1 Min	b_p Max	b_b Min	h_1 Min	h_1 Max	l_2 Min	b_2 Ma	b_3 Max	b_4 Max	b_7 Max.	N	F_u kN	N
208A	25.40	7.92	7.95	3.96	4.01	12.33	12.07	6.9	11.18	11.31	17.8	3.9	127.5	13.89	4.71
208B	25.40	8.51	7.75	4.45	4.50	12.07	11.81	6.9	11.30	11.43	17.0	3.0	127.4	17.85	4.31
210A	31.75	10.16	9.53	5.08	5.13	15.35	15.09	8.4	13.28	13.97	21.8	4.1	196.1	21.77	—
210B	31.75	10.16	9.65	5.08	5.13	14.99	14.73	8.4	13.28	13.41	19.6	4.1	196.1	22.26	5.88
212A	38.10	11.91	12.70	5.94	5.99	18.34	18.08	9.9	17.75	17.88	26.9	4.6	284.4	31.18	14.22
212B	38.10	12.07	11.68	5.72	5.77	16.39	16.13	9.9	15.62	15.75	22.7	4.6	284.4	28.93	7.35
216A	50.80	15.88	15.88	7.92	7.79	24.39	24.13	13.0	22.61	22.74	33.5	5.4	500.1	55.60	—
216B	50.80	15.88	17.02	8.28	8.33	21.34	21.08	13.0	25.25	25.58	36.1	5.4	500.1	42.27	—
220A	63.50	19.05	10.05	9.53	9.58	30.48	30.18	16.0	27.46	27.59	41.1	6.1	774.7	86.79	—
220B	63.50	19.05	19.56	10.19	10.24	26.68	26.42	16.0	29.41	29.14	43.2	6.1	774.7	64.53	—
224A	76.20	22.23	25.40	11.10	11.25	36.55	36.20	19.1	35.46	35.59	50.8	6.6	1108.1	124.54	—
224B	76.20	25.40	25.40	14.63	14.68	33.73	33.40	19.1	37.92	38.05	53.4	6.6	1108.1	97.87	—
228B	88.90	27.94	30.99	15.90	15.95	37.46	37.08	21.3	46.58	46.71	65.1	7.4	1510.2	129.05	—
232B	101.60	29.21	30.99	17.81	17.86	42.72	42.29	24.4	45.57	45.70	67.4	7.9	2000.5	169.07	—

IS : 3542—1980

Table 16.16: Pitch Circle Diameters for Chain Wheels. For transmission Steel Roffer Chains

The following table gives correct pitch circle diameters for wheels to suit a chain of unit pitch (1 mm). For chain of any other pitch the pitch circle diameter of the wheel is obtained by multiplying the figures given below by the pitch of that chain :

IS : 2403—1975

Number of teeth	*Pitch circle diameter*	*Number of teeth*	*Pitch circle diameter*	*Number of teeth*	*Pitch circle diameter*	*Number of teeth*	*Pitch circle diameter*
9	2.923 8	45	14.335 6	81	25.789 5	117	37.245 7
10	3.236 1	46	14.653 7	82	26.107 8	118	37.565 0
11	3.549 4	47	14.971 7	83	26.426 0	119	37.883 3
12	3.863 7	48	15.289 8	84	26.744 3	120	38.201 6
13	4.178 6	49	15.607 9	85	27.062 5	121	38.519 8
14	4.494 0	50	15.926 0	86	27.380 7	122	38.838 1
15	4.809 7	51	16.244 1	87	27.699 0	123	39.1564
16	5.125 8	52	16.562 2	88	28.017 2	124	39.474 6
17	5.442 2	53	16.880 3	89	28.335 S	125	39.792 9
18	5.758 8	54	17.198 4	90	28.653 7	126	40.111 2
19	6.075 5	55	17.516 6	91	28.971 9	127	40.429 5
20	6.392 5	56	17.834 7	92	29.290 2	128	40.747 8
21	6.709 5	57	18.152 9	93	29.608 4	129	41.0660
22	7.026 6	58	18.471 0	94	29.926 7	130	41.3843
23	7.343 9	59	18.789 0	95	30.244 9	131	41.702 6
24	7.661 3	60	19.107 3	96	30.563 2	132	42.020 9
25	7.978 7	61	19.425 5	97	30.881 5	133	42.339 1
26	8.296 2	62	19.743 7	98	31.1997	134	42.6574
27	8.613 8	63	20.061 9	99	31.5180	135	42.975 7
28	8.931 4	64	20.380 0	100	31.836 2	136	43.294 0
29	9.249 1	65	20.698 2	101	32.154 5	137	43.612 3
30	9.566 8	66	21.0164	102	32.472 7	138	43.930 6
31	9.884 S	67	21.334 5	103	32.791 0	139	44.248 8
32	10.202 3	68	21.6528	104	33.109 3	140	44.567 1
33	10.520 1	69	21.971 0	105	33.427 5	141	44.885 4
34	10.838 0	70	22.289 2	106	33.745 8	142	45.203 7
35	11.155 8	71	22.607 4	107	34.064 0	143	45.522 0
36	11.473 7	72	22.925 6	108	34.382 3	144	45.840 3
37	11.791 6	73	23.243 8	109	34.700 6	145	46.158 5
38	12.109 6	74	23 .562 0	110	35.018 8	146	46.476 8
39	12.427 5	75	23.880 2	111	35.337 1	147	46.795 1
40	J2.745 5	76	24.198 5	112	35.655 4	148	47.1134
41	13.063 5	77	24.516 7	113	35.973 7	149	47.401 7
42	13.381 5	78	24.834 0	114	36.291 9	150	47.750 0
43	13.699 5	79	25.153 1	115	36.610 2		
44	14.017 6	80	25.471 3	116	36.928 5		

Table 16.17: Pitch Cirde Diameter for Extended Pitch Precision Roller Chains

The following table gives correct pitch circle diameter, for wheels to suit a chian of unit pitch (for example, 1 mm). The pitch circle diameters for wheels to suit a chain of any other pitch are directly proportional to the pitch of the chain :

Number of teeth	*Pitch circle diameter*	*Number of teeth*	*Pitch circle diameter*	*Number of teeth*	*Pitch circle diameter*	*Number of teeth*	*Pitch circle diameter*
5	1.701 3	23	7.343 9	41	13.063 5	58½	18.630 1
5½	1.849 6	23½	7302 6	4I½	13.222 5	59	18.789 2
6	2.000 0	24	7.661 3	42	13.381 5	59½	18.948 2
6½	2.151 9	24½	7.820 0	42½	13.540 5	60	19.107 3
7	2.304 8	25	7.978 7	43	13.699 5	60½	19.266 5
7½	2.458 6	25½	8.137 5	43½	13.858 5	61	19.425 5
8	2.613 1	26	8.296 2	44	14.017 6	61½	19384 7
8½	2.768 2	26½	8.455 0	44V4	14.176 5	62	19.743 7
9	2.923 8	27	8.613 8	45	14.335 6	62½	19.902 9
9½	3.079 8	27½	8.772 6	45½	14.494 6	63	20.061 9
10	3.236 1	28	8.931 4	46	14.653 7	63½	20.221 0
10½	3.392 7	28½	9.090 2	46	14.812 7	64	20.380 0
11	3.549 4	29	9.249 1	47	14.971 7	64½	20.539 3
11½	3.706 5	29½	9.408 0	47½	15.130 8	65	20.098 2
12	3.863 7	30	9.566 8	48	15.289 8	65½	20.857 5
12½	4.021 1	30½	9.725 6	48½	15.448 8	66	21.0164
13	4.178 6	31	9.884 5	49	15.607 9	66½	21.175 7
13½	4.336 2	31½	10.043 4	49½	15.766 9	67	21.334 6
14	4.494 0	32	10.202 3	50	15.926 0	67½	21.493 9
14½	4.651 8	32½	10.361 2	50½	16.085 0	68	21.652 8
15	4.809 7	33	10.520 1	51	16.244 1	68½	21.812 1
15½	4.967 7	33½	10.679 0	51½	16.403 1	69	21.971 0
16	5.125 8	34	10.838 0	52	16362 2	69½	22.130 3
16½	5.284 0	34½	10.996 9	52½	16.721 2	70	22.289 2
17	5.442 2	35	11.155 8	53	16.880 3	70½	22.448 5
17½	5.600 5	35½	11.314 8	53½	17.039 3	71	22.607 4
18	5.758 8	36	11.473 7	54	17.198 4	71½	22.766 7
18½	5.917 1	36½	11.632 7	54½	17.357 5	72	22.925 6
19	6.075 5	37	11.791 6	55	17316 6	72½	23.084 9
19½	6.234 0	37½	11.950 6	55½	17.675 6	73	23.243 8
20	6.392 5	38	12.109 5	56	17.834 7	73½	23.403 1
20½	6.550 9	38½	12.268 5	56½	17.993 8	74	23362 0
21	6.709 5	39	12.427 5	57	18.152 9	74½	23.721 3
21½	6.868 1	39½	12.586 5	57½	18.311 9	75	23.880 2
22	7.026 6	40	12.745 5	58	18.471 0		
22½	7.185 3	40½	12.904 5				

IS : 3842—1966

Table 16.18: Common wire rope applications

Type of service	*Type of rope*	*Sheave diameter*	
		Recommended	*Minimum*
Haulage rope Mine haulage Factory yard haulage Inclined planes Tramway Power transmission Guy wires	6 × 7	72 d	42 d
Standard hoisting rope (most commonly used rope) Mine hoists Quarries Ore docks Cargo hoists Car pullers Cranes Derricks Dredges Tram ways Well drilling Elevators	6 × 19 60-100 20-30 d	45 d	30 d
Extra-flexible hoisting rope	8 × 19	31 d	21 d
Special flexible hoisting rope : Steel mill ladles Cranes High speed elevators Service where sheave Diameters are limited	6 × 37	27 d	18 d

Table 16.19: D_{min}/d as a functions of the number of bends in the systems

No. of bends	1	2	3	4	5	6	7	8	9	10	11	12	13	14	15	16
D_{min}/d	16	20	23	25	26.5	28.0	30	31	32	33	34	35	36	37	37.5	38

Table 16.20: Recommended facators of safety for wire ropes, K_f

Rope application	*Factor of safety*		
Mining ropes	*10 or other figure laid down by the statutory authority*		
	Class 1	Class 2 & 3	Class 4
Wire ropes used on the cranes and other hoisting equipment. Fixed guys Unreeved rope bridles of jib cranes or ancillary appliances, such as lifting beams. Ropes which are straight between terminal fittings	3.5	4.0	4.5
Hoisting, lifting and reeved bridle systems of inherently flexible cranes (for example, mobile crawler tower, guy derrick, stiff leg derrick, where jibs are supported by ropes or where equivalent shock absorbing devices are incorporated in jib supports.	4.0	4.5	5.5
Cranes and hoists in general hoist blocks.	4.5	5.0	6.0

IS: 3973—1967

Table 16-21: Approximate Wire Rope and Sheave Data

Rope Construction	*Weight N/m*	*Wire dia (d_w) (mm)*	*Area mm²*	*Recommended Sheave dia*	
				Average	*Minimum*
6 × 7	$0.0333d^2$	0.106d	$0.38d^2$	75d	42d
6 × 19	$0.0363d^2$	0.063d	$0.38d^2$	45d	30 d
6 × 37	$0.0343d^2$	0.045d	$0.38d^2$	27d	18d
8 × 19	$0.0339d^2$	0.050d	$0.38d^2$	31d	21d

Table 16.22: Numerical values of factor K and K′

Rope designation	*Ropes mass factor*			$\frac{K_2}{K_{1n}}$	$\frac{K_2}{K_{1p}}$	*Min. breaking load / actor*		$\frac{K_2'}{K_1'}$	E_r N/mm²
	Natural fibre cred ropes K_{1n}	*Polypropylene fibre cored ropes, K_{1p}*	*Steel cored ropes K_2*			*Fibre cored ropes K_1'*	*Steel cored ropes K_2'*		
	kg(100 m. mm²)								
6 × 7	0.3574	03485	03485	1.10	1.135	13322	03588	1.08	68600
6 × 19	0.3460	0.3374	0.3906	1.10	1.135	0.3073	0.3319	1.08	64700
6 × 37	0.3460	0.3374	—	—	—	0.2948	—	—	
6 × 19 Seale	0.3726	0.3633	0.4099	1.10	1.135	0.3309	0.3574	1.08	64700

6 × 19 Filler	0.3802	0.3707	0.4182	1.10	1.135	0.3377	0.3647	1.08	
6 × 26 Warrington —Seale									
6 31 Warrington —Seale	0.3802	0.3707	0.4182	1.10	1.135	0.3299	0.3563	1.08	
6 × 36 Warrington Seale									
6 × 41 Warrington Seale									
6 × 12	0.2506	0.2306	—	—	—	0.2086	—	—	
6 × 24	0.3184	0.3041	—	—	—	0.2802	—	—	
8 × 19 Seale	0.3484	0.3362	0.4252	1.22	1.28	0.2870	0.3386	1.18	68600
8 × 19 Filler	0.3655	9.3440	0.4349	1.22	1.28	0.2936	0.3464	1.18	
17 × 7 & 18 × 7	0.3828	0.3790	0.4019	1.05	1.06	0.3186	0.3281	1.03	53900
34 × 7 & 36 × 7	0.3902	0.3863	0.4019	1.03	1.035	0.3122	0.3184	1.02	

IS: 6494—1977

Table 16.23: Wire ropes for general engineering purposes

(*a*) 6 × 7 (6/1) construction

Nominal dia, mm	*Approximate mass kg / 100m*		*Minimum breaking load in kN corresponding to tensile designation of wires of*					
	Fibre core	*Steel core*	*1570*		*1770*		*1660*	
			Fibre core	*Steel core*	*Fibre core*	*Steel core*	*Fibre core*	*Steel core*
8	22.9	25.2	33	36	38	41	42	45
9	28.9	31.8	42	46	48	51	53	57
10	35.7	39.1	52	56	59	64	65	70
11	43.2	47.6	63	68	71	77	79	85
12	51.5	56.6	75	81	85	91	94	101

(*b*) 6 × 19 (12/6/1) construction (aggregate breaking load = 1.163 × tabulated load)

8	22.1	31	35	39
9	38.0	39	44	49
10	34.6	48	54	60
11	41.9	58	66	73
12	49.8	69	78	82
13	585	82	92	102
14	67.8	95	107	118
16	88.6	124	139	154
18	112	156	176	195
19	125	174	199	217
20	138	193	218	241
22	167	234	263	291
24	199	278	313	347
26	334	326	368	407
28	371	378	426	472
32	351	494	557	617
36	448	625	705	781
40	554	772	870	964
44	670	934	1053	1166
48	797	1112	1253	1388
52	936	1305	1471	1629

Table 16.23: Continued.

(c) 6 × 37 (18/12/6/D construction

8	22.1	24.4	30	32	33	36	37	40
9	28.0	30.8	37	41	42	46	47	51
10	34.6	38.1	46	50	52	56	58	62
11	41.9	46.1	56	60	63	68	70	76
12	49.8	54.8	67	72	75	81	83	90
13	58.5	64.3	78	84	88	95	98	106
14	67.8	74.6	91	98	102	110	113	122
16	88.6	97.4	118	128	133	144	148	160
18	112	123	150	162	169	183	187	202
19	125	137	167	180	188	203	209	224
20	138	152	185	200	209	225	231	250
22	167	184	224	242	252	273	280	302
24	199	219	267	289	300	225	333	359
26	234	257	313	338	333	381	390	422
28*	271	—	363	—	409	—	453	—
32*	354	—	474	—	531	—	592	—

36*	448	—	600	—	676	—	749	—
40*	554	—	740	—	835	—	924	—
44*	670	—	896	—	1010	—	1119	—
48*	797	—	1070	—	1200	—	1331	—
52*	936	—	1250	—	1410	—	1562	—
56*	1085-	—	1450	—	1640	—	1812	—

Aggregate load for fibre core = 1.21 × tabulated load

Aggregate load for steel core = 1316 × tabulated load

*Sizes not preferred.

(*d*) 6 × 17 (8/8/1) and 6 × 19 (9/9/1) construction

Nominal dia, mm	*Approximate mass kg / 100m*		*Minimum breaking load in kN corresponding to tensile designation of wires of*					
			1570		*1770*		*1660*	
	Fibre core	*Steel core*	*Fibre core*	*Steel core*	*Fibre core*	*Steel core*	*Fibre core*	*Steel core*
8	23.8	26.2	33	36	37	40	42	45
9	30.2	33.2	42	45	47	51	53	57
10	37.3	41.0	52	56	59	69	65	70
11	45.1	49.6	63	68	71	77	78	85
12	53.7	49.0	75	81	84	91	93	101
13	63.0	693	88	95	99	107	110	118
14	73.0	80.3	102	110	115	124	127	137
16	95.4	105	133	144	150	162	166	179
18	121	133	168	182	190	205	210	227
19	135	148	188	203	211	228	234	253
20	149	164	208	224	234	253	259	280
22	180	198	251	272	283	306	314	339
24	215	236	299	323	337	364	374	403
26	252	277	351	379	396	428	438	479
28	292	321	407	440	459	496	508	549
32	382	420	532	575	600	648	664	717
36	483	531	673	727	759	820	841	908
40	596	656	831	898	937	1012	1038	1121
44	721	794	1006	1086	1134	1225	1226	1356
48	858	944	1197	1293	1349	1458	1494	1614
52	1008	1108	1405	1517	1584	1711	1754	1894

Table 16.24: Specificatioas for steel wire suspension ropes for lifts, elevators and hoists

(*a*) 6 × 19 (12/6/1) construction

Nominal did, mm	*Approximate mass kg/100 m*	*Minimum breaking load in kN corresponding to tensile designation of wires of*		
		1230	*1420*	*1570*
6	12.5	13.6	15.7	17.4
7	17.0	185	21	24
8	22.1	24	28	31
9	28.0	31	35	39
10	346	38	44	48
11	41.9	46	53	58
12	49.8	54	63	69

(*b*) 6 × 19 (9/9/1) construction

6	13.4	14.7	16.9	18.7
7	18.3	19.9	23	25
8	23.8	6	30	33
9	30.2	33	38	42
10	37.3	41	47	52
11	45.1	49	57	63
12	53.7	59	68	75
13	63.0	69	79	88
14	73.0	80	92	102
16	95.4	104	120	133
18	121	132	152	168
19	135	147	170	189
20	149	162	188	208

(*c*) 6 × 19 (12/6/6F/1) construction

6	13.7	15.0	173	19.1
7	18.6	20.4	23	26
8	24.3	27	31	34
9	30.8	34	39	43
10	38.0	42	48	53
11	46.0	50	58	64
12	54.7	60	69	76
13	64.3	70	81	90
14	74.5	81	94	104
16	97.3	106	123	136
18	123	135	155	172
19	137	150	173	191
20	152	166	192	212

The nominal diameter 19 mm is non-preferred.

Table 16.25: Ratio of dram and sheave diameter to rope diameter

Purpose		*ConstructionMinimum ratio* upto v = 50 m/min.*		
Mining installations	*All*	100		
		Class I	Class 2 & 3	Class 4
Cranes and allied hoisting equipment	6 × 37 8 × 19 Filler wire	15	17	22
	8 × 19 8 × 19 Warrington	17	18	24
	8 × 19 Scale 37 × 7 Non-rotating 6 × 24	18	19	25
	6 × 19 Filler wire	18	20	23
	6 × 19 6 × 19 Warrington	19	23	27
	17 × 9 Non-rotating 18 × 7 Non-rotating 6 × 19 Scale	24	28	35

*Increase this ratio by 8% for each additional speed of 50 m/min. IS : 3973—1967

Table 16.26: Dimensions of sheave for steel wire ropes

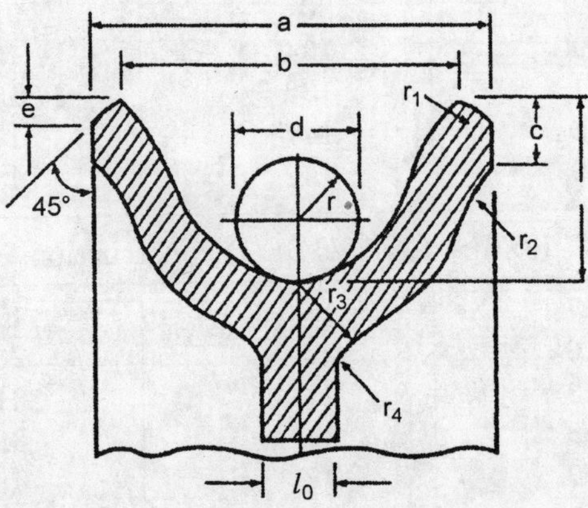

Fig. 16.13 Dimensions of sheave for steel wire ropes.

AD dimensions hi mm

Rope dia, d	*a*	*b*	*c*	*e*	*h*	l_0	*r*	r_1	r_2	r_3	r_4
4.8	22	15	5	0.5	125	8	4	25	2	8	6
6.2	22	15	5	0.5	125	8	4	25	2	8	6
8.7	28	20	6	1	15	8	5	3	2.5	9	6
11.0	40	30	7	1	25	10	85	4	3	12	K
13.0	40	30	7	1	25	10	85	4	3	12	8
15.0	40	30	7	1	25	10	85	4	3	12	K
19.5	55	40	10	1.5	30	15	12	5	5	17	10
24.0	65	50	10	1.5	37.5	18	14.5	5	5	20	15
28.0	80	60	12	3	45	20	17	6	7	25	15
34.5	90	70	15	2	55	22	20	7	8	28	20
39.0	110	85	18	27	65	22	25	9	10	40	30

Rudenko, N. 'Materials Handling Equipment', Mir Publishers, Moscow (1969), p. 86, Table 16.

Table 16.27: Tolerance on pulley or sheave groove diameters

Nominal dia of rope, mm	6 to 8	10 to 20	22 to 29	32 to 38	31 to 57	Above 57
Minimum clearance mm	0.80	1.60	2.40	3.20	4.00	4.80

IS : 3973—1967

The unit pressure should not exceed the values given in Table 16.28.

Table 16.28: Unit pressure for sheave bearings

$v_1 m/s$	0.1	0.2	0.3	0.4	0.5	0.6	0.7	0.8	0.9	1.0	1.1	1.2	1.3	1.4	1.5
$p, N/mm^2$	7.35	6.86	6.47	6.08	5.88	5.59	5.39	5.29	5.19	5.09	5.00	4.90	4.80	4.71	4.61

Table 16.29: Dimensions of drum grooves

All dimensions in mm

Rope dia, d	$r_1 = 0.53\,d$	*Standard groove*		r_2	*Deep groove*	
		$s_1 = 1.15d$	$c_1 = 0.25d$		s_2	c_2
44	35	7	2	1.0	9	45
63	4	8	2	1.5	11	45
8.7	5	11	3	1.5	13	65
11.0	7	13	3	1.5	17	85
13.0	8	15	4	15	19	9.5
15.0	9	17	5	2.0	22	11.0
195	115	22	5	2.0	27	135
24.0	135	27	6	25	31	16
28.0	155	31	8	25	36	18
345	19	38	10	3.0	41	22
39	21	42	12	35	50	245

Rndenko, N., "Materials Handling Equipment", Mir Publishers, Moscow (1969), p. 90, table 17.

Table 16.30 Clearance between adjacent ropes

Type of groove	*Rope diameter, mm*		*Clearance*
	From	*Upto and including*	
Concentric groove, CG	—	12	1.5
	12	28	2.5
	29	36	3.0
Helical groove, HG	—	32	1.5
	32	44	2.5
	44	51	3.0

IS : 3973— 1967

Table 16.31: Dimensions for Fastening Steel Wire Ropes on Drums

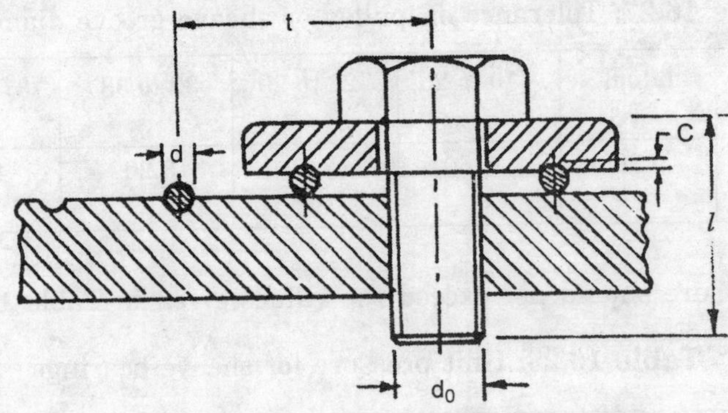

Fig. 16.16. Dimensions for fastening steel wire ropes on drums.

All dimensions in mm

Rope dia d	*Locating dimensions*		*Screw size*			*Plate*
	Pitch of screws	*t*	d_0	*l*	*c*	*Number of fastenings*
(1)	*(2)*	*(3)*	*(4)*	*(5)*	*(6)*	*(7)*
4.8 6.2	28	19	8	20	2	1
7.7 8.7	34	25	10	30	4	1
11 13	52	35	16	45	5	1
15 17.5 18	53	43	18	50	7	1

19.5 20.5 21.5	58	53	22	60	8	1
23 24 26	78	60	22	60	10	1
28 28.5 30 31	78	70	27	75	12	1
32.5 34.5 36	100	85	30	100	14	2

Rudenko, R, "Materials Handling Equipment", Mir Publishers, Moscow, (1969), p. 94, Table 18.

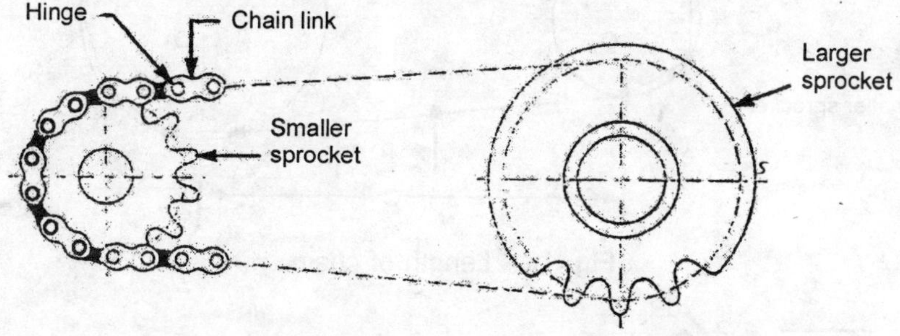

Fig. 16.1 Sprockets and chain.

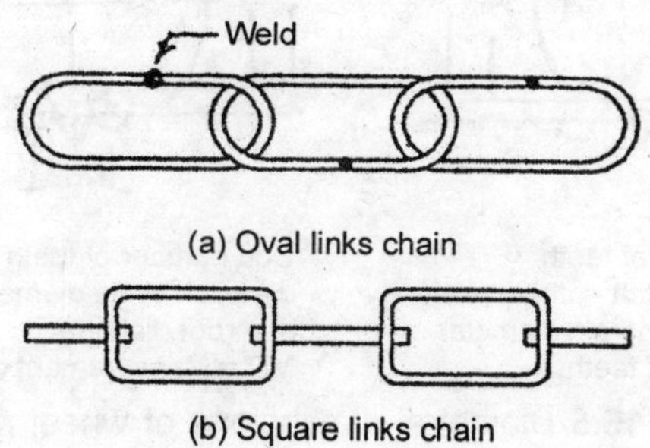

Fig. 16.2 Hoisting and Hauling chains.

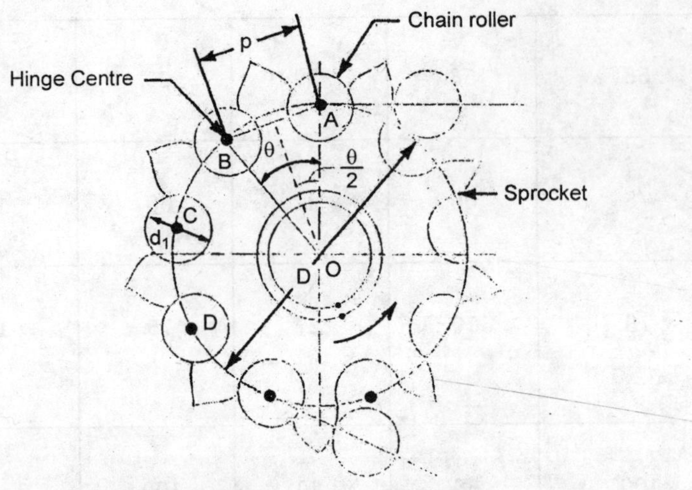

Fig. 16.3 Terms used in chain drive.

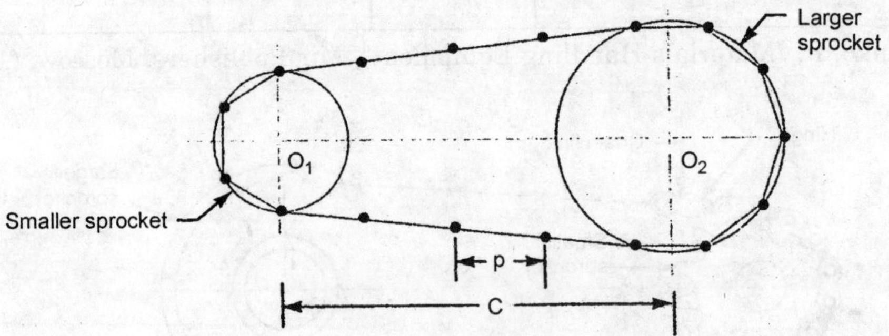

Fig. 16.4 Length of chain.

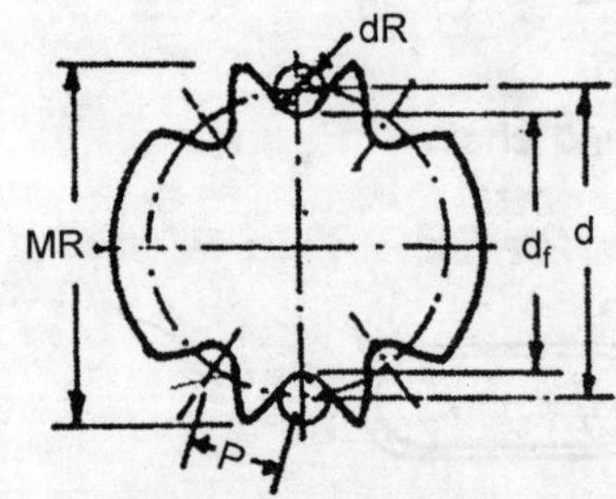

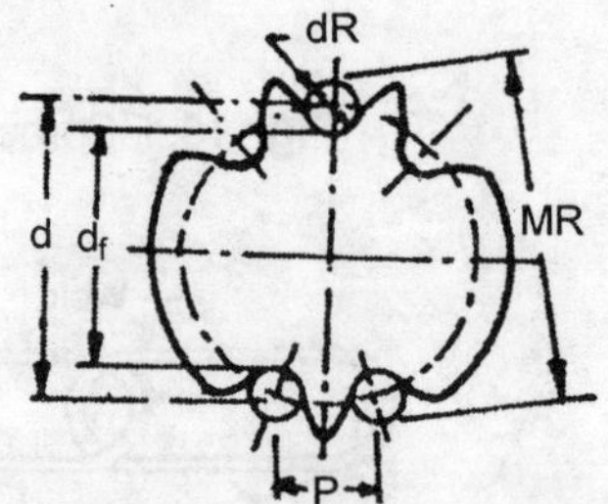

Even number of teeth	Odd number of teeth
p = chordal pitch = chain pitch	d = pitch circle diameter = p(sin 180°/z)
dR = measuring pin diameter	dt = root diameter
Z = number of teeth	MR = Measurement over pins.

Fig. 16.5 Diametral dimensions of wheel rims.

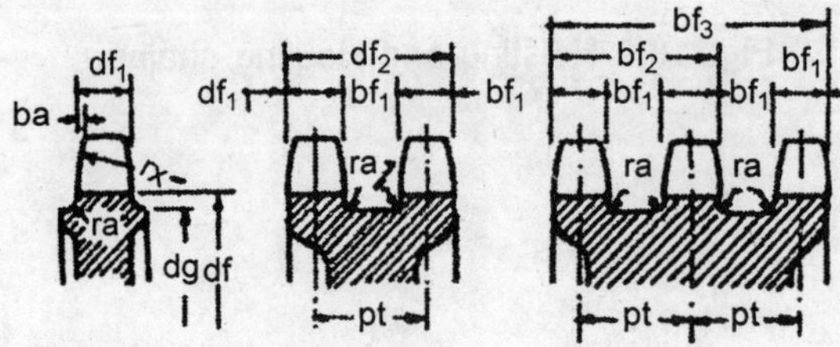

Fig. 16.6 Wheel rim profile.

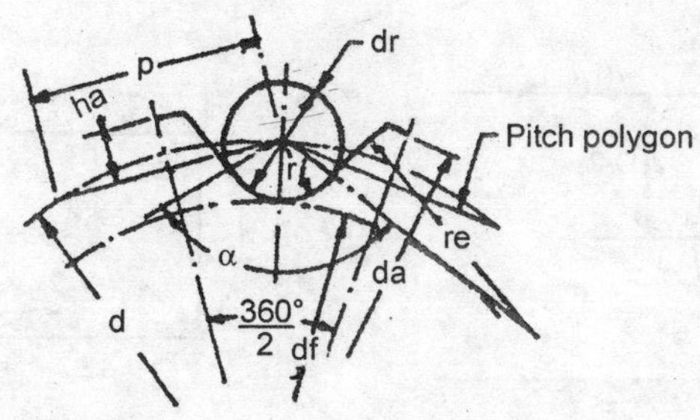

Fig. 16.7 Wheel tooth gap forms for steel roller chain.

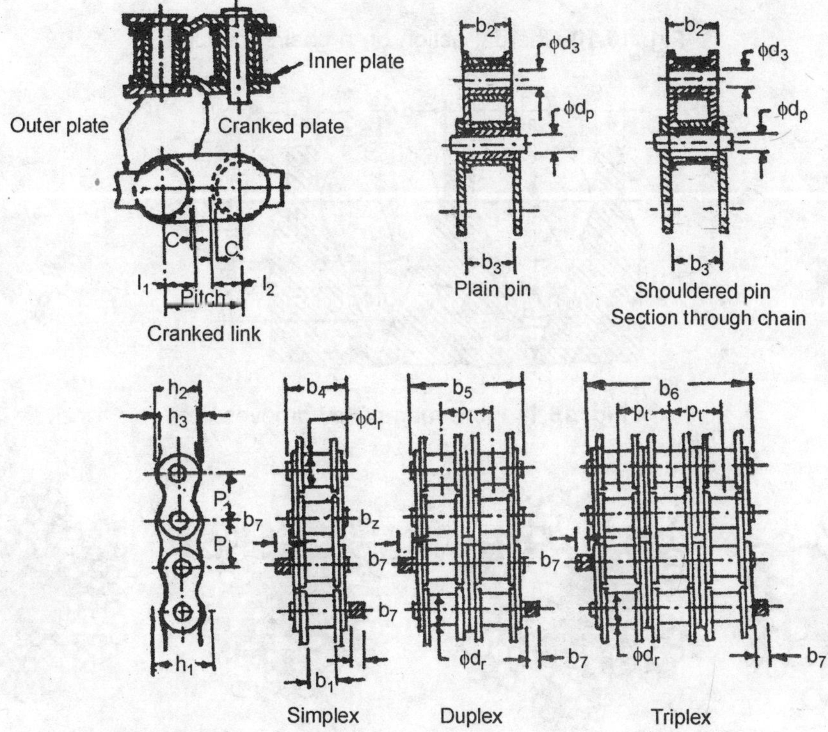

Fig. 16.8 Dimensions for transmission steel roller chains.

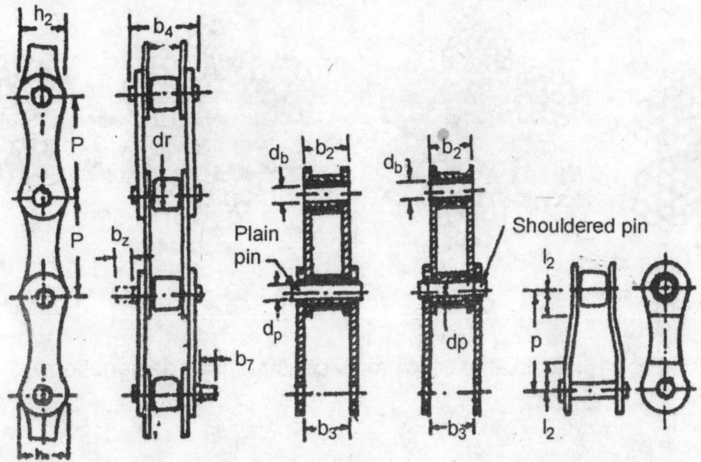

Fig. 16.9 Extended pitch precision transmission roller chain.

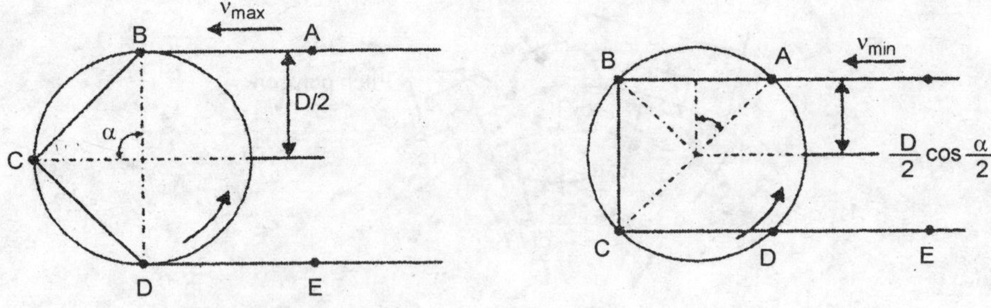

Fig. 16.10 Chordal action on a chain sprocket.

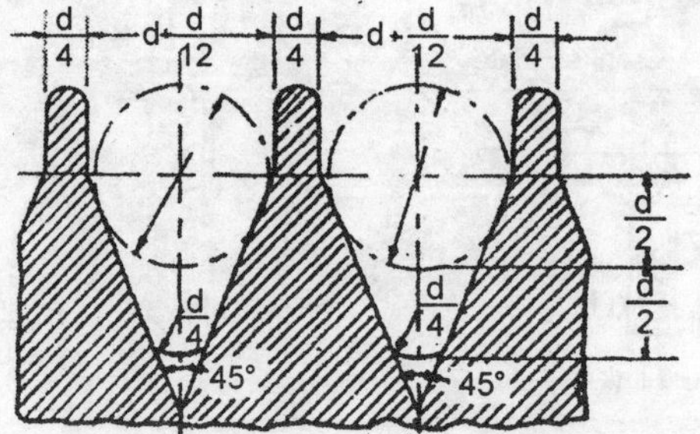

Fig. 16.11 Fibre rope pulley grooves.

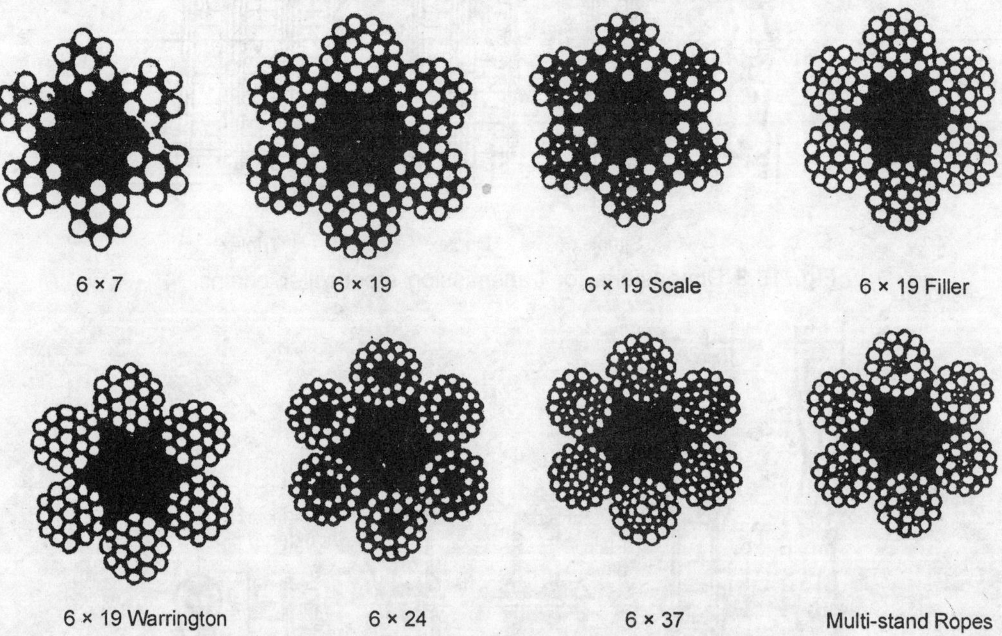

Fig. 16.12 Examples of rope construction, designation.

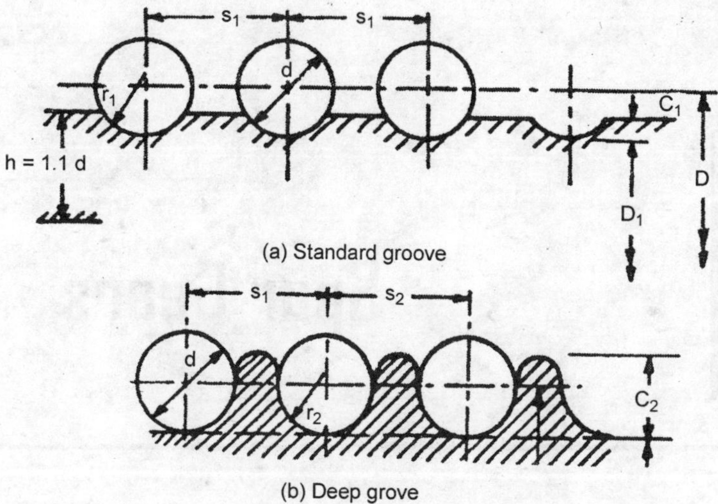

Fig. 16.14 For Table 16.29 Dimensions for drum groove.

CHAPTER

17

Spur Gears

1. Spur Gear Terminology (Fig. 17.1)

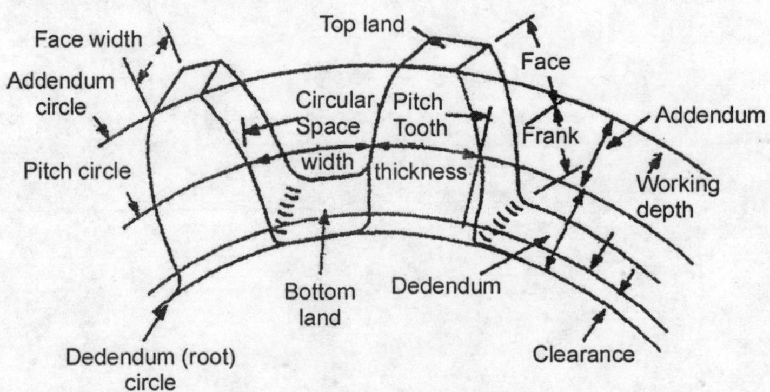

Fig. 17.1 Gear terminology.

1. Circular pitch, $p = \pi d/2$ ...(17.1)
2. Module, $m = d/z$ ...(17.2)
3. Diametral pitch, $P = 1/\text{m} = z/d$ ...(17.3)
4. Speed ratio, $i = z_g/z_p = d_a/d_p = \omega_p/\omega_g = n_p/n_g$ ...(17.4)
5. Centre distance, $a = (d_g + d_p)/2$ ...(17.5)
6. Pitch line velocity, $v = \pi d_g n_g/(60 \times 10^3)$ m/s ...(17.6*b*)

2. Interference in Gears (Fig. 17.2)

Radius of addendum circle of pinion,

$$r_{a1} = r_1 + h_{a1}$$

Base circle radius of pinion, $r_{b1} = r_1 \cos \phi$

Radius of addendum circle of gear,

$$r_{a2} = r_2 \cos \phi$$

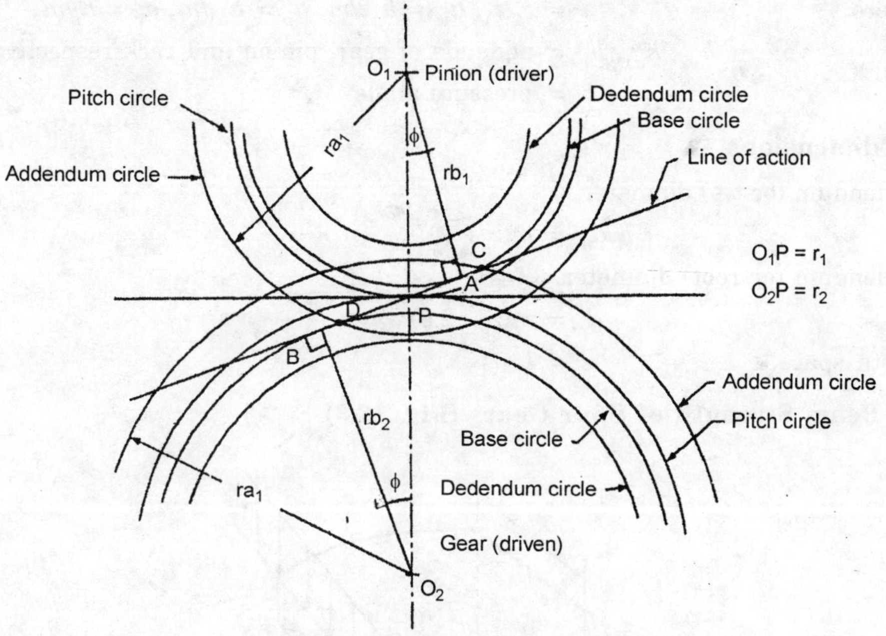

Fig. 17.2 Interference in gears.

Base circle radius of gear, $r_{b2} = r_2 \cos \phi$

Where r_1, r_2 = pitch circle radii of pinion and gear respectively

h_{a1}, h_{a2} = addenda of pinion and gear respectively

Length of path of approach,

$$L_a = [r^2_{a2} - r^2_{b2}]^{1/2} - r_2 \sin \phi \quad ...(17.7a)$$

Length of path of recess, $L_r = [r^2_{a1} - r^2_{b1}]^{1/2} - r_1 \sin \phi$ ...(17.7b)

Maximum possible length of path of approach,

$$(L_a)_{max} = r_1 \sin \phi \quad ...(17.8a)$$

Maximum possible length of path of recess,

$$(L_r)_{max} = r_2 \sin \phi \quad ...(17.8b)$$

Interference will occur, when

$$L_a < (L_a)_{max} \quad ...(17.9a)$$

and $L_r < (L_r)_{max}$ ...(17.9b)

3. Minimum Number of Teeth

The minimum number of teeth to avoid interference are:

Gear:

$$z_2 = 2a_w/[\{1 + (1/i^2)\sin^2 \phi + (2/i)\sin^2 \phi\}^{1/2} - 1]$$

Pinion: $z_1 = 2a_p/[\{1 + i^2 \sin^2 \phi + 2i \sin^2 \phi\}^{1/2} - 1]$ ...(17.10)

$$z_1 = 2a_p/[\{1 + i^2 \sin^2 \phi + 2i \sin^2 \phi\}^{1/2} - 1] \quad ...(17.11)$$

Rack: $z_{min} = 2a_r/\sin^2 \phi$ ...(17.12)

where $i = z_2/z_1,\ a_w = h_a2/m,\ a_p = h_{a1}/m,\ a_r = h_r/m$

h_{a2}, h_{a1}, h_r = addenda of gear, pinion and rack respectively.

ϕ = pressure angle

4. Tooth dimensions

Addendum (or tip) diameter,

$$d_a = m(z + 2) \quad ...(17.13)$$

Dedendum (or root) diameter,

$$d_f = m(z - 2.5) \quad ...(17.14)$$

Tooth space, $e = t$ ...(17.15)

5. Static Beam Strength of Spur Gears (Fig. 17.3)

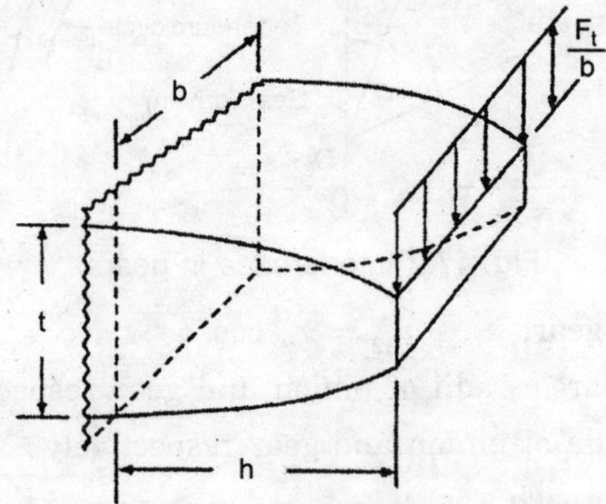

Fig. 17.3 Gear tooth as a cantilever.

Lewis equation:

$$F_b = \sigma_d \text{ bmY} \quad ...(17.16a)$$

$$= \sigma_d \text{ b}\pi\text{my} \quad ...(17.16b)$$

Where y, Y = Lewis form factors based upon circular pitch and diametral pitch respectively.

σ_d = allowable static bending stress of gear material

= S_{ut}/FOS = $\sigma_{ut}/3$

b = face width, which varies from 3πm to 4πm.

m = module

6. Lewis Form Factors (Fig. 17.4)

Lewis form factors based upon circular pitch are given by,

$$y = 0.124 - 0.684/z \text{ for } 14.5° \text{ involute FD} \quad ...(17.18a)$$

$$= 0.154 - 0.912/z \text{ for } 20° \text{ involute FD} \quad ...(17.18b)$$

$$= 0.170 - 0.950/z \text{ for } 20° \text{ involute stub} \quad ...(17.18c)$$

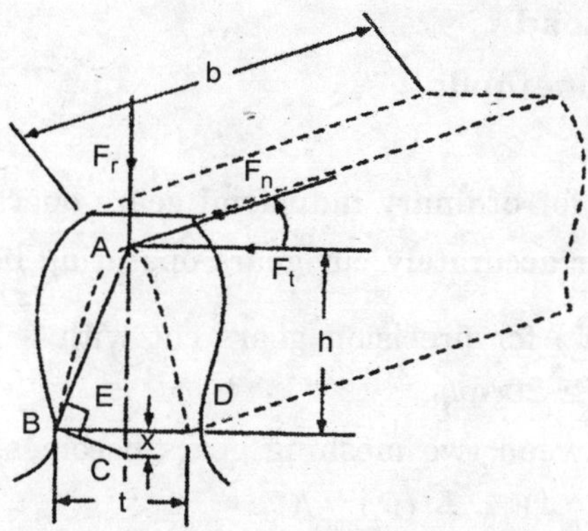

Fig. 17.4 Gear tooth as parabolic beam.

7. Tangential Load on Gear Tooth (Fig. 17.5)

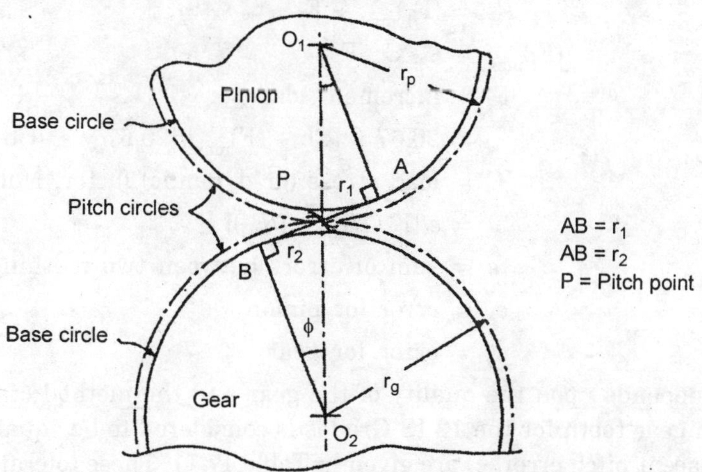

Fig. 17.5 Contact between meshing teeth.

$$F_t = 10^3 \times P/vN \qquad ...(17.19)$$

Where P = power transmitted in kW

v = mean pitch line velocity = $\pi dn/(60 \times 10^3)$ m/s

d = pitch circle diameter, mm

n = speed in rpm

8. Service Factor, C_s (Table 17.7)

$$C_s = \text{maximum torque/rated torque} \qquad ...(17.20)$$

9. Load Distribution Factor, C_m for Spur Gears (Table 17.9)

10. Maximum Tangential Load, $(F_t)_{max}$

$$(F_t)_{max} = C_s C_m F_t \qquad ...(17.21)$$

11. Design for Dynamic Load

(*a*) Velocity Factor method:

Velocity factor:

C_v = 3.05/(3.05 + v) for ordinary industrial gears operating upto 8 m/s. ...(17.22*a*)

= 6.1/(6.1 + v) for accurately cut gears operating between 6 to 20 m/s...(17.22*b*)

= 5.55/(5.55 + $\sqrt{v}$) for precision gears cut with a high degree of accuracy and operating at $v \geq 20$ m/s. ...(17.22*c*)

The effective load between two meshing gears becomes,

$$F_{eff} = (F_t)_{max}/C_v \quad ...(17.23)$$

Face width, $b = C_s C_m F_1/(C_v \alpha_d \pi y m)$ ...(17.24)

(*b*) Earle Buckingham's Increment Load Method

The dynamic load is given by,

$$F_d = (F_1)_{max} + F_i \quad ...(17.25)$$

Where $(F_t)_{max} = C_s\, C_m\, F_t$ ...(17.25)

F_i = increment load

= 20.67 v [Cb + $(F)_{max}$]/[20.67 v + {Cb + $(F_1)_{max}\}^{1/2}$}

C = load stress (or dynamic) factor, N/mm

= $e/[K_1(1/E_p + 1/E_g)]$

e = sum of errors between two meshing teeth, mm

e_p = error for pinion

e_g = error for gear

The error depends upon the quality of the gear and the method of manufacture. The expected error on gear tooth for the 12 IS Grades is considered to be equal to tolerance. The tolerances for adjacent pitch error (e) are given in Table 17.11. These tolerances are calculated by using the following basic equation:

$$\phi_p = m + 0.25\sqrt{d} \quad ...(17.27)$$

Where ϕ_0 = tolerance factor m = module, mm

d = pitch circle diameter, mm

K_1 = a constant

= 9.345 for 14.3° invoiute FD ...(17.28*a*)

= 9.000 for 20° involute FD ...(17.28*b*)

= 8.700 for 20° involute stub ...(17.28*c*)

E_p, E_q = moauin ot elasticity ot pinion and gear materials

Values or load stress factor are given in Table 17.12

(*c*) Prof, Sports Adjacent Pitch Error Method

According to Prot. bpotts, the dynamic load on gear tootn is given by,

$$F_\alpha = (F_t)_{max} + F_L' \quad ...(17.29)$$

where $F_i = e n_p z_p b_{rp} r_g / [K_2(r_p^2 + r_g^2)^{1/2}]$ ...(17.30)

K_2 = 2530 for steel pinion with steel gear

= 3785 for C.I. pinion with C.I. gear ...(17.31*a*)

= 3260 for steel pinion with C.I. gear.

Where n_p = rpm of pinion

r_p, r_g = pitch circle radii of pinion and gear respectively, mm

e = sum of errors between two meshing teeth, mm

= $e_p + e_g$

e_p = error for pinion

e_g = error for gear

Multiply r^2_g in the bracket by 0.92 for steel pinion with C.I. gear.

12. Endurance Strength of Gear

The endurance limit for steel gears is given by,

$$S_e = k_a k_b k_c k_d k_e k_f S'_e \qquad ...(17.32)$$

Where S'_e = uncorrected endurance strength

= $0.5S_{ut}$ for steel

k_a = surface finish factor (Fig. 4.7 or 4.8)

k_b = size factor (Table 4.3)

k_c = reliability factor (Table 4.4)

k_d = temperature factor (Eq. 4.15)

k_e = mean stress factor

= 1 .0 for idler gears

= 1 .4 for input and output gears

k_f = fatigue stress reduction factor (Table 17.13)

$$S_e = 1.75 \times (BHN)_{core} \text{ for steel} \qquad ...(17.33)$$

The endurance strength of gear teeth is given by

$$F_e = S_e b \pi y m \qquad ...(17.34)$$

Where S_e = corrected endurance limit of gear material

For a safe design,

$$F_e \geq k F_d \qquad ...(17.35)$$

Where k = factor of safety

= 1.25 for steady loads

= 1.35 for pulsating loads

= 1.5 for shock loads.

13. Design for Wear

Surface stress, $\sigma = \{F(1/r_1 - 1/r_2)/\{\pi L[(1 - \nu^2_1)/E_1 + (1 - \nu_2^2)/E_2\}]$ ...(17.36)

Where L = length of contact of cylinders

r_1, r_2 = radii of curvature of cylinders

ν_1, ν_2 = Poisson's ratio of cylinders

E_1, E_2 = modulii of elasticity of cylinders

Wear load:

$$F_w = (\pi S^2_{es}\, b \sin\phi \cos\phi/2)\, [d_p\, d_g/(d_p + d_g)]\, [(1 - \nu^2_p)/E_p + (1 - n^2_g/E_g] \quad ...(17.37)$$

For $\nu_p = \nu_g = 0.3$ so that $1 - \nu^2_p = 1 - \nu^2_g = 1 - (0.3)^2 = 0.91$

$$F_w = (\sigma^2_{es}\, b \sin\phi \cos\phi/1.4)\, [2d_p\, d_g/(d_p + d_g)]\, [1/E_p + 1/E_g] \quad ...(17.38)$$

Let $Q = 2d_g/(d_p + d_g) = 2z_g/(z_p + z_g)$ ...(17.39)

and $K = (\pi S^2_{es} \sin\phi \cos\phi/4)\, [(1 - \nu^2_p)/E_p + (1 - \nu^2_g)/E_g]$ ...(17.40)

Then $F_w = d_p b Q K$ ...(17.41)

Where K = load – stress factor.

$= c_1(\text{BHN}/100)^2$ ...(17.42)

Refer to Table 17.14 for values of c_1.

$K = 0.16\ (\text{BHN}/100)^2$ for steel pinion – steel gear and 20° pressure angle

$= 0.21\ (\text{BHN}/100)^2$ for cast iron pinion – cast iron gear and 20° pressure angle

$= 0.18\ (\text{BHN}/100)^2$ for steel pinion – cast iron gear and 20° pressure angle

According to G. Niemann, the approximate value of S_{es} for steel pinion is given by,

$S_{es} = 2.65\ (\text{BHN})_{case}$ MPa ...(17.43)

Where BHN = Brinell hardness number in MPa

For a safe design $F_w > F_d$ ...(17.44)

The pinion should invariably be designed for wear load. If $F_w < F_d$, then the hardness of the pinion should be increased. In that case put $F_w = F_d$, calculate the value of K from Eq. (17.41) and substitute that value in Eq.(17.42) to determine the hardness of pinion required.

14. Force Analysis (Fig. 17.6)

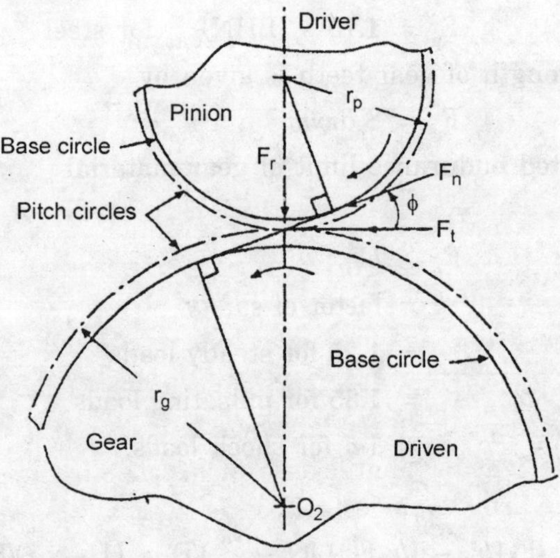

Fig. 17.6 Forces on a apur gear tooth.

Radial force, $F_r = F_t \tan \phi$ ...(17.45)

Where ϕ = pressure angle

Tangential force, $F_t = 2T/d_1$ ...(17.46)

Where d_1 = pitch circle diameter of pinion.

15. Dimensions of Pinion and Gear Wheel (Fig. 17.7)

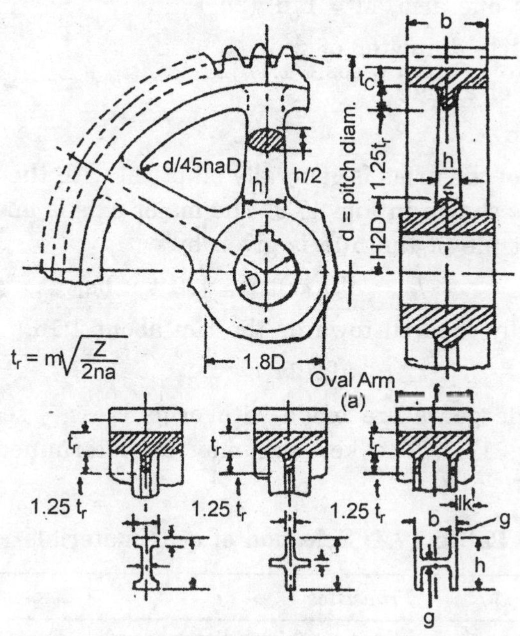

Fig. 17.7 Proportions of spur gear wheels.

When a solid pinion is keyed to the shaft, the minimum pitch diameter is given by:

$$(d_p)_{min} = 2 \times \text{bore} + 0.25 \times m \quad ...(17.47)$$

The minimum thickness of metal between the keyway and the root circle is:

$$t_{min} = m\sqrt{(z/5)} \quad ...(17.48)$$

Outside diameter of hub = 1.8 × bore for steel gear ...(17.49*a*)

= 2 × bore for cast iron gear ...(17.49*b*)

= 1.65 × bore for forged steel ...(17.49*c*)

Hub length = 1.75 × bore for light service ...(17.50*a*)

> face width of gear teeth ...(17.50*b*)

Web thickness for small pinions = (1.6 – 1.9)m ...(17.51)

Number of arms = 4 for split gears upto 1 metre diameter ...(17.52*a*)

= 6 for gears upto 3 metre diameter ...(17.52*b*)

= 8 for larger gears ...(17.52*c*)

The arms should be designed as cantilever beams loaded at the pitch line, with the load equally distributed to all the arms. The design load is the stalling load that shall develop in the arms at zero velocity.

Stalling load, $F_0 = \sigma b \pi y m$

Section modulus of arms, $z = I/c$

Where I = moment of inertia of arms about the neutral axis

c = distance of the outermost fibres of the arm cross – section.

Bending moment on the arms = $F_0 d/2$

Bending moment in one arm, $M = F_0 d/(2n_a)$

Where d = pitch circle diameter of gear

n_a = number of arms

$I/c = F_0 d/(2n_a \sigma)$ ...(17.53)

The cross – section of the arms is generally elliptical with the major axis of arms lying in the plane of rotation, as shown in Fig. 17.9. The major axis is made twice the minor axis. The major axis at the outside of the hub is given by:

$$h = 4[I/(\pi c)]^{1/3} = [20.4 \times I/c] \quad ...(17.54)$$

The arms are usually tapered towards the rim about 1:16.

Rim thickness, $t_r = m[z/(2n_a)]^{1/3}$ ...(17.55)

Very large and wide gears are made with arms cross – section of I, cross or H – section, as shown in Fig. 17.7. The key size may be determined by knowing the shaft diameter.

Table 17.1: Selection of qear materials

Material	*Properties*	*Applications*
1. Cast iron	Low wear, excellent machinability, ease of casting, low noise, shock absorbent, low tensile strength.	Suitable for complicated shapes.
2. Plain carbon Steels	High toughness, tensile strength and endurance strength.	Used for industrial gears.
3. Cast steels	Can with stand high stresses.	Used for gears difficult to fabricate.
4. Alloy steels	High tooth .strength and low wear.	Used for industrial and automobile gears.
5. Heat treated Steels	High toughness and tensile strength, long life.	Used for gears subjected to excessive wear and high loads.
6. Case hardened Steels	Hard shell and soft core.	Used for gears subjected to varying loads and shocks.
7. Surface hardened steels	Gives hard surface and soft core loads.	Suitable for transmitting shock
8. Non - ferrous Metals	High corrosion resistance, low inertia of rotating parts.	Used in instruments and appliances.
9. Non-metallics	Quiet operation, internal lubrication, damping of shock and vibrations, manufacturing economy, low load carrying capacity.	Used for toys, instruments and appliances.

Table 17.2: Mechanical properties of cast irons used for gears

Material	*Condition (Finished Gear)*	*Ultimate tensile strength, S_{ut} MPa*	*Brinell Hardness Number, BHN MPa*
Grey Cast Iron			
FG200	As cast	200	179 (min)
FG220	As cast	220	197(min)
FG350	As cast	350	207 (min)
FG350	Heat treated	350	300 (min)
Malleable Cast Iron			
White heart Grade B	-	275	21 7 (max)
Black heart Grade B	-	314	149 (max)
Nodular (Ductile) Cast Iron			
SG 800 / 2	As cast	800	245 - 350
SG 700 / 2	As cast	700	225 - 300
SG 600 / 3	As cast	600	190-270
SG 500 / 7	As cast	500	170-240
SG 400/12	As cast	400	200 (max)

Table 17.3: Mechanical properties of steels used for gears

Material	*Heat treatment*	*Ultimate tensile strength MPa*	*Brinell Hardness Number BHN*
Cast Steel:			
Grade 27-54	-	540	145
Grade 1	Annealed	620	182 (min)
	H + T	790	229 (min)
	Surface Hardened	620	165 (core, min) 460 (case, min)
High Tensile Steel Casting:			
Grade 1 CS 640	-	640	190 (min)
Grade 2 CS 700	-	700	207 (min)
Grade 3 CS 840	-	840	248(min)
Grade 4 CS 1030	-	1030	305(min)
Grade 5 CS 1230	-	1230	355(min)
Forged Steel:			
30C8	N	490	143
30C8	H+T	590	152

40C8	N	570	152
40C8	H + T	590	179
55C8	N	706	201
55C8	H + T	686	223
55C8		784	248
Carbon Chromium Steels:			
55Cr3	N	883	225 (min)
55Cr3	H+T	981	285 (min)
40Cr4	H + T	785	229 (min)
Carbon Manganese Steels:			
27Mn6	N	540	
27Mn6	H + T	590	170 min
27Mn6	H + T	687	201 min
37Mn6	-	687	201 min
37Mn6	-	785	229 min
Manganese Molybdenum Steels:			
35 Mn 6 Mo 3	N	687	201 min
35 Mn 6 Mo 3	H + T	785	229 min
35 Mn 6 Mo 4	N	785	229 min
35 Mn 6 Mo 4	H + T	883	255 min
35Mn2Mo28	H+T	700	200 min
35Ni1Cr60	H + T	700	200 min
Chromium Molybdenum Steels:			
40Cr4Mo3	H + T	687	201 min
40 Cr 5 Mo 6	H + T	883	248 min
15 Cr 13 Mo 6	H + T	883	255 min
25 Cr 13 Mo 6	H + T	883	255 min
Nickel steels:			
40IMM4	H + T	785	229 min
Nickel Chromium Steels:			
30 Ni 6 Cr 5	H+T	1510	444 min
Nickel Chromium Molybdenum Steels:			
40 Ni 6 Cr 4 Mo 3	H + T	883	255 min
31 Ni 10 Cr 3 Mo 6	H + T	883	255 min
40 Ni 10 Cr 3 Mo 6	H + T	980	285 min

Surface Hardened Steels:			
40 C 8	-	540	145 core 460 case
55 C 6	-	695	200 core 520 case
55 C 3	-	850	250 core 500 case
40Cr4	-	695	500 case
3.5 % Ni Steels	-	695	200 core 300 case
3.5 % Ni Cr Steels	-	850	250 core 500 case
1.5% Nil % Cr Steels	-	850	250 core 500 case
Case Hardened Steels:			
0.12-0.22% C steels	-	495	650 case
0.15 %C steels	-	495	140 core
0.20 % C steels	-	495	140 core 640 case
3 % Ni steels	-	695	200 core 300 case
3.5 % Ni steels	-	695	200 core 620 case
5 % Ni steels	-	850	600 case

H = Hardened, T – Tempered, N = normalized

Table 17.4: Recommended combination of steels for pinion and gear

Wheel	*Surface hardness, BHN*	*Gear ratio, i*	*Materials*					
Pinion	≤350		45C8	50C4	50C4 55C8	40Cr1	35Ni1Cr60	40Ni1CMMo15
Gear	≤350	i < 8	35C8 St 50 Cast steel	35C8 Cast Steel	45C8 St63 Cast steel	50C4 55C8 Cast steel	40Cr1 Cast steel	35Ni1Cr60
Pinion	>350	i < 4	45C8 50C4	55C8	40Cr1	35Ni1Cr60	40Ni1Cr1Mo15	40Ni2Cr1Mo28
Gear	>350		35C8	45C8 St63	50C4 55C8	40Cr1	35Ni1Cr60	15Ni2Cr1Mo15
Pinion	>350		45C8	50C4	40Cr1	15Ni2Cr1Mo15	40NiCr1Mo28	
Gear	≤350	i ≥ 4	Cast steel	35Ni1 Cr60	Cast steel 35Ni1 Cr60	45C8	55C8	

Table 17.5: Standard tooth forms of involute profile

Particulars	*14.5° full-depth*	*20° full – depth*	*20° stub*
Addendum, ha	m	m	0.8m
Dedendum, hf	1.157m	1.250m	m
Total depth, h	2.157m	2.125m	1.8m
Working depth, hw	2m	2m	1.6m
Tooth thickness, t	1.5708m	1.5708m	1.5708m
Clearance, c	0.157m	0.250m	0.200m
Fillet radius, r	0.209m	0.300m	0.304m
Minimum number of teeth	32	18	14

Table 17.6: Standard modules

First choice	1.0, 1.25, 1.5, 2, 2.5, 3, 4, 5, 6, 8, 10, 12, 16, 20, 25, 32, 40 and 50
Second choice	1, 125, 1.375, 1.75, 2.25, 2.75, 3.5, 4.5, 5.5, 7, 9, 11, 14, 18, 22, 28, 36 and 45

Table 17.7: Service factor, C_s

Type of load	*Type of service*		
	Intermittent 3 hours per day	*8 - 10 hours per day*	*Continuous 24 hours per day*
Steady	1.25	1.00	0.80
Light shock	1.00	0.80	0.65
Medium shock	0.80	0.65	0.55
Heavy shock	0.65	0.55	0.50

Table 17.8: Type of loads for different machinery

SI. No.	*Type of load*	*Driven machinejy*
1.	Steady	Generators, liquid mixing agitators, centrifugal blowers, compressors and pumps, medium duty conveyors, fans, etc.
2.	Moderate shock	Liquid and sold mixing agitators, multi – cylinder reciprocating compressors and pumps, heavy duty conveyors, sugar cane crushers, hoists, concrete mixers, textile machinery, etc.
3.	Heavy shock	Single – cylinder reciprocating compressors and pumps, stone crushers, passenger lifts, heavy duty hoists, hammer mills, presses, rolling mills, tumbling barrels, printing presses, etc.

Table 17.9: Load distribution factor for spur gears, C_m

SI.No	*Characteristics of support*	*Face width upto, mm*			
		50	150	225	400
1.	Accurate mountings, small bearing clearances, minimum deflection, precision gears	1.3	1.4	1.5	1.8
2.	Less rigid mountings and less accurate gears, contact across the full face	1.6	1.7	1.8	2.0
3.	Accuracy and mounting such that less full face contact exists.	More than 2.0			

Table 17.10: Standard Gear Ratios for Spar Gears

Single	1.25, 1.4, 1.6, 1.8, 2.0, 2.24, 2.5, 2.8, 3.15, 3.55, 4.0, 4.5, 5.0, 5.6, 7.1, 8.0, 9.0 and 10.
Double	8, 9, 10, 11, 12, 12.5, 14, 16, 18, 20, 22.4, 25, 28, 31.5, 35.5, 40, 45 and 50.
Triple	40, 45, 50, 56, 63, 71, 80, 90, 100, 112, 125, 140, 160, 180, 200, 250, 280, 315, 355 and 400

Table 17.11: Number of Teeth and Sums of numbers of Teeth for Standard Transmission Ratios

Thmsmission	*1*	*1.12*	*1.25*	*1.4*	*1.6*	*1.8*	*2.0*	*224*	*2.5*	*2.5*	*5.15*	*3.55*	*4.0*
Ratio Sum of number of teeth, $z_1 + z_2$	*Number of teeth-pinion, $z_1$lgear, z_2*												
100	50/50	41/53	44/56	—	39/61	36/64	—	31/69	—	26/79	24/76	22/78	20/80
101	15/50	—	45/56	42/54	39/62	36/65	34/67	31/72	—	24/77	22/79	20/81	
102	52/51	45/57	42/60	40/63	37/66	—	32/71	—	27/76	25/78	—	21/82	
103	52/51	—	—	43/60	40/63	37/66	—	32/71	—	27/76	—	21/82	
104	52/52	49/55	46/58	43/61	40/69	—	35/69	32/72	—	—	25/79	23/81	21/83
105	53/52	—	—	—	41/64	38/67	35/70	—	30/75	—	25/80	25/82	21/84
106	53/53	50/56	47/59	44/62	41/65	38/68	—	33/73	30/76	28/78	—	23/83	21/85
107	54/53	—	47/60	44/63	41/66	—	36/71	33/74	—	28/79	26/81	—	—
108	54/54	51/57	44/60	45/63	42/66	39/69	36/72	33/75	31/77	28/80	26/82	24/84	22/86
109	55/54	51/58	48/61	45/64	42/67	39/70	—	34/75	31/78	—	26/85	24/85	22/87

Table 17.12 BIS Gear Grades.

IS Grades	*Accuracy level*	*Manufacturing methods*	*Typical applications*	*Permissible pitch line velocity, v m/s*
1, 2 or 3	Ultra – high (AA)	Finish grinding followed by honing.	Used for master gears.	$40 \leq v < 200$
4 or 5	High (A)	Shaving and finish grinding with first rate machine tools and skilled operators.	Used for turbine gearing, aerospace gearing, and sometimes critical industrial gears.	$v \geq 30$
6 or 7	Medium high (B)	Shaving and grinding with emphasis on production rate rather than highest quality.	Used for medium speed industrial gears and critical vehicle gears.	$25 \leq v < 30$
8 or 9	Medium (C)	Find hobbling or hobbing followed by shaving with first rate machine tools and skilled operators	Used for vehicle gears, geared motors, and low speed industrial gears.	$20 \leq v < 25$
10 or 11	Low (D)	Hobbing, shaping or milling.	Used for low gears.	$v < 20$
12	Very low (E)	Casting or moulding	Used for low speeds and light loads.	$v < 10$

Table 17.13: Tolerances on the adjacent pitch error

IS Grade	*Adjacent pirch error e, microns*
1	$0.80 + 0.06\phi_p$
2	$1.25 + 0.10\phi_p$
3	$2.00 + 0.16\phi_p$
4	$3.20 + 0.25\phi_p$
5	$5.00 + 0.40\phi_p$
6	$8.00 + 0.63\phi_p$
7	$11.00 + 0.90\phi_p$
8	$16.00 + 1.25\phi_p$
9	$22.00 + 1.80\phi_p$
10	$32.00 + 2.50\phi_p$
11	$45.00 + 3.55\phi_p$
12	$63.00 + 5.00\phi_p$

Table 17.14: Values of load stress factor, C in N/mm for 20° pressure angle FD involute teeth.

Material		*Poisson's ratio, ν GPa*		*Modulus of elasticity, E GPa*		*Load stress factor, C*
Pinion	*Gear*	*Pinion*	*Gear*	*Pinion*	*Gear*	*N/mm*
Steel	Steel	0.292	0.292	207	207	11500e
Cast iron	Cast iron	0.211	0.211	100	100	8900e
Steel	Cast iron	0.292	0.211	207	100	7500e
Steel	Cast iron	0.292	0.211	207	160	10000e

Table 17.15: Fatigue strength reduction factor, k_f

z	20	24	30	40	50	80	125	250
k_f	0.944	0.941	0.924	0.915	0.901	0.896	0.886	0.861

Table 17.1.6: Values of load – stress factor, K for ϕ = 20° FD involute
K = d(BHN/100)2 MPa

Material		*Poisson's ratio*		*Modulus of elasticity, GPa*		c_1
Pinion	*Gear*	ν_p	ν_q	E_p	E_q	
Steel	Steel	0.30	0.30	210	210	0.1536
Cast Iron	Cast Iron	0.28	0.28	100	100	0.3267
Steel	Cast Iron	0.30	0.28	210	100	0.2402
Steel	Steel	0.30	0.30	207	207	0.15585
Steel	Cast Iron	0.30	0.28	207	100	0.2413
Steel	Steel	0.292	0.292	207	207	0.16
Cast iron	Cast iron	0.211	0.211	100	100	0.21
Steel	Cast iron	0.292	0.211	207	160	0.18

CHAPTER 18

Helical Gears

1. Nomenclature

β = Helix angle

ϕ_n = normal pressure angle

p_n = normal circular pitch

p_t = transverse circular pitch

p_x = axial pitch

P_t = transverse diametral pitch

P_n = normal diametral pitch

m_n = normal module

m_t = transverse module

2. Various relationships (Fig. 18.1)

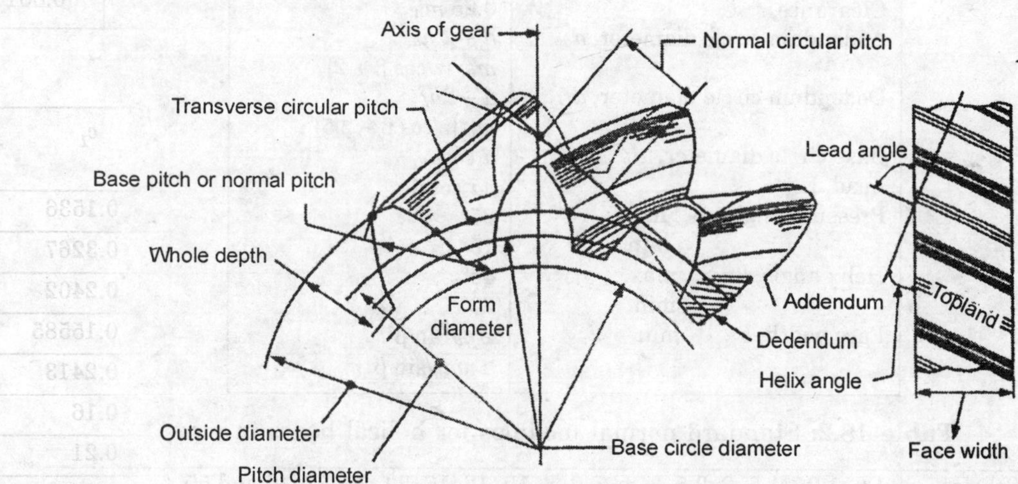

Fig. 18.1 Helical gear terminology.

$$p_n = p_t \cos \beta \qquad ...(18.1)$$

$$p_t = \pi \text{ d/z} \qquad ...(18.2)$$

$$P_t = \text{z/d}$$

$$m_t = 1/P_t = \text{d/z}$$

$$p_t P_t = \pi = p_n P_n$$

$$P_n = P_t / \cos \beta = z/(d \cos \beta) \qquad ...(18.3)$$

$$m_n = m_t \cos \beta \qquad ...(18.4)$$

$$p_x = p_t / \tan \beta = (\pi d/z) \cot \beta \qquad ...(18.5)$$

$$\cos \beta = \tan \phi_n / \tan \phi_t \qquad ...(18.6)$$

$$d = z p_t / \pi = z m_t = z m_n / \cos \beta \qquad ...(18.7)$$

Centre distance,

$$a = (1/2)\,(d_1 + d_2) \qquad ...(18.8a)$$

$$= [m_n / (2 \cos \beta)]\,(z_1\ z_2) \qquad ...(18.8b)$$

Speed ratio,

$$i = z_2 / z_1 = n_1 / n_2 \qquad ...(18.9)$$

Virtual number of teeth,

$$z_v = z/\cos^3 \beta \qquad ...(18.10)$$

Pitch circle diameter of virtual gear,

$$d_c = d/\cos^2 \beta \qquad ...(18.11)$$

3. Tooth Proportions - Refer to Tabte 18.1.

Table 18.1: Proportions of helical gear tooth.

Item		*Proportions*
Addendum, h_a		m_n
Dedendum, h_f		$1.25\ m_n$
Total depth, h		$18.25\ m_n$
Working depth, h_w		$2\ m_n$
Clearance, c		$0.25\ m_n$
Addendum circle diameter, d_a		$d + 2\ h_a$
		m_n [z/cos β + 2]
Dedendum circle diameter, d_f		$d - 2h_f$
		m_n [z/cos β – 2.5]
Base circle diameter, d_b		$d \cos \phi_t$
Lead, L		$\pi d \cot \beta$
Pressure angle, ϕ_t	max.	25°
	min	15°
Helix angle, β	max.	45°
	min.	20°
Face width, b	min	$\geq p_t/\tan \beta$
		$\geq \pi m_n/\sin \beta$

Table 18.2: Standard normal modules for helical hears in mm.

First choice	1.0, 1.25, 1.5, 2, 2.5, 3, 4, 5, 6, 8, 10, 12, 16, 20, 25, 32, 40 and 50
Second choice	1.125, 1.375, 1.75, 2.25, 2.75, 3.5, 4.5, 5.5, 7, 9, 11, 14, 18, 22, 28, 36, and 45

4. Force on helical gear tooth (Fig. 18.2)

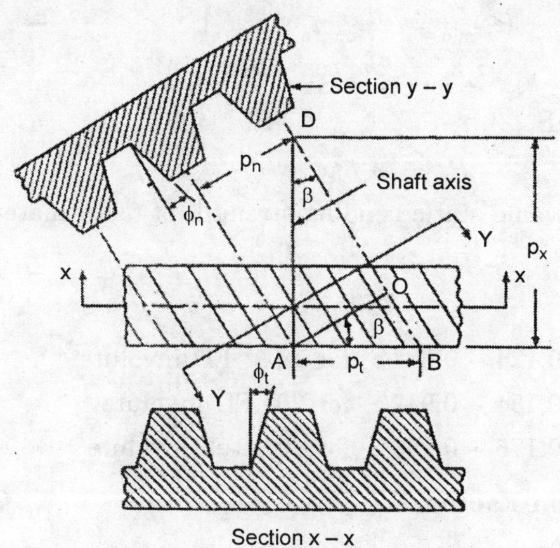

Fig. 18.2 Tooth relationships.

Tangential force,

$$F_t = 10^3 \times P/v \qquad ...(18.12)$$

Where P = power transmitted, kW

v = pitch line velocity, m / s

$= \pi\, dn/(60 \times 10^3)$

where n = rpm,

d = pitch diameter, mm

Radial force, $F_r = F_t \tan \phi_n / \cos \beta$ ...(18.13)

Axial force, $F_a = F_t \tan \beta$ ...(18.14)

5. Thrust Directions - Refer to Fig. 18.3.

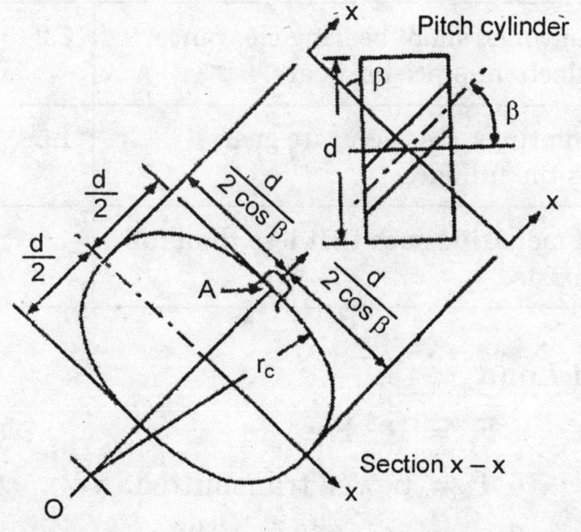

Fig. 18.3 Concept of virtual number of teeth.

6. Minimum Number of Teeth on Helical Gears

$$(z_v)_{min} = 2h_a/[m_n \sin^2 \phi_n] \quad ...(18.15)$$

$$z_P = 2h_a \cos^3 \beta/[m_n \sin^2 \phi_n] \quad ...(18.16)$$

7. Static Beam Strength

$$F_b = m_n b \pi y \sigma_d \quad ...(18.17)$$

Where σ_d = allowable static bending strength of tooth material

$= S_{ut}/3$

8. Lewis Form Factor

$$y = 0.124 - 0.684/z_v \text{ for } 14.5° \text{ FD involute} \quad ...(18.18a)$$

$$= 0.154 - 0.912/z_v \text{ for } 20° \text{ FD involute} \quad ...(18.18b)$$

$$= 0.175 - 0.950/z_v \text{ for } 20° \text{ stub involute} \quad ...(18.18c)$$

9. Wear and Lubrication Factor, C_w

Refer to Table 18.3.

Table 18.3: Wear and lubrication factor , C_w.

Nature of lubrication	C_w
Enclosed gears continuously lubricated with oil of the proper viscosity and character	1.15
Scant lubrication but regular with frequent inspection	1.25
Indifferent lubrication	1.35

10. Load Distribution Factor, C_m

Refer to Table 18.4.

Table 18.4: Load distribution factor for helical gears, C_m

Characteristics of support	*Face width upto, mm*			
	50	*150*	*225*	*400*
1. Accurate mountings, small bearing clearance, minimum deflection, precision gears.	1.2	1.3	1.4	1.7
2. Less rigid mountings, less accurate gears, contact across the full face.	1.5	1.6	1.7	2.0
3. Accuracy and mounting such that less than full face contact exists.	Over 2.0			

11. Maximum Tangential Load

Tangential force, $F_t = 10^3 P/v$

P = power transmitted, kW

Where $v = \pi dn/(60 \times 10^3)$, m/s

$$N = \text{rpm},$$
$$d = \text{pitch circle diameter, mm}$$

Maximum load, $(F_t)_{max} = C_w\, C_m\, F_t$ ...(18.19)

Where C_w = wear and lubrication factor

C_m = load distribution factor

12. Velocity Factor, C_v.

The velocity factors are given by,

$$C_v = 4.58/(4.58 + v) \text{ for } v \le 5 \text{ m/s} \qquad ...(18.20a)$$
$$= 6.1/(6.1 + v) \text{ for } 5 \le v \le 10 \text{ m/s} \qquad ...(18.20b)$$
$$= 15.25/(15.25 + v) \text{ for } 10 \le v \le 20 \text{ m/s} \qquad ...(18.20c)$$
$$= 5.55/(5.55 + \sqrt{v}) \text{ for } v > 20 \text{ m/s} \qquad ...(18.20d)$$
$$= 0.7625/(1.0167 + v) \text{ for non-metallic gears} \qquad ...(18.20e)$$

13. Effective Tangential Load

Effective load, $F_{eff} = (F_t)_{max}/C_v$ ...(18.21)

14. Face Width

$$b = C_w C_m F_t/(C_v\, m_n\, \pi y \sigma_d) \qquad ...(18.22)$$
$$b_{min} = 1.15\, \pi m_n/\sin\beta \qquad ...(18.23)$$

For helical gears:

$$12.5\, m_n \le b \le 20\, m_n \qquad ...(18.24)$$

For herringbone gears:

$$20\, m_n \le b \le 30\, m_n \qquad ...(18.25)$$

15. Dynamic Load

(a) Earle Buckingham's Method

The dynamic load is given by,

$$F_d = (F_t)_{max} + F_i \qquad ...(18.26)$$

Where F_i = increment load

$$= 20.67\, v\, [Cb \cos^2\beta + (F_t)_{max}] \cos\beta/[20.67\, v + \{Cb\cos^2\beta + (F_t)_{max}\}^{1/2}] \qquad ...(18.27)$$

C = load stress (or dynamic) factor, N/mm (see Table 17.12)

$= e/[K_1(1/E_p + 1/E_g)]$

e = probable error in action between gears, mm

$= e_p + e_g$

Where e_p = pitch error for pinion

e_g = pitch error for gear

Tolerance factor,

$$\phi = m_n + 0.25\sqrt{d}$$

Where m_n = normal module, mm

d = pitch circle diameter, mm

K_1 = a constant

= 9.345 for 14.5° involute FD ...(18.28*a*)

= 9.000 for 20° involute FD ...(18.28*b*)

= 8.700 for 20° involute stub ...(18.28*c*)

E_p, E_g = modulii of elasticity of pinion and gear materials

C = 11,500e N/mm, for steel pinion and steel gear ...(18.29a)

= 8900e N/mm, for cast iron pinion and cast iron gear ...(18.29b)

= 10,000e N/mm, for steel pinion and cast iron gear ...(18.29c)

(b) Prof. Spotts Method

Increment load,

$$F_i = en_p z_p b r_1 r_2/[2530\ (r^2{}_1 + r^2{}_2)^{1/2}] \quad ...(18.28)$$

The increment load is assumed to be inclined at an angle 4>$_n$ to the tangent plane. The tangential component of F_i is $F_i \cos \phi_n \cos \beta$. Effective load is given by,

$$F_{eff} = C_s C_w F_t + F_i \cos \phi_n \cos \beta \quad ...(18.29)$$

16. Endurance Strength of Gear

Endurance strength of gear is given by,

$$F_e = S_e b \pi y m_n \quad ...(18.30)$$

Where S_e = endurance limit of gear material

For gears made of steel,

$$S_e = 1.75\ (BHN)_{core}$$

17. Check for Safe Design

For a safe design,

$$F_e \geq kF_d \quad ...(18.31)$$

Where k = factor of safety

= 1.25 for steady loads

= 1.35 for pulsating loads

= 1.5 for shock loads

18. Wear Load

$$F_w = d_p b QK/\cos^2 \beta \quad ...(18.32)$$

$$K = [\pi S^2{}_{es} \sin \phi \cos \phi/4]\ [(1 - \upsilon^2{}_p)/E_p + (1 - \upsilon^2{}_g)/E_g] \quad ...(18.33a)$$

$$= [S^2{}_{es} \sin \phi \cos \phi/1.4]\ [1/E_p + 1/E_g] \text{ for } \nu_p = \nu_g = 0.30 \quad ...(18.33b)$$

$$= c_1(BHN/100)^2$$

For values of c_1, refer to Table 17.14.

$K = 0.16\ (BHN/100)^2$ for steel pinion - steel gear and 20° normal pressure angle

$= 0.21\ (BHN/100)^2$ for cast iron pinion - cast iron gear and 20° normal pressure angle

$= 0.18\ (BHN/100)^2$ for steel pinion - cast iron gear and 20° normal pressure angle

S_{es} = surface endurance limit

$= 2.75\ (BHN) - 70$ MPa for steel ...(18.34a)

$= 2.65$ BHN ...(18.34b)

$\tan \phi_n = \tan \phi_t \cos \beta$

$Q = 2d_g/(d_p + d_g) = 2Z_g/(Z_p + Z_g)$ ...(18.35)

For safe design,

$F_w \geq F_d$ ...(18.36)

19. Factor of safety

$FOS = F_b/F_{eff}$ against bending failuro ...(18.37*a*)

$= F_w/E_{eff}$ against wear failure ...(18.37*b*)

Table 18.5: Tolerances on the adjacent pitch error.

IS Grade	*Adjacent pitch error e, microns*
1	$0.80 + 0.06\ \phi_p$
2	$1.25 + 0.10\phi_p$
3	$2.00 + 0.16\phi_p$
4	$3.20 + 0.25\phi_p$
5	$5.00 + 0.40\phi_p$
6	$8.00 + 0.63\phi_p$
7	$11.00 + 0.90\phi_p$
8	$16.00 + 1.25\phi_p$
9	$22.00 + 1.80\phi_p$
10	$32.00 + 2.50\phi_p$
11	$45.00 + 3.55\phi_p$
12	$63.00 + 5.00\phi_p$

A
B
E
F_r
F
F_n
F_a
Centre lines of teeth
β
D
F_t
C
d
Pitch cylinder
(a)
A
F_r
F_n
ϕ_n
B
$F_n \cos \phi_n$
C
(b)
B
F_a
$F_n \cos \phi_n$
β
D
F_t
C
(c)

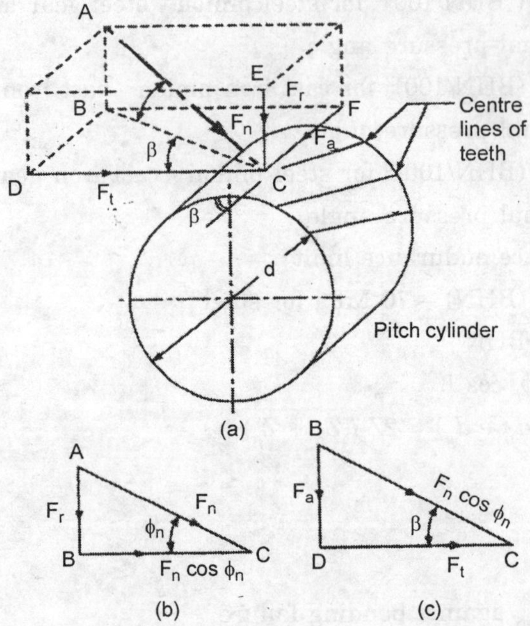

Fig. 18.4 Tooth forces in a helical gear.

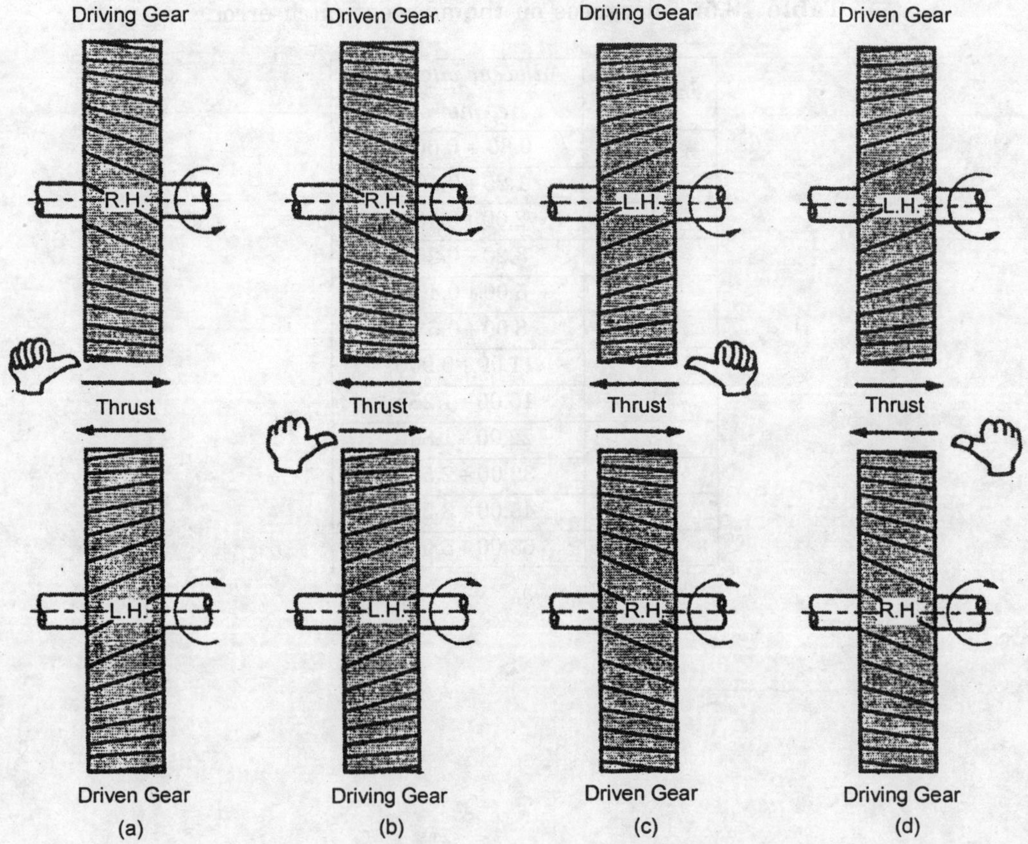

Driving Gear
Driven Gear
Driving Gear
Driven Gear
R.H.
R.H.
L.H.
L.H.
Thrust
Thrust
Thrust
Thrust
L.H.
L.H.
R.H.
R.H.
Driven Gear
Driving Gear
Driven Gear
Driving Gear
(a)
(b)
(c)
(d)

Fig. 18.5 Thrust directions in helical gears.

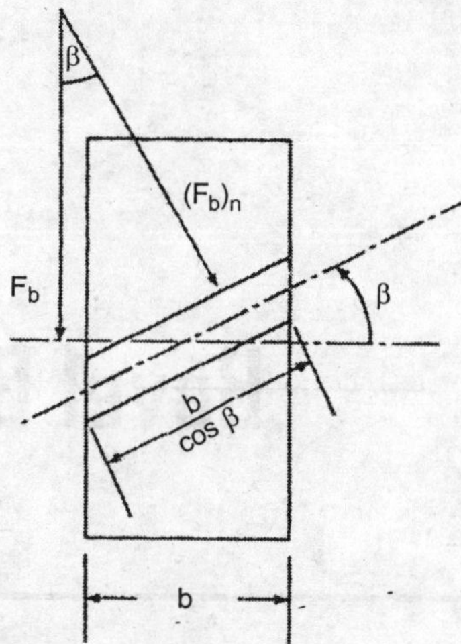

Fig. 18.6 Beam strength of helical gear tooth.

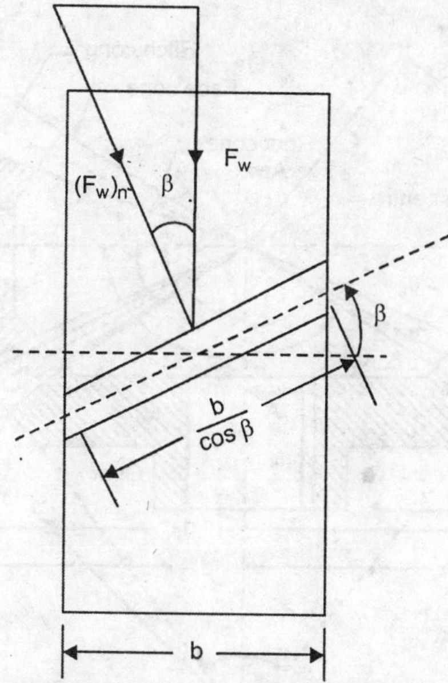

Fig. 18.7 Wear strength of helical gear tooth.

CHAPTER

19

Bevel Gears

1. Terminology (Fig. 19.1):

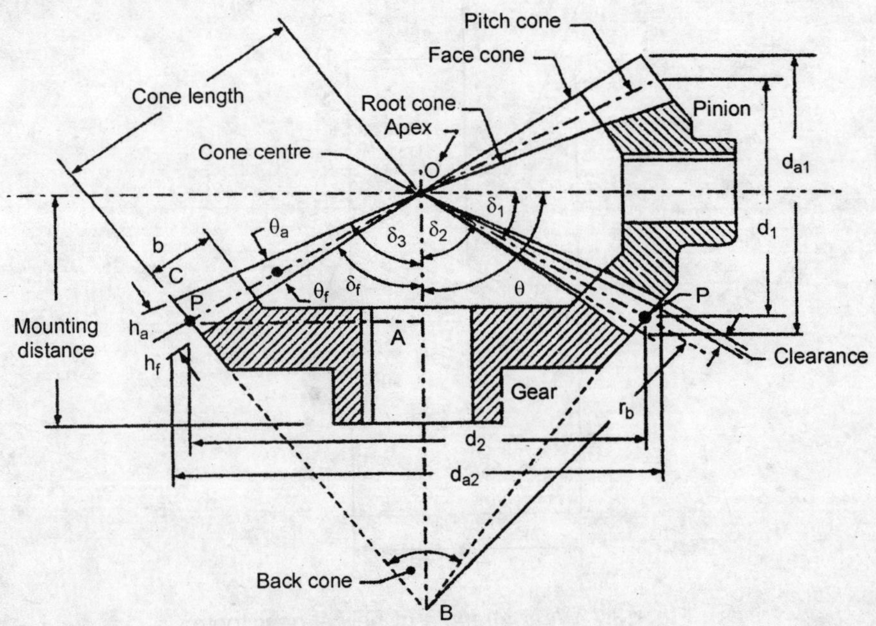

Fig. 19.1 Bevel gear terminology.

$$\delta = \text{pitch angle}$$

$$\delta_a = \text{face angle}$$

$$\delta_f = \text{root angle}$$

$$\theta_a = \text{addendum angle}$$

$$\delta_f = \text{dedendum angle}$$

d = pitch circle diameter (large end)
d_a = addendum diameter
d_f = dedendum diameter
b = face width
r_b = back cone radius
L = OP = cone length or pitch cone distance
OP = pitch cone

Pitch cone distance, $L = [r^2{}_1 + r^2{}_2]^{1/2}$ ...(19.1)

Where r_1, r_2 = pitch circle radii of pinion and gear respectively

Module, $m = d/z$ ...(19.2)

Where d = pitch circle diameter at the larger end of the teeth
z = number of teeth

Addendum, $h_a = 1.0\text{m}$ ...(19.3*a*)

Dedendum, $h_f = 1.2\text{ m}$ ...(19.3*b*)

Speed ratio, $i = n_1/n_2 = z_2/z_1 = d_2/d_1$ ...(19.4)

Back cone radius, $r_{b1} = r_1/\cos\delta_1$ ...(19.5*a*)

Mean radius (r_m) – It is the pitch radius at the midpoint along the race width of the teeth.

$$r_{m1} = r_1 - (b \sin\delta_1)/2 \quad ...(19.6a)$$
$$r_{m2} = r_2 - (b \sin\delta_2)/2 \quad ...(19.6b)$$

2. Angle Relationship

Let θ = angle between the axes of the pinion and gear

(*i*) For θ < 90°, we have

$$\tan\delta_1 = d_1 \sin\theta/(d_2 + d_1\cos\theta) = \sin\theta/(z_2/z_1 + \cos\theta) \quad ...(19.7a)$$
$$\delta_2 = q - \delta_1 \tan\theta_a = 2h_a \sin\delta_1/d_1 = 2h_a \sin\delta_2/d_2 \quad ...(19.7b)$$
$$\tan\theta_f = 2h_f \sin\delta_1/d_1 = 2h_f \sin\delta_2/d_2 \quad ...(19.8)$$
$$d_{a1} = d_1 + 2h_a \cos\delta_1 \quad ...(19.9a)$$
$$d_{a2} = d_2 + 2h_a \cos\delta_2 \quad ...(19.9b)$$

(*ii*) For θ > 90°

$$\tan\delta_1 = d_1 \sin(180° - \theta)/(d_2 + d_1\cos(180° - \theta) \quad ...(19.10a)$$
$$= \sin(180° - \theta)/[z_2/z_1 + \cos(180° - \theta)] \quad ...(19.10b)$$

Radius of imaginary spur gear,

$$r_b = BC = D/(2\cos\delta)$$

(*iii*) For θ = 90°

$$\tan\delta_1 = z_1/z_2 = 1/i \quad ...(19.11a)$$
$$\tan\delta_2 = z_2/z_1 = i \quad ...(19.11b)$$

3. Virtual Number of Teeth

Back cone radius, $r_b = d/(2\cos\delta)$ ...(19.12)

Virtual number of teeth, $z_y = z/\cos\delta$ ...(19.13)

4. Minimum Number of Teeth

$$z_v \geq 2h_a/(\mathrm{m}\ \sin^2 \phi)$$

$$z_1 > 2h_a \cos \delta_1/(\mathrm{m}\ \sin^2 \phi) \quad ...(19.14)$$

5. Static Beam Strength

Static beam strength, $F_b = (\sigma_d\ \mathrm{bm}\pi\mathrm{y})\ [1 - b/\mathrm{L}]$ for $b \leq \mathrm{L}/3$ ...(19.15)

Where $\mathrm{L} = [r^2{}_1 + r^2{}_2]^{1/2}$, $r = d/2$, and $(1 - \mathrm{b/L})$ = bevel factor.

σ_d = design stress

6. Velocity Factor

The velocity factor for straight bevel gears is given by,

$$C_v = 3.05/(3.05 + v), \text{ for teeth cut by form cutter} \quad ...(19.16a)$$

$$= 6.1/(6.1 + v), \text{ for generated tooth} \quad ...(19.16b)$$

7. Maximum Load

Tangential load, $F_t = 10^3 \times \mathrm{P}/v$ ...(19.17)

Maximum tangential load,

$$(F_t)_{max} = C_s C_m F_t \quad ...(19.18)$$

Where C_s = service factor (see Table 17.7)

C_m = load distribution factor (see Table 19.1)

P = power transmitted

8. Effective Load

$$F_{eff} = (F_t)_{max}/C_v \quad ...(19.19)$$

$$\mathrm{FOS} = F_b/F_{eff} \quad ...(19.20)$$

9. Face Width

$$F_{eff} = (\sigma_d\ \mathrm{bm}\pi\mathrm{y})\ [1 - b/\mathrm{L}] \quad ...(19.21)$$

The face width lies between 6 m to 10 m or $b \leq \mathrm{L}/3$, whichever is less.

10. Dynamic Load

(*a*) Earle Buckingham's Method

Dynamic load, $F_d = (F_t)_{max} + F_i$

$$F_i = 20.67\ \mathrm{v}\ [\mathrm{Cb} + (F_t)_{max}]/[20.67\ \mathrm{v} + \{\mathrm{Cb} + (F_t)_{max}\}^{1/2}$$

...(19.22)

Where C = dynamic factor

$$= K_1 \cdot e[E_p E_g/(E_p + E_g)] \quad ...(19.23)$$

e = sum of errors on the meshing teeth

$= e_p + e_g$

$$\phi = \mathrm{m} + 0.25\sqrt{d_m}$$

d_m = mean diameter at the larger end of the teeth.

Refer to Table 17.16 for values of tolerances for adjacent pitch error.

K_1 = tooth form factor

= 0.111 for 20°FD involute

= 0.107 for 14.5° FD involute

E_p, E_g = modulii of elasticity for pinion and gear respectively

The values of C are as follows:

C = 11500 e N/mm for steel pinion and steel gear ...(19.24*a*)

= 8900 eN/mm for cast iron pinion and cast iron gear

= 10000 eN/mm for steel pinion and cast iron gear

(*b*) Prof. Spotts Method

For bevel gears made of steel the incremental load is given by,

$$F_i = e n_p z_p b r_{mg}/[2530\,(r^2_{mp} + r^2_{mg})^{1/2}] \quad ...(19.25)$$

Where

$$b' = b/\cos\delta_p$$

$$r_{mp} = (d_p - b\sin\delta_p)/2$$

$$r_{mg} = (d_g - b\sin\delta_g)/2$$

$$F_d = (F_t)_{max} + F_i \quad ...(19.26)$$

11. Endurance Strength

$$F_e = S_e b\pi y m\,(1 - b/L) \quad ...(19.27)$$

$$FOS = F_e/F_d \quad ...(19.28)$$

12. Wear Load

$$F_w = bQd_1\,K/\cos d_1 \quad ...(19.29)$$

Ratio factor, $Q = 2\,z_{2v}/(z_{1v} + z_{2v})$

Where $z_v = z/\cos\delta$

Load stress factor, $K = (\pi S^2_{es}\sin\phi\cos\phi/4)\,[(1 - \nu^2_1)/E_1 + (1 - \nu^2_2)/E_2]$

Where S_{es} = surface endurance strength of pinion teeth material

The values of *k* are as given below:

K = 0.16 (BHN/100)^2 N/mm^2 for steel pinion and steel gear ...(19.30*a*)

= 0.21 (BHN/100)^2 N/mm^2 for cast iron pinion and gear ...(19.30*b*)

= 0.18 (BHN/100)^2 N /mm^2 for steel pinion and cast iron gear ...(19.30*c*)

$FOS = F_w/F_{eff}$ ...(19.31)

The recommended factor of safety for bevel gears is 1.5 – 2.0.

13. Force Analysis (Fig. 19.3)

Mean radius, $r_m = d_1/2 - (b\sin\delta_1)/2$

Tangential force, $F_t = T/r_m$ ...(19.32)

Pinion:

Radial force, $F_r = F_s\cos\delta_1 = F_t\tan\phi\cos\delta_1$ ...(19.33)

Axial force, $F_a = F_s\sin\delta_1 = F_t\tan\phi\sin\delta_1$ ...(19.34)

Gear:

Radial force, $F_r = F_s\cos\delta_2 = F_t, \tan\phi\cos\delta_2$ ...(19.35)

Axial force, $F_a = F_s\sin\delta_2 = F_t\tan\phi\sin\delta_2$ ...(19.36)

Table 19.1: Standard normal modules for helical gears in mm

First choice	1.0, 1.25, 1.5, 2, 2.5, 3, 4, 5, 6, 8, 10, 12, 16, 20, 25, 32, 40 and 50
Second choice	1.125, 1.375, 1.75,2.25,2.75,3.5,4.5, 5.5,7, 9, 11, 14, 18, 22, 28, 36 and 45

Table 19.2: Service factor, Cs

Type of load	*Type of service*		
	Intermittent 3 hours per day	*8 - 10 hours per day*	*Continuous 24 hours per day*
Steady	1.25	1.00	0.80
Light shock	1.00	0.80	0.65
Medium shock	0.80	0.65	0.55
Heavy shock	0.65	0.55	0.50

Table 19.3: Type of loads for different machinery

SI. No.	*Type of load*	*Driven machinery*
1.	Steady	Generators, liquid mixing agitators, centrifugal blowers, compressor and pumps, medium duty conveyors, fans, etc.
2.	Moderate shock	Liquid and sold mixing agitators, multi – cylinder reciprocating compressors and pumps, heavy duty conveyors, sugar cane crushers, hoists, concrete mixers, textile machinery, etc.
3.	Heavy shock	Single – cylinder reciprocating compressors and pumps, stone crushers, passenger lifts, heavy duty hoists, hammer mills, presses, rolling mills, tumbling barrels, printing presses, etc.

Table 19.4: Load distribution factor, C_m for bevel gears

Type of application	*General commercial gears*	*High quality gears*
Type of mounting		
1. Both gears straddle mounted	1.25	1.00
2. One gear straddle mounted and other overhung mounted	1.4	1.1
3. Both gears overhung mounted	1.5	1.25

Table 19.5 Tolerances on the adjacent pitch error.

IS Grade	*Adjacent pirch error e, microns*
1	$0.80 + 0.06\phi_p$
2	$1.25 + 0.10\phi_p$
3	$2.00 + 0.16\phi_p$
4	$3.20 + 0.25\phi_p$
5	$5.00 + 0.40\phi_p$
6	$8.00 + 0.63\phi_p$
7	$11.00 + 0.90\phi_p$
8	$16.00 + 1.25\phi_p$
9	$22.00 + 1.80\phi_p$
10	$32.00 + 2.50\phi_p$
11	$45.00 + 3.55\phi_p$
12	$63.00 + 5.00\phi_p$

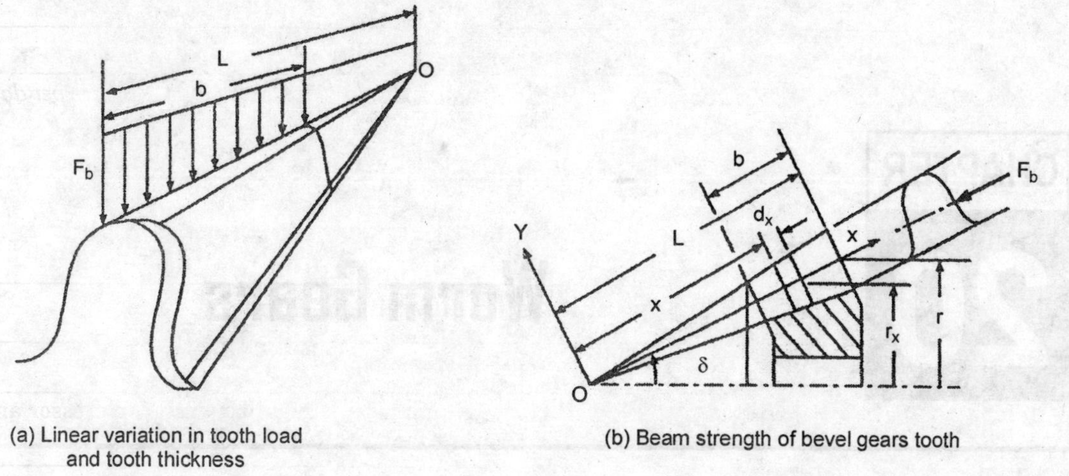

Fig. 19.2

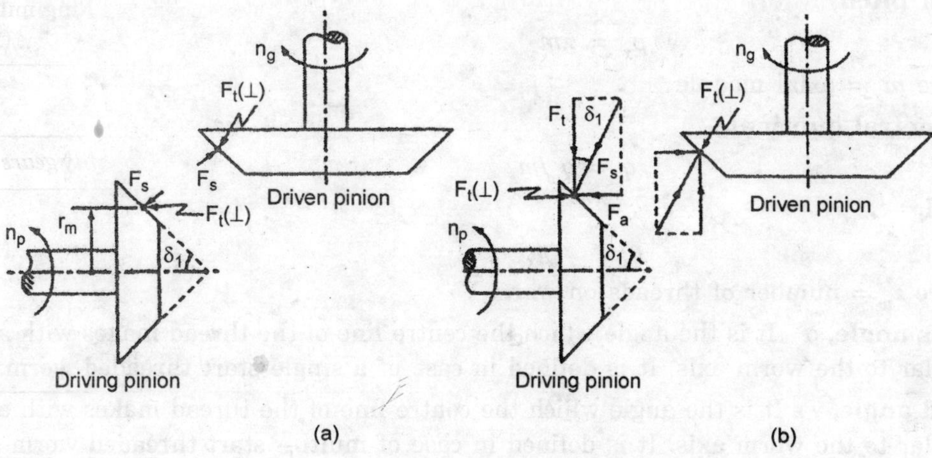

Fig. 19.3 Force analysis in Bevel gears.

CHAPTER

20

Worm Gears

1. Terminology (Fig. 20.1)

Axial pitch,

$$p_x = \pi m_x \quad ...(20.1)$$

where m_x = axial module

Diametral quotient,

$$q = q_w/m_x$$

Lead,

$$p_z = z_w p_x \quad ...(20.2)$$

where z_w = number of threads on worm

Helix angle, α - It is the angle which the centre line of the thread makes with a line perpendicular to the worm axis. It is defined in case of a single start threaded worm.

Lead angle, γ - It is the angle which the centre line of the thread makes with a line perpendicular to the worm axis. It is defined in case of multi – start threaded worm.

$$\tan \gamma = p_z/(\pi d_w)_{13} = \pi m_x z_w/(\pi q m_x) = z_w/q = m_x z_w/d_w \quad ...(20.3a)$$

$$= (n_g/n_w), \text{ for a compact drive} \quad ...(20.3b)$$

Speed ratio,

$$i = n_w/n_g = z_g/z_w \quad ...(20.4)$$

Centre distance,

$$a = (1/2)\,(d_w + d_g) = (m_x/2)\,(q + z_g) \quad ...(20.5a)$$

$$a \approx 8[(i + 5)\text{ kW}]^{0.588} \text{ mm} \quad ...(20.5b)$$

Normal circular pitch,

$$p_n = p_x \cos \gamma \quad ...(20.6)$$

Axial module,

$$m_x = m_n/\tan \gamma \quad ...(20.7)$$

Width (or face) angle, 2ψ : It is the angle at the centre included between the points of intersection of the generating circle with the lateral faces of the teeth.

$$\tan \psi = \tan \phi_n/\tan \gamma \quad ...(20.8)$$

Normal pressure angle, ϕ_n

$$\tan \phi_n = \tan \phi_t \cos \gamma \quad ...(20.9)$$

Where ϕ_t = transverse pressure angle

Normal module,

$$m_n = m_x \cos \gamma$$

Designation of worm drive

$$z_1/z_2/q/m_x \quad ...(20.10)$$

2. Number of Starts

The number of starts on the worm should not normally exceed 6. The recommended values of number of starts are given in Table 20.1.

3. Standard Axial Modules on Worm

The standard axial modules for the worm are given in Table 20.2.

4. Normal Pressure Angle, ϕ_n

The normal pressure angle in relation to lead angle is given in Table 20.3.

5. Proportions of Worm and Worm Gear

The proportions of worm and worm gear are given in Table 20.4.

6. Effective Face with of Worm Wheel (Refer to Fig. 20.3)

$$b^2/4 = d^2_{a1}/4 - d^2_1/4 = [m_x (q + 2)/2]^2 - (qm_x/2)^2$$

$$\sin \psi = b/(d_{a1} + 2c)$$

$$\tan \psi = \tan \phi_n/\tan \psi$$

where c = clearance = $0.2\ m_x \cos \gamma$

Length of root of worm gear teeth,

$$I_r = [2\psi/(2\pi)]\ \pi(d_{a1} + 2c)$$

$$= (d_{a1} + 2c)\ \psi = (d_{a1} + 2c) \sin^{-1} [b/(d_{a1} + 2c)] \quad ...(20.11a)$$

Length of worm,

$$L_w = \pi m_x(4.5 + z_g/50) \quad ...(20.11b)$$

7. Friction in Worm Gears

The variation of coefficient of friction μ v's sliding velocity v_s between the worm and worm gear is shown in Fig. 20.4.

Velocity of worm,

$$v_w = \pi \phi_w n_w/(60 \times 10^3) \text{ m/s}$$

Where d_w = pitch diameter of worm, mm

n_w = rpm of worm

Rubbing (or sliding) velocity,

$$v_s = v_w/\cos \gamma = v_g/\cos \gamma$$

$= \pi d_w n_w/(60 \times 10^3 \times \cos\gamma)$ m/s ...(20.12)

Coefficient of friction,

$\mu = 0.0422/(v_s)^{0.28}$ for 0.2 m/s $< v_r < 2.8$ m/s ...(20.13*a*)

$= 0.025 + 3.281\, v_s/10^3$ for $v_r > 2.8$ m/s ...(20.13*b*)

Also $\mu = 0.051 \times S/(v_s + 0.4)^{1/2}$ ...(20.14)

Where S = a factor which depends upon the material combination and surface finish of mating teeth of worm and worm gear. The values of factor S are given in Table 20.5.

8. Efficiency of Worm and Worm Gear Drive

Efficiency, η = Power output to the gear/Power input to the worm

$= P_g/P_w = F_{aw}\, v_g/(F_{tw}\, v_w)$ (Because $F_{tg} = F_{aw}$)

$= (\cos\phi_n - \mu\tan\gamma)/(\cos\phi_n + \mu\cot\gamma)$ when worm is the driver ...(20.15*a*)

$= (\cos\phi_n - \mu\cot\gamma)/(\cos\phi_n + \mu\tan\gamma)$ when worm gear is the driver (20.15*b*)

If $\eta < 50\%$ then the drive is self – locking type.

For maximum efficiency,

$\tan\gamma = (1 + \mu^2)^{1/2} - \mu$ ...(2.16)

For a compact design,

$\gamma = 45° - 0.5\tan^{-1}(\mu/\cos\phi_n)$ ...(20.17)

9. Thermal Capacity of Worm Gears

Heat generated,

$H_g = (1 - \eta)\,P$ kW ...(20.18*a*)

$= (1 - \eta)\,F_{tw} v_w/10^3$ kW ...(20.18*b*)

Heat dissipated,

$H_d = h_{cr} A_c \cdot \Delta t/10^3$ kW ...(20.19)

Where P = input power, kW

η = efficiency pf worm gear drive

h_{cr} = combined heat transfer coefficient, W/m^2 · °C

= 10 to 18 W/m^2 · °C with natural air circulation

= 20 to 28 W/m^2 · °C with forced circculation with a fan

A_c = effective surface area of housing, m^2

$= 1.14 \times 10^{-4} \times a^{1.7}$ m^2

a = centre distance, mm

Δt = rise in temperature of lubricant, °C $\le$ 50°C

For a safe design,

$$H_d > H_g$$

For $H_g = H_d$, we have

$(1 - \eta)\,P = h_{cr} A_c \cdot \Delta t/10^3$

$P = h_{cr} A_c \cdot \Delta t/[1 - \eta) \times 10^3]$ kW ...(20.20)

10. Static Beam Strength of Worm Gears

Select normal module and calculate the axial module,

$$m_x = m_n/\cos\gamma$$

Approximate diameter of worm,

$$d_w = a^{0.875}/2.2 \approx 3 \times \text{circular pitch of worm gear} \quad ...(20.21a)$$

$$d_w = m_x\, z_w/\tan\gamma \quad ...(20.21b)$$

$$d_g = m_x z_g$$

$$v_g = \pi d_g n_g/(60 \times 10^3)\ \text{m/s}$$

Velocity factor,

$$C_v = 3.05/(3.05 + v) \text{ for teeth cut using a form cutter} \quad ...(20.22a)$$

$$= 6.1/(6.1 + v) \quad \text{for generated teeth by hobbing} \quad ...(20.22b)$$

Virtual number of teeth,

$$z_{vg} = z_g \cos^3\gamma \quad ...(20.23)$$

Lewis form factor,

$$y = 0.154 - 0.912/z_v \quad ...(20.24)$$

Tangential load,

$$F_t = 10^3/v_g \quad ...(20.25)$$

Maximum load,

$$(F_t)_{max} = C_s F_t \quad ...(20.26)$$

Effective load,

$$(F_t)_{eff} = C_s F_t/C_v \quad ...(20.27)$$

Where C_s = Service factor, which depends on type of load and machinery (Table 20.16)
= 1.25 to 2.0

Lewis equation for beam strength of gear teeth,

$$F_b = \sigma_{dg}\, \pi m_n b y \quad ...(20.28)$$

Where all the terms have their usual meanings.

σ_{dg} = allowable static bending strength of gear tooth material
= $S_{ut}/3$

Calculate face width from,

$$F_b = F_{eff} \quad ...(20.29)$$

The face width of worm gear is taken as the larger of the following two values:

$$b = 2m(q + 1)^{1/2} \quad ...(20.30a)$$

or

$$= 0.73\, d_w \quad ...(20.30b)$$

which ever is less.

Factor of safety against bending,

$$FOS = F_b/F_{eff} \quad ...(20.31)$$

11. Materials for Worm and Worm Gear

The materials used for worm are given in Table 20.12.

Phosphor bronze is widely used for worm gear. The worm gears are either sand cast or sand cast and chilled. The mechanical properties of worm gear materials are given in Table 20.13.

12. Wear Load

$$F_w = d_g b K \quad ...(20.32)$$

Where K = load stress factor for wear of worm gear (Table 20.14)

Also $$F_w = A \cos\gamma \times C_v\, \sigma_c / C_s \quad ...(20.33)$$

Where A = projected area of tooth,

$$\text{mm}^2 = h_w d_w \psi / 57.3 \quad (20.34)$$

h_w = 2.16 m

ψ = half the face angle, deg

σ_c = compressive strength, MPa (Table 20.15)

For a safe design,

$$F_w > F_d$$

The value of K depends upon the combination of materials for worm and worm gear and lead angle. Table 20.14 gives the values of K.

Factor of safety against wear,

$$\text{FOS} = F_w / F_{eff} \quad ...(20.35)$$

13. Force Analysis (Refer to Fig. 20.6 and 20.7)

$$F_{tg} = -F_{aw};\ F_{ag} = -F_{tw};\ F_{rg} = -F_{rw}$$

$$F_{tw} = 2T/d_w \quad ...(20.36)$$

$$F_{aw} = F_{tw}\ [(\cos\phi_n \cos\gamma - \mu \sin\gamma)/(\cos\phi_n \sin\gamma + \mu\cos\gamma)] \quad ...(20.37)$$

$$F_{rw} = F_{tw}\ [\sin\phi_n/(\cos\phi_n \sin\gamma + \mu\cos\gamma)] \quad ...(20.38)$$

Table 20.1: Number of starts on worm.

Speed	*Number of starts on worm, z_w*
20 and over	1
12 – 36	2
8 – 12	3
6 – 12	4
4 – 10	6

Table 20.2: Standard axial modules for worm, m_x (mm)

First choice	1.0, 1.25, 1.5, 2, 2.5, 3, 4, 5, 6, 8, 10, 12, 16, 20, 25, 32, 40 & 50
Second choice	1.125, 1.375, 1.75, 2.25, 2.75, 3.5, 4.5, 5.5, 7, 9, 11, 14, 18, 22, 28, 36 & 45.

Table 20.3: Recommended normal pressure angle in relation to lead angle

Lead angle, γ, deg	0 – 15	15 – 30	30 – 35	35 – 40	40 – 50
Normal pressure angle, ϕ_n	14.5°	20°	25°	25°	30°

Table 20.4: Proportions of worm and worm gear.

Worm	*Worm wheel*
Pressure angle, ϕ = 20° Addendum, $h_{a1} = m_x$ Dedendum, $h_{f1} = 1.25\ m_x$ Tooth depth, $h = 2.25\ m_x$ Bottom clearance, c = (0.2 to 0.3) m_x Tooth thickness: Axial = 1.57 m_x Normal = 1.57 $m_x \cos\beta$ Reference (or pitch) diameter, $d_1 = qm_x$ Tip diameter, $d_{a1} = d_1 + 2m_x$ Root diameter, $d_{f1} = d_1 - 2m_x - 20$ Addendum modification coefficient, $x - a/m_x - 0.5\ (q + \sigma_2)$ Base diameter, $d_{b1} = m_x(q + 2x)$	Pitch diameter, $d_2 = m_x\ z_2$ Bade circle diameter, $d_{b2} = d_2$ Tip diameter, $d_{a2} = d_2 + 2m_x = d_2 + 2h_{a2}$ Throat diameter, $d_{f2} = d_2 - 2m_x - 2c$ $= d_2 - 2h_2$

Table 20.5: Values of factor S

Material of worm	*Material of worm gear*	*S*
1. Steel – Hardened and Ground	Cu – Sn Bronze Al – Alloy Cast iron	1.0 1.15 1.25
2. Steel – Heat Treated but not Ground	Cu – Sc Bronze Al – Alloy Cast Iron	1.5 1.73 1.83

Table 20.6: Dimension of worms

Module m	*Axial pitch p_x*	*No. of states z*	*Diametral Quotient q*	*Reference dia d_1*	*Tip dia d_{a1}*	*Root dia d_{f1}*	*Lead angle r*
(1)	*(2)*	*(3)*	*(4)*	*(5)*	*(6)*	*(7)*	*(8)*
1	3.142	1	16	16.00	18	13.60	3° 35′
1.25	3.927	1	(18)	22.50	25	19.50	3° 11′
1.50	4.712	1	16	24	27	20.40	3°35′
			(18)	27	30	23.40	3°11′
2.0	6.283	1	(11)	22	26	17.50	5°12′
2.0	6.283	1	(18)	36	40	31.20	3°11′
		2	(11)	22	26	17.20	10°18′
		4	(11)	22	26	17.20	19°59′
		1	(11)	27.50	32.50	21.50	5°12′
		1	16	40.00	45.00	34.00	3°35′
2.5	7.854	2	(11)	33	39	25.80	10°18′
		4	(11)	33	39	25.80	19°59′
		6	(11)	34	52	36.555	28°27′

		1	(11)	33	39	25.80	5°11′
		1	(18)	54	60	46.80	3°11′
3	9.425	2	(11)	33	39	25.80	10°18′
		4	(11)	33	39	25.80	19°59′
		6	(11)	33	39	27.41	28°27′
		1	10	40	48	30.40	5°43′
		1	(18)	79	80	62.40	3°11′
4	12.566	2	10	40	48	30.40	11°19′
		4	10	40	48	30.40	21°48′
		6	(11)	44	52	36.55	28°37′
		1	10	50	60	38.00	5°43′
		1	(18)	90	100	78.00	3°11′
5	15.708	2	10	50	60	38.00	11°19′
		4	10	50	60	38.00	21°48′
		6	(11)	55	65	45.69	28°37′
		1	10	60	72	45.60	5°43′
		1	(18)	108	120	93.60	3°11′
6	18.860	2	10	60	72	45.60	11°19′
		4	10	60	72	45.60	21°48′
		6	10	60	72	49.36	30°58′
		1	10	80	96	60.80	5°43′
		1	(18)	144	160	124.80	3°11′
		2	10	80	96	60.80	11°19′
8	25.133	4	10	80	96	60.80	21°48′
		4	8	80	80	48.52	26°34′
		6	(9)	72	88	58.71	33°41′
		8	(11)	88	104	75.33	36°02′
		1	10	100	120	76.00	5°43′
		1	(18)	180	200	156.00	3°11′
10	31.416	2	10	100	120	76.00	11°19′
		4	10	100	120	76.00	21°48′
		8	(11)	110	130	94.42	36°02′
		1	(9)	008	132	79.20	6°20′
		1	(18)	216	240	187.20	3°11′
12	37.699	2	(9)	108	132	79.20	12°32′
		4	(9)	108	132	79.20	23°58′
		8	(11)	132	156	113.30	36°02′
		1	(9)	144	176	105.60	6°20′
		2	(9)	144	176	105.60	12°32′
16	50.266	4	(9)	144	176	105.60	23°58′
		8	(11)	176	208	151.06	36°02
		1	(9)	180	220	132.00	6°20′
		2	(9)	180	220	132.00	12°32′
20	62.832	4	(9)	180	220	132.00	23°58′
		8	(11)	220	260	188.83	36°02′

IS : 3734—1966

Note: The values given within brackets are non-preferred. Pressure angle = 20 deg.

Table 20.7: Recommended values of transmission ratios, centre, distance * $z_1/z_2/q/m$ for general purpose worm gears All dimensions in millimetres

Centre distance	Transmission Ratio					
	5	7.5	10	12.5	15	20
25	2/10/(11)/2.5	1/7/(18)/2	—	—	1/15/(18)/1.5	1/21/(18) 1.25
31.5	4/21/(11)/2	2/15/(11)/2.5	2/21/(11)/2	1/13/(18)/2.5	1/15/(11)/2.5	1/10/(11)2
40	4/21(11)/2.5	—	2/21/(11)/2/5	—	1/16/16/2.5	1/21/(11)/2.5
50	9/29/(11)/2.5	4/29/(11)/2.5	4/38/(11)/2	—	2/29/(11)/2.5	2/38/(11)/2
63	6/31/(11)/3	4/31/(11)/3	4/39/(11)/2.5	4/51/(11)/2	2/31/(11)/3	2/39/(11)/2.5
80	6/29/(11)/4	4/30/(10)/4	4/42/(11)/3	4/53/(11)/2.5	2/30/10/4	2/42/(11)/3
100	6/29/11/(5)	4/30/10/5	4/40/10/4	4/49/(18)/3	2/30/10/5	2/40/10/4
125	6/32/10/6	4/31/10/6	4/40/10/5	4/52/10/4	2/31/10/6	2/40/10/5
160	6/31/(9)/8	40/30/10/8	4/43/10/6	4/54/10/5	20/30/10/8	—
200	8/39/(11)/8	4/30/10/10	4/40/10/8	—	2/30/10/10	2/40/10/8
250	8/39/(11)/10	—	4/40/10/10	4/52/8/8	—	2/40/10/10
315	8/41/(11)/12	4/30/(9)/16	4/43/(9)/12	4/53/10/10	7/30/(9)/16	—
400	8/39/(11)/16	4/31/(9)/20	40/41/(9)/16	—	2/31/(9)/20	2/41/(19)/16
500	9/39/(11)/20	—	4/41/(9)/20	4/53/(9)/16	—	2/41/(9)/20

Centre distance	Transmission Ratio				
	25	30	40	50	60
25	—	1/33/16/1	—	—	—
31.5	1/26/16/1.5	1/32/(18)/1/25	—	1/48/16/1	—
40	—	1/29/(11)/2	—	—	1/16/61/1
50	—	1/29/(11)/2.5	1/38/(11)/2	—	1/62/(18)/1.25
63	2/51/(11)/2	1/31/(11)/3	1/39/(11)/2.5	1/51/(11)/2	—
80	2/53/(11)/2.5	1/30/10/4	1/42/(11)/3	1/53/(11)/2.5	1/62/(18)/2
100	2/54/(11)/3	1/30/10/5	1/40/10/4	1/55/(11)/3	1/64/16/2.5
125	2/52/(10)/4	1/31/10/6	1/40/10/5	1/52/10/4	—
160	2/52/10/5	1/30/10/8	—	1/54/10/5	1/62/(18)/4
200	—	1/30/10/10	1/40/10/8	—	—
250	2/52/10/8	—	1/40/10/10	1/52/10/8	—
315	2/53/10/10	1/30/(9)/16	—	1/53/13/10	1/60/(18)/8
400	—	1/31/(9)/20	1/41/(9)/16	1/49/(49)/16	1/58/(9)/12
500	2/53/(9)/16	—	1/41/(9)/20	1/53/(9)/16	—

IS : 3734—1966

Note: The values within brackets are non-preferred.

*6/32/10/6 denotes, 6 starts on the worm, 32 teeth on the wheel, q = 10 and m = 6.

Table 20.8: Selection of number of teeth, z_2 on the worm wheel †

m_x, mm	4	5	5	6	6	8	8	10	10	12	12	16	20
$q \rightarrow$	11	10	12	9	11	8	11	8	11	8	9	11	8
a, mm ↓													
100	39	30	28	—	—	—	—	—	—	—	—	—	—
122	49	38	36	31	29	—	—	—	—	—	—	—	—
150	64	50	48	41	39	29*	—	—	—	—	—	—	—
180	79	62	60	51	49	37	34	28	—	—	—	—	—
210	—	74	72	61	59	44*	41*	34	31	—	—	—	—
240	—	—	—	71	69	52	49	40	37	32	29	—	—
270	—	—	—	81	79	59*	56*	46	43	37	34	—	—
300	—	—	—	—	—	67	64	52	49	42	39	28*	—
360	—	—	—	—	—	82	79	64	61	52	49	36	28
420	—	—	—	—	—	—	—	76	73	62	59	43*	34
480	—	—	—	—	—	—	—	—	—	72	69	51	40
540	—	—	—	—	—	—	—	—	—	82	79	58*	46
600	—	—	—	—	—	—	—	—	—	—	—	66	52

† Coould be ± 2 teeth from values found from the table above except numbers marked with* where the deviation permissible is –1 or +2 teeth.

Table 20.9: Approximate efficiencies of worm gear drives

Number of starts	*Efficiency, %*
1	70 – 75
2	70 – 85
3 and 4	80 – 92

Table 20.10: Values of Lewis form factor for worm gears

Normal pressure angle, deg	14.5	20	25	30
Lewis form factor, y	0.100	0.125	0.150	0.175

Table 20.11: Face width of worm wheel, *b*

Number of starts, z_w	*Face width, b*
1	0.75 d_w
2 or 3	0.75 d_w
4	0.67 d_w

Table 20.12: Mechanical properties of materials used for worm

Material	*Heat treatment*	*Ultimate tensile Strength, S_{ut} MPa*	*Brinella hardness Number, BHN MPa*	
			Core	*Case*
40C8	N	580	217	217
55C8	N	720	265	265
10C4	CH	340	140	400
14C6	CH	370	140	400
16Ni80Cr60	CH	700	250	500
20Ni2Mo25	CH	750	250	500
15Ni2Cr1Mo15	CH	800	250	500
40Ni2Cr1Mo28	CH	800	250	600
13Ni3Cr18	CH	900	250	600
15Ni4Cr1	CH	1000	250	600

N = Normalized, CH = Case Hardened

Table 20.13: Mechanical properties of materials for worm gear

Sl. No.	*Material*	*Ultimate tensile Strength, S_{ut} MPa*	*Brinell Hardness Number, BHN MPa*	*Applications*
1.	Grey Cast Iron	200 – 350	179 – 241	Used for light duty and low speed drive, $v_s < 2$ m/s
2.	Gun Metal	220	80	Used for light to medium duty and low speed drive, $v_s < 5$ m/s
3.	Phosphor Bronze	245	95	Used for medium to heavy duty and high speed drive, $5 < v_s < 25$ m/s
4.	Nickel Bronze	340	95	Used for medium to heavy duty and high speed drive, $5 < v_s < 25$ m/s
5.	Aluminium Bronze	550	120	Used for heavy duty and low speed drive, $v_s < 5$ m/s

Table 20.14: Worm gear load stress factor, K

Material		*Load stress factor, K (MPa)*		
Worm	Worm Gear	$\gamma < 10°$	$10° \le \gamma \le 25°$	$\gamma > 25°$
Hardened Steel	Cast Iron	0.345	0.430	0.515
Steel, 250 BHN	Phosphor Bronze	0.415	0.515	0.620
Hardened Steel	Phosphor Bronze	0.550	0.685	0.825
Hardened Steel	Chilled Phosphor Bronze	0.825	1.030	1.235
Cast Iron	Phosphor Bronze	1.030	1.285	1.545

Table 20.15: Permissible compressive stress σ_c*, N/mm²

Material of Worm	*Material of Gear*	*Number of teeth in the gear*							
		10	*20*	*30*	*40*	*50*	*60*	*70*	*80 and over*
20 *C* steel, Untreated	Cast iron	0.49	1.85	3.23	5.20	6.18	7.46	8.63	9.32
40 *C* steel, untreated	Bronze, sand cast	0.79	2.35	4.61	7.36	9.32	12.47	31.05	13.83
40 *C* steel, heat treated ground	Bronze, sand cost	1.18	3.53	6.47	10.99	13.83	18.54	19.62	29.70
10 *C* alloy steel, carburized	Bronze, and cost	1.86	5.00	8.63	14.81	18.54	22.37	26.19	27.57
Hardened ground	Bronze, chill cast	2.16	6.38	11.87	20.30	25.60	31.00	36.20	37.97
	Nickel Bronze, sand cast	2.94	7.75	20.50	23.05	31.00	37.57	43.75	46.20
	Nickel Bronze, chill cast	3.14	9.32	12.27	29.63	37.18	44.83	53.39	55.13

*Values tabulated are for 14.5° pressure angles. Multiply by 1.05 for 20° and 1.10 for 30° pressure angles

Table 20.16: Service factor for worm gears, C_s

Types of services	*Service factor, C_s*
1. Intermittent with light shock	1– 1.5
2. Continuous with medium shock, line shafts, crushers, etc.	1.5 – 2
3. Continuous with heavy shock, reciprocating pumps, paper and rubber mills, etc.	2 – 2.5
4. Continuous with frequent and heavy shocks, main line drives, steel mills, etc.	2.5 – 3

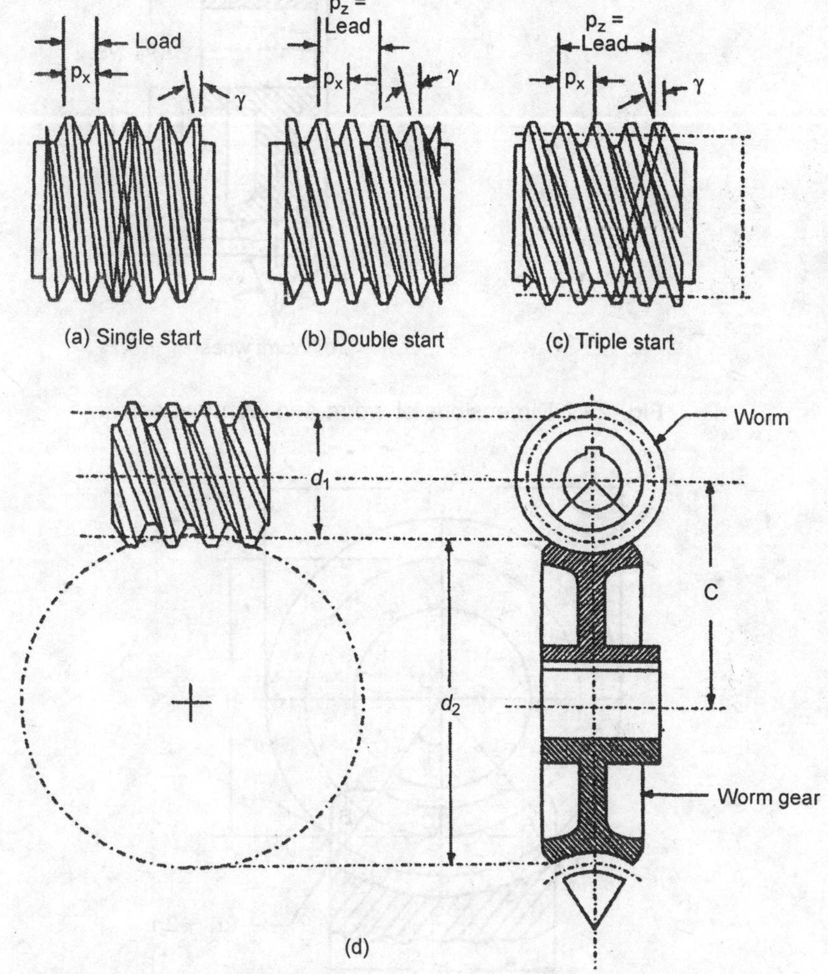

Fig. 20.1 Worms and worm wheel.

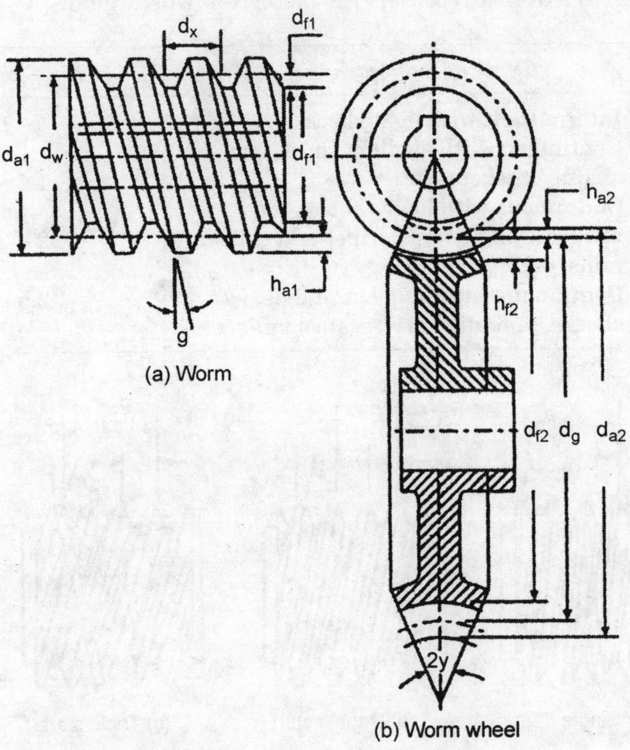

Fig. 20.2 Dimensions of worm and worm wheel.

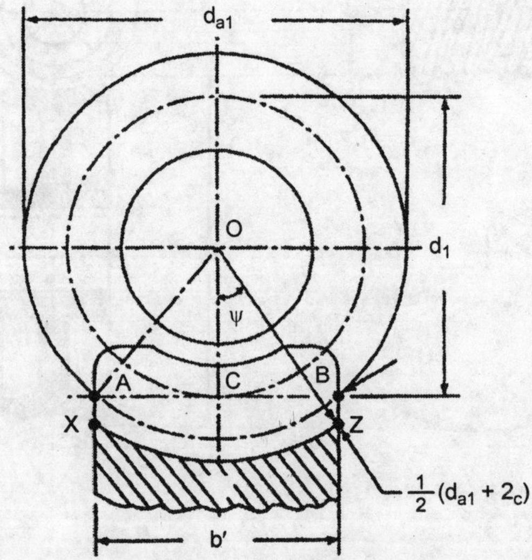

Fig. 20.3 Face width of worm wheel.

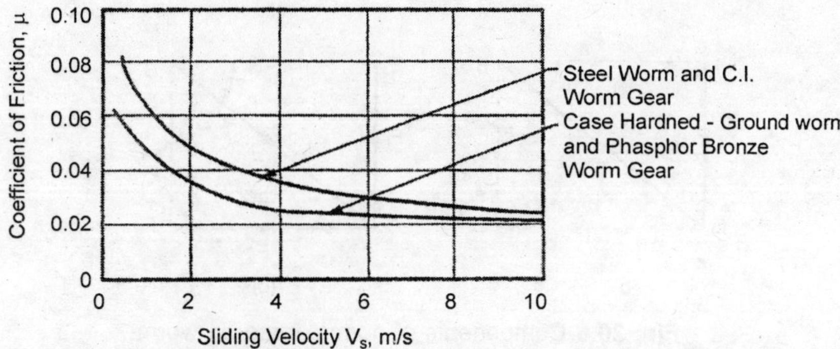

Fig. 20.4 Plot of coefficient of friction versus sliding velocity.

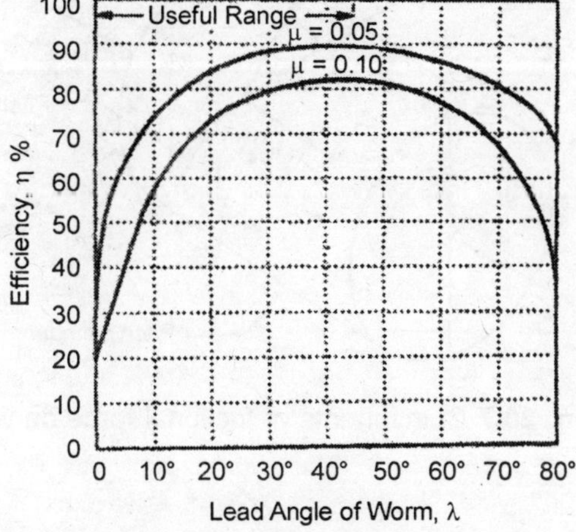

Fig. 20.5 Efficiency of worm gear pair.

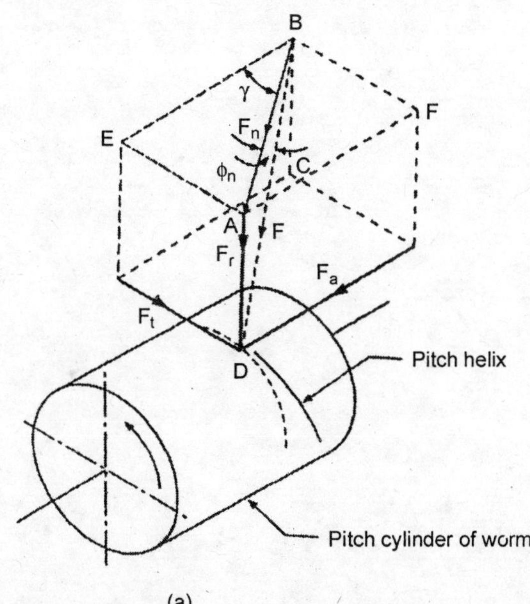

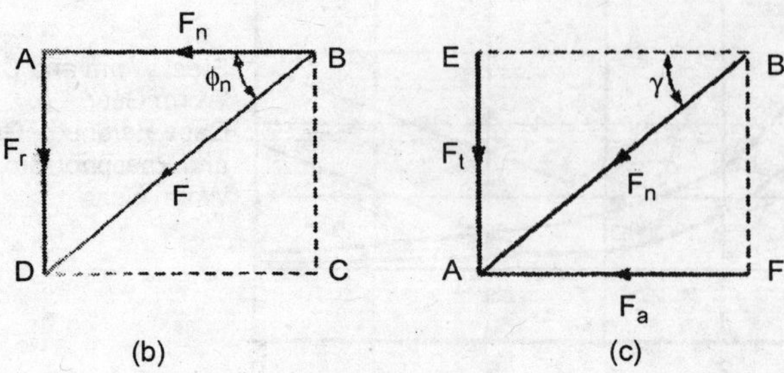

Fig. 20.6 Components of normal force on worm.

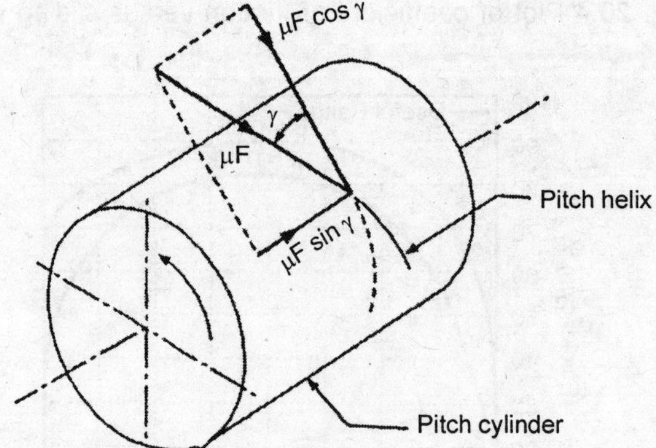

Fig. 20.7 Components of frictional force on worm.

CHAPTER

21

Sliding Bearings

Nomenclature

c = $r_1 - r$, radial clearance

C_p = specific heat of lubricant, J/kg. °C

d = journal diameter, mm

h = film thickness

H_d = $mCp \cdot \Delta t$, heat dissipated

H_g = heat generated

l = length of bearing, mm

m = ρQ, mass of lubricating oil passing through the bearing, kg/s

N = journal speed in rpm

n = journal speed, rps

p = W/Id, unit bearing pressure, N/mm²

r = journal radius, mm

r_1 = bearing radius, mm

W = radial load on the journal, N

ρ = density of oil, kg/m³

ϕ = attitude angle, deg

μ = dynamic viscosity of lubricant, N.s/mm²

ε = e/c, eccentricity ratio

ε = eccentricity

Δt = temperature rise of lubricant, °C

1. Dynamic Viscosity

Dynamic viscosity of lubricant,

$$\mu = (0.22 \text{ SUS} - 180/\text{SUS})\ \rho \times 10^{-6} \text{ N.s/m}^2 \text{ (Pa.s)} \quad ...(21.1)$$

Where SUS = viscosity in Saybolt's Universal Seconds

ρ = density, kg/m^3

$= \rho_{15.6} - 0.63\ (t_f - 15.6)$ kg/m^3 ...(21.2)

t_f = oil film temperature, °C

1cP = 10^{-2} P = 10^{-3} Pa.s or mPa.s

1 kg/m.s = 1 Pa.s

The properties of lubricants are given in Table 21.1. The specific gravity of oils at 15.6°C is given in Table 21.2. The effect of temperature on dynamic viscosity is shown in Fig. 21.1 for a range of SAE oils.

2. Hydrodynamic Theory of Lubrication

(*i*) Reynold's Equation (Fig. 21.3)

$$(\partial/\partial x)\ (h^3 \partial p/\partial x) + (\partial/\partial z)\ (h^3 \partial p/\partial z) = 5\mu U\ (\partial h/\partial x) \quad ...(21.3)$$

(*ii*) Short Bearing (*l*/d < 1)

$$h_{min} = c(1 - \varepsilon) \quad ...(21.4)$$

$$\rho = [\mu U/(rc^2)]\ [3\varepsilon \sin\theta/(1 + \varepsilon\cos\theta)^3]\ [l^2/4 - z^2] \quad ...(21.5)$$

$$W_x = (mUl^3/c^2)\ [\varepsilon^2/(1 - \varepsilon^2)^{3/2}] \quad ...(21.6)$$

$$W_y = (\mu Ul^3/c^2)\ [\pi\varepsilon^2/4(1 - \varepsilon^2)^{3/2}] \quad ...(21.7)$$

$$W = [4\mu Ul^3\ \varepsilon/\{c^2(1 - \varepsilon^2)\}]\ [16\ \varepsilon^2 + \pi^2(1 - \varepsilon^2)]^{1/2} \quad ...(21.8)$$

Or $[W/(l\mu U)]\ (c/r)\ (c/l)^2 = \varepsilon[16\ e^2 + \pi^2(1 - \varepsilon^2)]^{1/2}/(1 - \varepsilon^2)^2$ ...(21.9)

Somerfeld Number, $S = [\mu n/p]\ (r/c)^2$ ...(21.10)

$$S(l/d)^2 = (1 - \varepsilon^2)^2/[\pi\varepsilon\{16\ \varepsilon^2 + \pi^2(1 - \varepsilon^2)\}] \quad ...(21.11)$$

The distribution of load on the bearing surface is shown in Fig. 21.4

Attitude angle, $\tan\phi = W_y/W_x = -\pi(1 - \varepsilon^2)^{1/2}/(4\ \varepsilon)$ ...(21.12)

Frictionalforce: $F = (\mu Uld/c)\ [2\pi/(1 - \varepsilon^2)^{3/2}]$ ...(21.13)

(*iii*) Long Bearing (*l*/*d* > 1)

$$\partial^2 p/\partial x^2 = (6\mu U/h^3)\ \partial h/\partial x \quad ...(21.14)$$

Film thickness h_1 where the pressure is maximum is given by,

$$h_1 = c(1 - \varepsilon^2)/(2 + \varepsilon^2) \quad ...(21.15)$$

The resultant force is given by,

$$[W/(\mu Ul)]\ (c/r)^2 = 12\pi\varepsilon/[(2 + \varepsilon^2)\ (1 - \varepsilon^2)^{1/2}] \quad ...(21.16)$$

Sommerfeld Number, $S = (2 + e^2)\ (1 - \varepsilon^2)^{1/2}/(12\ \pi^2\ \varepsilon)$ ...(21.17)

Tangential frictional force,

$$F = (\mu Uld/c)\ [4\pi(1 + 2\varepsilon^2)/\{(2 + \varepsilon^2)\ (1 - \varepsilon^2)^{1/2}\}] \quad ...(21.18)$$

Coefficient of friction, $f = F/W = [(1 + 2\varepsilon^2\ )/(3\varepsilon)]\ (c/r)$ ...(21.19)

Angle θ for maximum pressure is given by, $\cos\theta = -3\varepsilon/(2 + \varepsilon^2)$ ...(21.20)

Film pressure at any angle θ is,

$$p_\theta = (3\mu Ud/c^2)\ [\varepsilon \sin\theta(2 + \varepsilon\cos\theta)/\{(2 + \varepsilon^2)\ (1 + \varepsilon\cos\theta)^2\}] \quad ...(21.21)$$

3. Bearing Pressure

$$p = W/(ld) \text{N/mm}^2 \quad ...(21.22)$$

4. Rubbing Velocity

Rubbing velocity, $U = \pi dN/(60 \times 10^3)$ m/s ...(21.23)

5. Bearing Modulus (Fig. 21.5)

Bearing modulus, $C = [\mu N/(60p)] \times 10^{-6}$ ...(21.24)

$(\mu N/p) \geq 3C$ for minimum coefficient of friction ...(21.24*a*)

$> 15C$ for severe load ing ...(21.24*b*)

The values of bearing modulus for various journal – bearing combinations are given in Table 21.3.

6. Clearance Ratio

$c/r = 0.001$ for plastic bearing ...(21.25)

$= 0.0015$ for hard bearing

Design data for bearings is given in Table 21.4.

7. Sommerfeld Number

$S = [\mu n/p]\,(r/c)^2 \times 10^{-6}$ without side leakage ...(21.26)

Where p = pressure, MPa, n = rps, μ = Pa.s, c = clearance, mm, r = journal radius, mm

8. Coefficient of Friction

(*i*) Petroffs equation for S > 150 and lightly loaded bearings,

$$f = (2\pi^2/10^6)[\mu n/p](r/c) \quad ...(21.27)$$

Where
p = pressure, Pa
r = journal radius, mm
c = radial clearance, mm
n = rps
μ = dynamic viscosity of lubricating oil, Pa.s

(*ii*) McKee's equation for well lubricated and lightly loaded bearings, S > 150,

$$f = (0.326/10^6)[\mu N/(60p)](r/c) + \Delta f \quad ...(21.28)$$

Where Δf = a correction factor for end leakage (Table 21.5)

$= 0.002$ for $0.75 < l/d < 2.8$

(*iii*) For medium to heavily loaded bearings,

$$f = [(1 + 2\varepsilon^2)/(3\varepsilon)](c/r) \quad ...(21.29)$$

(*iv*) For partially lubricated bearings having incomplete fluid film, Louis Illmer equation for coefficient of friction,

$$f = (C_1 C_2/8533)\,(p_a/U)^{1/4} \quad ...(21.30)$$

Where C_1 and C_2 = constants (Table 21.6 and 21.7)

U = rubbing velocity, m/s

p_a = average pressure, Pa

= $p_{max}/2$

9. Operating Pressure

(*i*) Victor Tatarinoff equation for permissible pressure for film lubrication,

$$p = (1.33\mu n/10^8)(d/c)^2 [l/(l + d)] \quad ...(21.31)$$

where p = pressure, MPa

μ = dynamic viscosity, mPas

n = speed of journal, rps

(*ii*) General Electric Company's formula (for motor and generator bearings),

$$p = 0.622\ U^{1/3}\ \text{MPa} \quad ...(21.32)$$

where U = journal speed, m/s

(*iii*) Moore's equation,

$$p = 0.726\ U^{1/2}\ \text{MPa} \quad ...(21.33)$$

(*iv*) Louis Illmer equation,

$$p = 4.65[U/(t - 121.5)^{1/3}\ \text{MPa} \quad ...(21.34)$$

where U = velocity, m/s

t = temperature, °C

Product of pressure and velocity is given in Table 21.8.

Recommended speed of some machinery is given in Table 21.9.

10. Minimum Oil Film Thickness

h_{min} = 0.01875 mm for ordinary babbitted bearings

= 0.0025 mm for fine - finished bronze bearings

= 0.075 to 0.125 mm for large steel shafts of fans, turbo - generators, etc.

= 0.00025D, D in mm ...(21.35*a*)

$$h_{min} = c(1 - e) \quad ...(21.35b)$$

$$h_{min} \geq C_p(R_{ab} + R_{aj}) \quad ...(21.36)$$

where R_{ab}, R_{aj} = roughness of bearing and journal respectively (Table 21.10)

C_p = 4.5 for ground finish

= 7.0 for super finish

= 10.0 for loose abrasive lapped

The Sommerfeld number for various h_{min}/c and l/d ratios are given in Table 21.11.

11. Heat Generated and Dissipated,

(*a*) Heat generated in bearing,

$$H_g = fpAU\ \text{Watts} \quad ...(21.37)$$

(*b*) Heat dissipated through bearing.

(*i*) Surface exposed in terms of projected area of journal

$$H_d = k_b A(t_b - t_a) \times 10^{-6} \text{ Watts} \quad ...(21.38)$$

Where t_b = bearing temperature, °C

$= t_a + 0.5(t_f - t_a)$

t_f = oil film temperature

t_a = surrounding air temperature, °C

k_b = energy dissipation coefficient, W/m^2.°C (Table 21.12)

A = Id = projected area of journal, mm^2

(*ii*) Based on area of housing,

$$H_d = K_1 A(t_b - t_a) \quad ...(21.39)$$

where $K_1 = (t_b - t_a)$ = heat radiating capacity of bearing, W/mm^2 (Table 21.13)

(*iii*) Based on projected area of bearing

$$H_d = K_2 ld(t_b - t_a) \text{ J/s} \quad ...(21.40)$$

where $K_2\ (t_b - t_a)$ = heat radiating capacity of bearing, W/mm^2 (Table 21.14)

Heat dissipated is also given by,

$$H_d = K_2 ld(t_b - t_a) \text{ J/s} \quad ...(21.41)$$

$K_3 = 2.6223 \times 10^6$ for well ventilated bearings

$= 4.6513 \times 10^6$ for unventilated bearings

12. Volume Flow Rate of Coolant

Heat carried away by lubricating oil,

$$H_c = \rho Q C_p(t_o - t_i) \quad ...(21.42)$$

Volume flow required,

$$Q = (H_g - H_d) \times 10^3/[\rho C_p(t_o - t_i)]\text{m}^3/\text{s} \quad ...(21.43)$$

ρ = density of coolant, kg/m^3

= 860 kg/m^3 for most of the oils

C_p = thermal capacity of coolant, J/kg.°C

= 1.76 kJ/kg.°C for most of the oils

t_i, t_o = inlet and outlet temperature of coolant

13. Permissible unit bearing pressure

Refer to Table 21.15 and 21.16.

14. Radial clearance ratio (c/r)

15. Minimum oil film thickness, h_{min}

$$h_{min} = (0,0002 - 0.0003)\ r \quad ...(21.44)$$

16. Bearing Materials

Refer to Table 21.19.

17. Bearing Cap

The thickness of the bearing cap, shown in Fig. 21.8, is given by,

$$t = [3WL/(2al)]^{1/2} \quad ...(21.45)$$

where L = distance between bolt centres

l = bearing length – oil hole diameter

σ = permissible bending stress for the cap material

Based upon deflection, the cap thickness is given by,

$$t = [WL^3/(4EI\delta)]^{10} \quad ...(21.46)$$

where I = moment of inertia of cap cross – section

δ = deflection of cap ≈ 0.0254 mm for medium size cap

= 0.0127 mm for small size cap

Empirically, t = 0.08 + 2.5 mm for small size brass and bronze bearing caps ...(21.47*a*)

= 0.2 d mm for medium size C.I. bearing caps...(21.47*b*)

= 0.202d + 3.2 mm for babbit bearing cap ...(21.47*c*)

18. Raimondi and BOYD method

Dimensionless parameters:

The geometry of the bearing is shown in Fig. 21.7.

h_0 = minimum film thickness, mm

$\underline{\varepsilon}$ = $e/c = 1 - h_0/c$, eccentricity ratio

h = h_0/c = minimum film thickness variable

l/d = length to diameter ratio

S = $(\mu n/p)\,(r/c)^2$, Sommerfeld number

f = $(r/c)\,f$ = coefficient of friction variable (CFV)

P_f = $f\,WU/10^3$, frictional power, kW

U = $\pi dN/(60 \times 10^3)$, journal tangential speed, m/s

p/p_{max} = maximum film pressure ratio

Q = Q/(rcnl), flow variable (FV)

Q = flow of lubricant, mm³/s

Q_s = side leakage of lubricant, mm³/s

Q_s/Q = flow ratio

p/p_{max} = maximum film pressure ratio

$\rho\, C_p \Delta t/p$ = temperature rise variable

Dimensionless parameters for full journal bearing with side flow are given in Table 21.20.

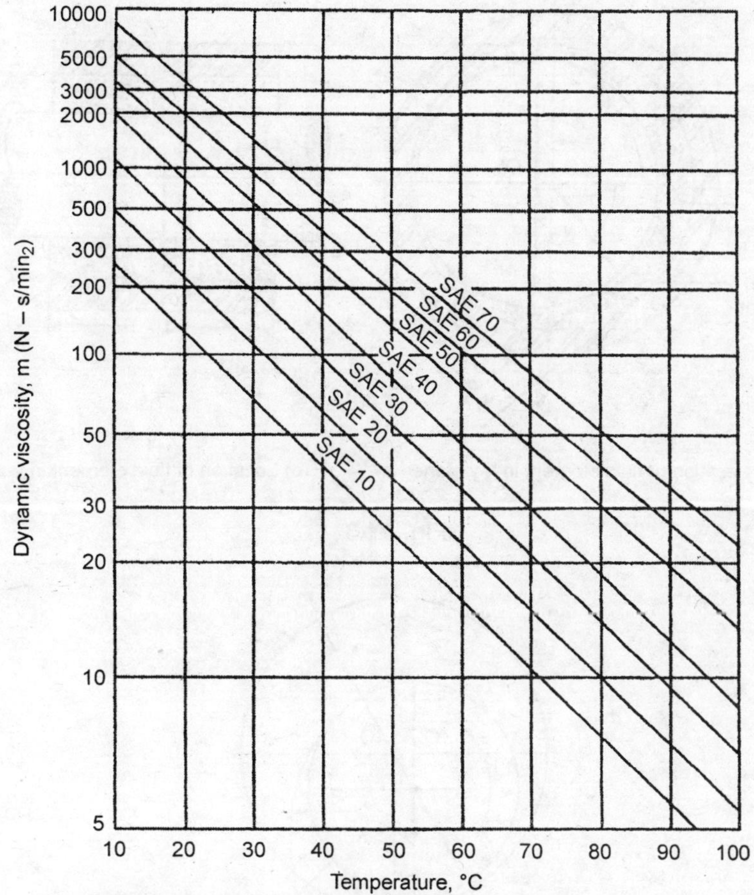

Fig. 21.1 Viscosity-temperature relationship.

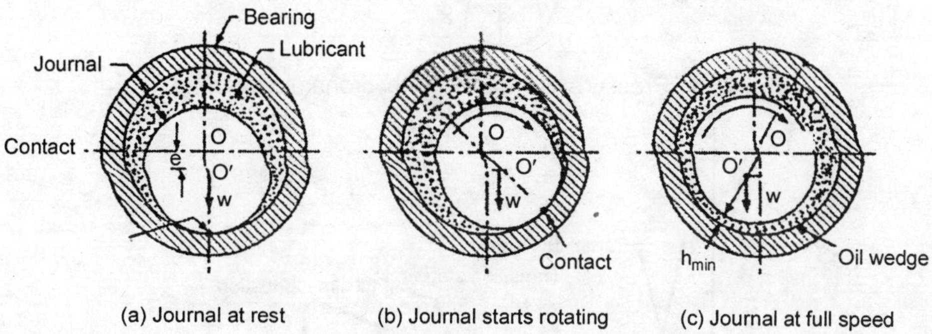

Fig. 21.2 Hydrodynamic Lubrication.

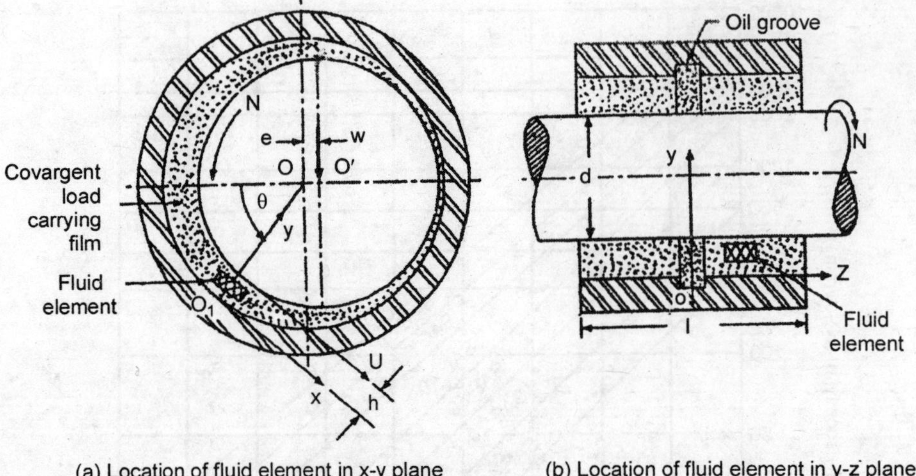

(a) Location of fluid element in x-y plane (b) Location of fluid element in y-z plane

Fig. 21.3

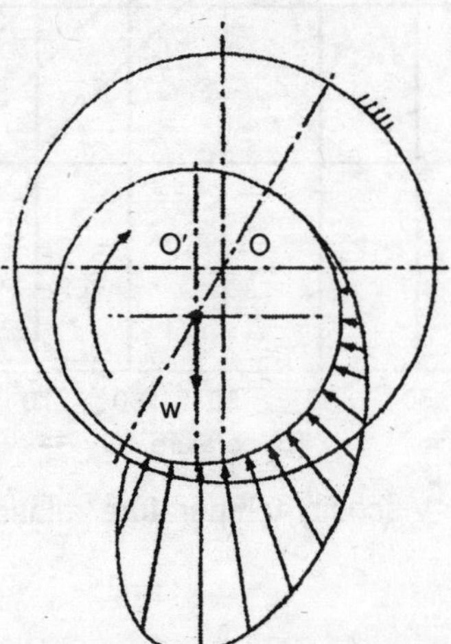

Fig. 21.4 Pressure distribution in hydronamic bearing.

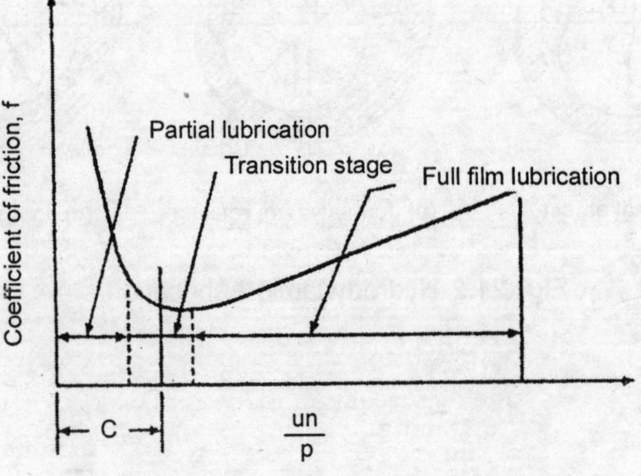

Fig. 21.5 Coefficient of friction *V's* $\frac{\mu n}{p}$.

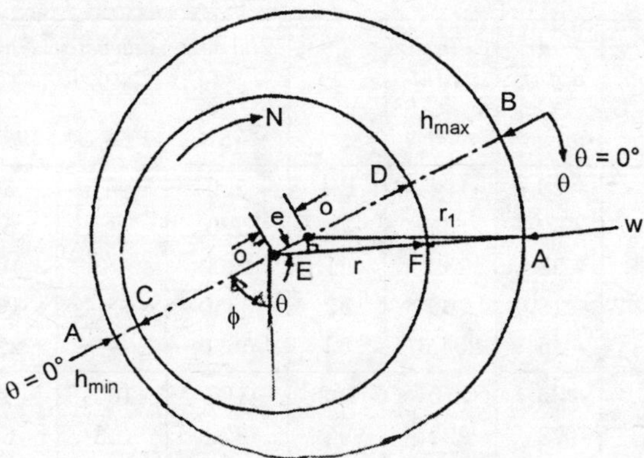

Fig. 21.6 Geometry of bearing.

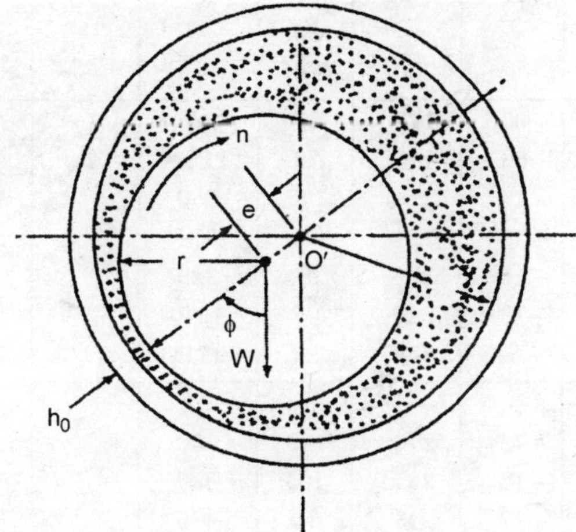

Fig. 21.7 Bearing geometry for Raimondi and Boyd meth.

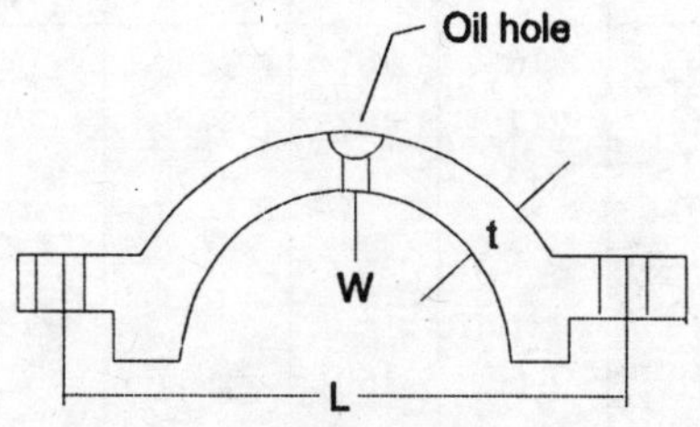

Fig. 21.8 Bearing cap.

Table 21.1: Properties of Lubricants

Puprpose	*SAE Number*	*Density kglm* at 15.5°C*	*Pour point °C*	*Flash point °C*	*Viscosity,Viscosity, Saybolt indexseconds at*				*Uses*
						38°C	*54°C*	*99°C*	
Transmis-	75	900	– 23	193	121	220	—	50	Combination
sion gear	80	934	– 32	185	78	320	—	52	pinion, reduction
oil	90	930	– 23	232	91	1300	—	100	gear units,
	140	937	– 18	260	82	3350	—	160	enclosed reduc-
	250	—	– 15	254.5	83	5660	—	220	tion gear sets.
Automo-	10	870	– 36	210	102	190	105	46	Automotive, truck
live oil	20	885	– 23	227	96	330	153	54	and marine
	30	891	– 20	238	92	530	220	64	reciprocating en-
	40	890	– 18	2403	90	800	255	77	gines, very heavy
	50	902	– 12	254.5	90	1250	—	97	duty oil used in
	60	—	—	—	80	—	—	115	diesel engines.
	70	—	—	260	80	—	—	137	
Aircraft engine oil		858	– 65	111	87	43		33	Turbojet engines.
		864	– 62	146	79	58		35	Various
		876	– 18	215.5	106	350		57	reciprocating
		884	– 18	224	96	514		64	aircraft engines.
		887	– 18	232	95	829		80	
		892	– 18	249	95	1240		99	
		892	– 7	318	96	1711		120	
Turbine grade oil								44	Direct connected
Light		872	– 18	210	109	150		53	turbines, electric
Medium		877	– 12	235	105	300		62	motors. Land
Heavy		885	– 12	243	10	460			geared turbines,
									electric motors.
									Marine propul-
									sion geared tur-
									bines.
Steam cylinder oil		895	1.4	210	101	1800		130	Rail road and sta-
		910	1.5	211	107	3750		210	tionary steam en-
		940	15.5	343	103	6470		300	gine cylinders,
									enclosed gears.
Hydraulic oils									Hydraulic fluids
Light	887	– 42	181	64	150			42	for most indoor in-
Medium	805	– 26	207	66	310			50	dustrial hydraulic
Heavy	901	– 12	267	70	910			74	equipments.
Extra-low temperature	844	– 28	110	226	74			43	Heavier loads,
									higher tempera-
									ture. Aircraft
									hydraulic system.

Refrigerating machine oil	895	– 45.5	146	53	72		36	Ammonia compressor
	898	– 37	165.5	22	195		43	
	909	– 29	182	34	235		45	
	902	– 23	1903	35	335		49	
Machine tools and general purpose oil	881	– 4	177	80	105		39	All general propose lubrication, machine tools.
	898	– 4	199	80	205		46	
	915	– 12	185	83	305		49	
	915	– 15	199	25	520		59	
	890	– 9	235	80	930		80	

Table 21.2: Specific gravity of oils at 15.6°C

Oil	*Type of oil*	*Specific gravity at 15.6°C*
SAE10	Turbine oil (oil ring) light service and high speed	0.889
SAE20	Light transmission for gears, automobile oil	0.925
SAE30	Automobile oil	0.926
SAE40	Medium transmission oil for large generators, motors, steam turbines, heavy speed gears, hoavy motor oil, automobile oil	0.927
SAE50	Light cylinder oil	0.890
SAE60	Airplane oil	0.893
SAE70	Gear oil	0.915

Table 21.3: Values of bearing modulus for various journal – bearing combinations

Journal material	*Bearing material*	$C = nN / p$	$C = \mu n / p$
Hardened and ground steel	Babbit	2.902	48.36×10^{-3}
Machined soft steel	Babbit	3.631	60.51×10^{-3}
Hardened and ground steel	Plastic Bronze	4.355	72.58×10^{-3}
Machined soft steel	Plastic Bronze	5.079	84.65×10^{-3}
Hardened and ground steel	Rigid Bronze	5.803	96.71×10^{-3}
Machined soft steel	Rigid Bronze	7.251	120.85×10^{-3}

μ in Pa.s, N in rpm, p in MPa, n in rps

Table 21.4: Journal bearing design practice

Machinery	*Bearing*	*Maximum pressure* p N/mm^2	*Diametral clearance ratio* c/d	*Ratio* l/d	*Dynamic viscosity* μ $N.s/m^2$	*Bearing modulus (minimum)* $C = \mu N/p$
Automobile and aircraft engines	Main	5.5 – 12	-	0.8 – 1.8		2.21
	Crank pin	10 – 24	-	0.7 – 1.4	0.007 – 0.008	1.45
	Wrist pin	16 – 34	-	1.5 – 2.2		1.16
Gas and oil engines (four stroke)	Main	4.8 – 8.2	0.001	0.6 – 2.0		2.90
	Crank pin	9.6 – 12.4	< 0.001	0.5 – 1.5	0.020 – 0.065	1.45
	Wrist pin	12.4 – 15.2	<	0.001	1.5 – 2.0	0.72

Gas and oil engines (two stroke)	Main	3.4 – 5.5	0.001	0.6 – 2.0	0.020 – 0.060	3.62
	Crank pin	6.9 – 10	< 0.001	0.6 = 1.5		1.74
	Wrist pin	8.2 – 12.4	< 0.001	1.5 – 2.0		1.45
Marine steam engine	Main	3.4	< 0.001	0.7 – 1.5	0.030	2.90
	Crank pin	4.1	< 0.001	0.7 – 1.2	0.040	2.17
	Wrist pin	10.3	< 0.001	1.2 – 1.7	0.030	1.45
Stationary slow speed steam engines	Main	2.8	< 0.001	1.0 – 2.0	0.060	2.90
	Crank pin	10.3	< 0.001	0.9 – 1.3	0.080	0.87
	Wrist pin	12.4	< 0.001	1.2 – 1.5	0.060	0.72
Stationary high speed steam engines	Main	1.8	< 0.001	1.5 – 3.0	0.015	3.50
	Crank pin	4.1	< 0.001	0.9 – 1.5	0.030	0.84
	Wrist pin	12.4	< 0.001	1.3 – 1.7	0.025	0.70
Steam locomotives	Driving axle	3.8	0.001	1.6 – 1.8	0.100	4.35
	Crank pin	13.8	< 0.001	0.7 – 1.1	0.040	0.72
	Wrist pin	27.6	< 0.001	0.8 – 1.3	0.030	0.72
Reciprocating pumps and compressors	Main	1.8	< 0.001	1.0 – 2.2	0.030 – 0.080	4.35
	Crank pin	4.1	< 0.001	0.9 – 1.7		2.90
	Wrist pin	6.9	< 0.001	1.5 – 2.0		1.45
Railway cars	Axle	3.0	0.001	1.8 – 2.0	0.100	7.25
Steam turbines	Main	0.7 – 1.9	0.001	1.0 – 2.0	0.002 – 0.016	14.50
Generators, motors, centri-fugal pumps	Rotor	0.7 – 1.4	0.0013	1.0 – 2.0	0.025	29.00
Cyroscope	Rotor	5.9	0.0013	-	0.030	7.97
Tansmission Shafting	Light, fixed	0.2	0.001	2.0 – 3.0		14.50
	Self-aligning	1.0	0.001	2.5 – 4.0	0.025 – 0.060	4.35
	Heavy	1.0	0.001	2.0 – 3.0		
Cotton mill	Spindle	0.007	0.005	-	0.002	14.50
Machine tools	Main	2.1	0.001	1.0 – 1.4	0.040	5.80
Punching and shearing machines	Main	27.5	0.001	1.0 – 2.0	0.100	-
	Crank pin	55.0	0.001	1.0 – 2.0	0.100	
Rolling mills	Main	20.6	0.0015	1.1 – 1.5	0.050	1.45

* m in Pa.s, N in rpm and p in N/mm^2

Table 21.5: Correction factor for end leakage, Δf

l/d	Δf	l/d	Δf
0.20	0.150	0.80	0.0018
0.30	0.012	1.00	0.0020
0.40	0.010	1.25	0.0022
0.50	0.005	1.50	0.0025
0.60	0.004	2.00	0.0027
0.70	0.003	2.50	0.0030

Table 21.6: Values of circumstance constant C_1

Lubrication	*Workmanship*	*Attendance*	*Operating condition*	*Value of C_1*
Oil bath or flooded	High grade	First class	Clean and protected	1
Oil, free drop (constant feed)	Good	Fairly good	Favourable (ordinary condition)	2
Oil cup or grease (intermediate feed)	Fair	Poor	Exposed to dirt, grit or other unfavourable conditions	4

Table 21.7: Values of type constant C_2

Type of bearing	*Value of C_2*
Rotating journals, such as rigid bearings and crank pins.	1
Oscillating journals, such as rigid wrist pins and pintle blocks.	1
Rotating bearings lacking ample rigidity, such as eccentrics and the like	2
Rotating flat surfaces lubricated from the centre to the circumference, such as annular step or pivot bearings.	2
Sliding flat surfaces wiping over the guide ends, such as reciprocating crosshead shoes. Long guides Short guides	 2 3
Sliding or wiping surfaces lubricated from the periphery or outer wiping edge, such as marine thrust bearings and worm gears.	3 – 4
Long power screw nuts and similar wiping parts over which it is difficult to effect a uniform distribution of lubricant or load	4 – 6

Table 21.8: Permissible pv values

Class of bearing or journal	*pv value*	*Class of bearing or journal*	*pv value*
Axles, locomotives	4.20	Generators, motors	1.75
Railway cars	4.20	Line shaft	0.88
Aircraft crank pins	87.60	Reducing gears	3.50
Gas engine crank pins	14.00	Machine tools	0.35
High speed steam engine crank pins	14.00	Steam turbines	35.00
Air craft crank shaft and main bearings	70.10	Mill shafting	0.43
Diesel engine crank shaft and main bearing	35.00		

* p in N / mm and v in m / s

Tabte 21.9: Recommended speed of some machinery

Location of bearing	*Speed, rpm*	*Location of bearing*	*Speed, rpm*
1. Automobile crankshaft	900 – 1400	9. Marine steam turbine	2000
2. Aeronautic engine Crankshaft	1800 – 2000	10. Stationary steam turbine	2000
3. Stationary gas engine:		11. De-Laval 5 kW steam	30,000
Main	250 – 800	turbine	
Crankpin	250 – 800		
Cross head	250 – 800		
4. Diesel engine:		12. Railway car axle	300
Main	180		
Crankpin	180		
5. Marine steam engine:		13. Generator and motor	150 – 500
Main			
Crankpin			
6. Stationary slow speed		14. rolling mill main	60
engine:			
Main	40 – 80		
Crankpin	40 – 80		
Cross head	40 – 80		
7. Cotton mill spindle	8000 – 12000	15. Gyroscope	800 – 1500
8. Stationary high speed		16. Locomotive drive wheel:	
engine:		Crankpin	250
Main	360	Cross head	250
Crankpin	360		
Cross head			

Table 21.10: Bearing and journal surface roughness R_{ab} and R_{aj}, μn

Type of surface	*Surface roughness*
High precision journal	0.125 – 0.2
High precision bearing	0.125 – 0.5
Medium precision journal	0.4
Medium precision bearing	0.8
Commercial grinding	0.4 – 1.6
Broaching, reaming, boring, finished turning	0.8 – 1.6

Tabte 21.11: Values of Sommerfeld number for various h_{min}/c and l/d ratios

l/d	h_{min}/C								
	1.0	0.2	0.1	0.067	0.05	0.04	0.033	0.0286	0.025
0.1	0	43.5	8.76	4.32	2.50	1.79	1.37	1.10	0.962
0.2	0	18.9	4.39	2.20	1.32	0.970	0.775	0.649	0.581
0.3	0	11.6	2.96	1.52	0.952	0.714	0.585	0.503	0.452
0.4	0	8.27	2.22	1.18	0.770	0.585	0.495	0.438	0.396
0.5	0	6.21	1.81	1.00	0.680	0.537	0.452	0.406	0.369
0.6	0	5.12	1.60	0.909	0.640	0.510	0.436	0.390	0.357
0.8	0	3.98	1.40	0.833	0.599	0.483	0.416	0.377	0.346
1.0	0	3.41	1.29	0.787	0.568	0.467	0.403	0.366	0.338
1.2	0	3.12	1.21	0.752	0.540	0.450	0.394	0.357	0.330
1.5	0	2.86	1.12	0.715	0.526	0.435	0.380	0.347	0.321
2.0	0	2.53	1.06	0.681	0.505	0.420	0.370	0.339	0.314
00	0	2.07	0.922	0.611	0.411	0.394	0.352	0.322	0.301

Table 21.12: Energy dissipation coefficient, k_b, W/m². °C

Bearing condition	k_b
1 Simple pillow block bearing	
(*a*) Quiet air	145 – 245
(*b*) air at 2.5 m/s	490 – 770
2. Heavy duty self – aligning ring oil lubricated	
(*a*) Quiet air	290 – 390
(*b*) Air at 2.5 m / s	880 – 1230

Table 21.13: Heat radiating capacity of bearing, K_1 $(t_b - t_a)$, W/mm2

	Δt, °C					
	10	15	20	25	30	35
$K_1(t_b - t_a)$ Ventilated	142×10^{-6}	214×10^{-6}	286×10^{-6}	361×10^{-6}	443×10^{-6}	531×10^{-6}
$K_1(t_b - t_a)$ Unventilated	69×10^{-6}	116×10^{-6}	163×10^{-6}	204×10^{-6}	252×10^{-6}	301×10^{-6}

Table 21.14: Heat radiating capacity of bearing, K_2 $(t_b - t_a)$, W/mm²

	Δt, °C					
	10	15	20	25	30	35
$K_2(t_b - t_a)$ Ventilated	284×10^{-5}	428×10^{-5}	567×10^{-5}	672×10^{-5}	886×10^{-5}	1060×10^{-5}
$K_2(t_b - t_a)$ Unventilated	137×10^{-5}	232×10^{-5}	325×10^{-5}	417×10^{-5}	502×10^{-5}	600×10^{-5}

Table 21.15: Permissible unit bearing pressure for different applications

Application	*Unit bearing pressure, p, MPa*
(*a*) Uniform loads:	
1. Electric motors	0.8 – 1.5
2. Steam turbines	0.8 – 1.5
3. Gear reducers	0.8 – 1.5
4. Centrifugal pumps	0.6 – 1.2
(*b*) Moderate shock loads:	
1 . Air compressors	
• Main bearings	1 – 2
• Connecting rod bearings	2 – 4
(*c*) Heavy shock loads:	
1. Diesel engines	
• Main bearings	6 – 12
• Connecting rod bearings	8 – 15
2. Automotive gasoline engines	
• Main bearings	4 – 5
• Connecting rod bearings	10 – 15

Table 21.16: Permissible unit bearing pressure for different bearing materials

Bearing materials	*Unit bearing pressure, p.MPa*
1. Lead base Babbit	4.2 – 5.6
2. Tin base Babbit	5.6 – 7.0
3. Copper – Lead (55 : 45)	14.0 – 21.0
4. Copper – Lead – Tin (72:25:3)	21.0 – 28.0
5. Silver	35.0
6. Bronzes	70.0

Table 21.17: Radial clearance for bearings

Bearings	*Material*	*Radial clearance ratio, c/r*
1 . Precision bearings	Babbits	0.0010 – 0.0015
2. General machinery bearings	Copper – Lead	0.0015 – 0.0025
3. Rough service bearings	Aluminium alloy	0.03 – 0.004

Table 21.18: Optimum values of h_{min}/c for full journal bearing

l/d	h_{min}/c *for maximum load*	h_{min}/c *for minimum friction or minimum power loss*
∞	0.66	0.6
1	0.53	0.3
0.5	0.43	0.12
0.25	0.27	0.03

Table 21.19: Journal bearing materials

Material	*Characteristics*	*Limitations*	*Applications*
1. Wood (maple lignum, vital)	Self – lubricating, low cost, long life	Lighter loads at high speeds under 65°C	Conveyors
2. Cast iron	Low friction, low cost	Not over 3.4 MPa and 40 m/min speed	Refrigerator, compressors, cam – shafts, light transmission
3. Steel	Low cost	Light loads, upto 45 m/min speeds	Guides
4. Bronze bushing	Low cost, simple construction	Loads upto 20 MPa. Speeds upto 270 m/min	All equipment
5. Heavy babbit linear on steel or cast iron	Long life, low friction, good lubrication	Steady loads upder 6.7 MPa	Motor, turbine shafting
6. Light liner on steel or bronze backing	Heavy duty, general purpose, good for dynamic loads	About 3.4 MPa loads and speeds upto 1200 m/min	Gas and diesel engines, compressors
7. Rubber	Low friction, resists abrasion, shock absorbent, long life	About 0.5 MPa, needs water lubrication	Marine propellers, pumps and turbines
8. Carbon graphite	No lubrication needed, light duty applications	Under 455°C, 3.2 MPa at low speeds	Electric motors, meters, conveyors
9. Moulded plastic	Low friction, clean	Low loads at high speeds	Dairy, textile and food machinery
10. Sintered bearings, steel bronze	Self lubricating	Low loads, high speeds	All inaccessible places, typewrite electric fans, instruments, piston rings.

Table 21.20: Dimensionless static performance parameters for full journal bearing with side flow

$\frac{l}{d}$	ε	$\bar{h} = \frac{h_a}{c}$	S	φ	$\bar{f} = f\left(\frac{r}{c}\right)$	$\bar{Q} = \frac{Q}{rcnl}$	$\frac{Q_s}{Q}$	$\frac{pc_p\Delta t}{H_g}$	$\frac{p}{p_{\max}}$
∞	0	1.0	∞	70.92	∞	∞	0	∞	-
	0.1	0.9	0.240	69.10	4.80	3.03	0	19.9	0.826
	0.2	0.8	0.123	67.26	2.57	2.83	0	11.4	0.814
	0.4	0.6	0.0626	61.94	1.52	2.26	0	8.47	0.764
	0.6	0.4	0.0389	54.31	1.20	1.56	0	9.73	0.667
	0.8	0.2	0.021	42.22	0.961	0.760	0	15.9	0.495
	0.9	0.1	0.115	31.63	0.756	0.411	0	23.1	0.358
	0.97	0.03	-	-	-	-	0	-	-
	1.0	0	0	0	0	0	0		0

1	0	1.0	∞	85	∞	π	0	∞	-
	0.1	0.9	1.33	79.5	26.4	3.37	0.150	106	0.540
	0.2	0.8	0.631	74.02	12.8	3.59	0.280	52.1	0.529
	0.4	0.6	0.264	63.10	5.79	3.99	0.497	24.3	0.484
	0.6	0.4	0.121	50.58	3.22	4.33	0.680	14.2	0.415
	0.8	0.2	0.0446	63.24	1.70	4.62	0.842	8.00	0.313
	0.9	0.1	0.0188	26.45	1.05	4.74	0.919	5.16	0.247
	0.97	0.03	0.00474	15.47	0.514	4.82	0.973	2.61	0.152
	1.0	0	0	0	0	-	1.0	0	-
0.5	0	1.0	∞	88.5	∞	π	0	∞	
	0.1	0.9	4.31	81.62	85.6	3.43	0.173	343.0	0.523
	0.2	0.8	2.03	74.94	40.9	3.72	0.318	164.0	0.506
	0.4	0.6	0.779	61.45	17.0	4.29	0.552	68.6	0.441
	0.6	0.4	0.319	48.14	8.10	4.85	0.730	33.0	0.365
	0.8	0.2	0.0923	33.31	3.26	5.41	0.874	13.4	0.267
	0.9	0.1	0.0313	23.66	1.60	5.69	0.939	6.66	0.206
	0.97	0.03	0.00609	13.75	0.610	5.88	0.980	2.56	0.126
	1.0	0	0	0	0	-	1.0	0	0
0.25	0	1.0	∞	89.5	∞	π	0	∞	-
	0.1	0.9	16.2	82.31	322	3.45	0.180	1287.0	0.515
	0.2	0.8	7.57	75.18	153.0	3.76	0.330	611.0	0.489
	0.4	0.6	2.83	60.89	61.1	4.37	0.567	245.0	0.415
	0.6	0.4	1.07	46.72	26.7	4.99	0.746	107.0	0.334
	0.8	0.2	0.261	31.04	8.80	5.60	0.884	35.4	0.240
	0.9	0.1	0.0736	21.85	3.50	5.91	0.945	14.1	0.180
	0.97	0.03	0.0101	12.22	0.922	6.12	0.984	3.73	0.108
	1.0	0	0	0	0	0	1.0	0	0

Rolling Contact Bearings

1. Basic Static Load Ratings

(i) Radial Ball Bearings

Equivalent static load,

$$P_0 = X_0 F_r + Y_0 F_a \qquad ...(22.1a)$$

and

$$P_0 = F_r \qquad ...(22.1b)$$

whichever is larger.

Where
X_0 = a radial factor
Y_0 = a thrust factor
F_r = radial load, N
F_a = axial load, N

For values of X_0 and Y_0, refer to Table 22. 5.

(ii) Radial Roller Bearings

$$P_0 = X_0 F_r + Y_0 F_a \qquad ...(22.2a)$$

and

$$P_0 = F_r \qquad ...(22.2b)$$

whichever is larger.

Where
X_0 = a radial factor
Y_0 = a thrust factor
F_r = radial load, N
F_a = axial load, N

For values of Xo and Y_0, refer to Table 22.5.

(iii) Thrust Ball Bearings

$$P_{oa} = F_a + 2.3\ F_r \tan \alpha \qquad ...(22.3)$$

Where α = contact angle

(iv) Thrust Roller Bearings

$$P_{oa} = F_a + 2.3\ F_r \tan \alpha \qquad ...(22.4)$$

2. Basic Dynamic Load and Life Ratings

(i) Radial Ball Bearings

Equivalent dynamic load,

$$P = XVF_r + YF_a \quad ...(22.5)$$

Where X = radial factor

V = rotation factor

Y = thrust factor

For values of X, V and Y, refer to Table 22.6.

The approximate value of rating life L_{10} for ball bearings, except filling slot bearings,

$$L_{10} = (C/P)^3 \text{ millions of revolutions} \quad ...(22.6)$$

(ii) Radial Roller Bearings

The equivalent dynamic load P in N for self – aligning and tapered roller bearings,

$$P = XVF_r + YF_a \quad ...(22.7)$$

Where X = radial factor

V = rotation factor

Y = thrust factor

For values of X, V and Y, refer to Table 22.6.

The approximate value of rating life L_{10} for radial roller bearings,

$$L_{10} = (C/P) \text{ millions of revolutions} \quad ...(22.8)$$

(iii) Thrust Ball Bearings

$$P_a = XF_r + YF_a \quad ...(22.9)$$

Where X = radial factor

Y = thrust factor

For values of X, Y, refer to Table 22.7.

The rating life L_{10} of thrust ball bearings is given by,

$$L_{10} = (C_a/P_a)^3 \text{ millions of revolutions} \quad ...(22.10)$$

(iv) Thrust Roller Bearings

$$P_a = XF_r + YF_a \quad ...(22.11)$$

Where X = radial factor

Y = thrust factor

For values of X, Y, refer to Table 22.7.

The rating life L_{10} of thrust ball bearings is given by,

$$L_{10} = (C_a/P_a)^{10/3} \text{ millions of revolutions.} \quad ...(22.12)$$

3. Service Factor

See Table 224.

4. Variable Loads

(a) Ball bearings,

$$C^3 = (60L_{10}/10^6) \sum_{i=1}^{m} r_i n_i (C_s P)_i^3 \qquad ...(22.13)$$

$$\sum_{i=1}^{m} r_i = 1$$

Equivalent steady load under fluctuating load,

$$P = (2P_{max} + P_{min})/3 \qquad ...(22.14)$$

(a) Roller bearings,

$$C^{10/3} = (60L_{10}/10^6) \sum_{i=1}^{m} r_i n_i (C_s P)_i^{10/3} \qquad ...(22.15)$$

5. Bearing Life

Bearing life in hours, $L_h = 10^6 L/(60n)$

Where n = rpm and L = life in millions of revolutions.

6. Bearing Life For Higher Reliability (Fig. 22. 3)

$$\text{Median life} = 5L_{10}$$

The reliability R is defined as,

$$R = \frac{\text{Number of bearings which have successfully completed L million revolutions}}{\text{Total number of bearings under test}}$$

$$(L/L_{10}) = [ln\,(1/R)/ln\,(1/R_{90})^{1/b} \qquad ...(22.18)$$

$$= K_L$$

The values of 'a' and b are:

$$a = 6.84,\ b = 1.17 \text{ for,}$$

Where R_s = probability of one out of n bearings failing during its life time.

7. Life Adjustment Factor for Operating Temperature

Actual bearing life, $L_a = K_L\, K_T\, L_{10}$

Where K_L = life adjustment factor for reliability (Table 22.2)

K_T = life adjustment factor for temperature (Table 22.3)

Table 22.1: Service factor for rolling contact bearings, C_s

Type of load	C_s
Constant or steady	1.0
Light shocks	1.5
Moderate shocks	2.0
Heavy shocks	3.0

Table 22.2: Life adjustment factor for reliability, K_L

Realibility	0.90	0.91	0.92	0.93	0.94	9.95	9.96	9.97	0.98	9.99	0.995	0.999
K_L	1.0	0.928	0.855	0.779	0.701	0.618	0.531	0.437	0.332	0.208	0.131	0.0448

Table 22.3 Life adjustment factor for operating temperatures, K_T

Temperature, °C	100	125	150	175	200	225	250
K_T	0.98	0.95	0.91	0.87	0.80	0.74	0.71

Table 22.4: Bearing life for vehicle applications

Type of vehicle	*Life in million revolutions*
1. Automobile cars	50
2. Trucks	100
3. Rail road cars	1000

Table 22.5: Values for Factors X_o and Y_o for Radial Ball and Roller Bearings

Type of Bearing	*Single Row Bearing*[1]		*Double Rmv Bearings*[2]	
	X_0	Y_0	X_0	Y_0
Radial Ball Bearings : Radial contact groove ball bearings[3]	0.6	0.5	0.6	0.5
Angular contact groove ball bearing[4] α = 20°	0.5	0.42	1	0.84
α = 25°	0.5	0.38	1	0.76
α = 30°	0.5	0.38	1	0.66
α = 35°	0.5	0.29	1	0.58
α = 40°	0.5	0.26	1	0.52
Self-aligning ball bearings	0.5	0.22 cot α	1	0.44 cot α
Radial roller bearings : Self-aligning and tapered roller bearings,	0.5	0.22 cot α	1	0.44 cot α

IS : 3823 (Part-1, II)–1966

Note:

1. P_0 is always $\geq F_r$
2. Double row bearings are presumed to be symmetrical.
3 Permissible maximum value of F_a/C_0 depends on the bearing design (groove depth and internal clearance).
4. For two similar single row-angular contact ball bearings, mounted 'face-to-face' or 'back-to-back' use the value of X_0 and Y_0 which apply to a double-row angular contact ball bearings. For two or more similar single row angular contact bearings mounted 'in tandem', use the values of X_0 and Y_0 which apply to single row angular contact ball bearing.

Table 22.6: Values of Factors V, X and Y

Type of Bearing	$\frac{iF_a}{C_a}$	$\frac{F_d}{Zd_{br}^2}$	$\frac{F_a}{C_0}$	$\frac{F_a}{izd_{br}^2}$	e[1]	*V* In relation to the load, the inner ring is		X[2] Single Row[2]	X[2] Double Row[3]		Y[2] Single Row[2]	Y[2] Double Row[3]	
						Rotating	Stationary	$\frac{F_a}{VF_r} > e$	$\frac{F_a}{VF_r} \le e$	$\frac{F_a}{VF_r} > e$	$\frac{F_a}{VF_r} > e$	$\frac{F_a}{VF_r} \le e$	$\frac{F_a}{VF_r} > e$
(1)	(2)	(3)	(4)	(5)	(6)	(7)	(8)	(9)	(10)	(11)	(12)	(13)	(14)
Ball		0.014	0.018	0.19						2.30		230	
Bearings:			0.028	0.035	0.22						1.99		1.99
			0.056	0.070	0.26						1.71		1.71
Non													
Filling	—	—	0.084	0.110	0.28		+	0.56	1	0.56	1.55	0	1.55
slot			0.110	0.140	0.30	1.0	1.2				1.45		1.45
assembly,			0.170	0.210	0.34						1.31		1.31
Radial			0.280	0.350	0.38						1.15		1.15
contact			0.420	0.530	0.42						1.04		1.04
groove			0.560	0.700	0.44						1.00		1.00
ball													
bearing													

0.014		0.018	—	—	0.23				For this type,			For this type	2.78	3.74
		0.028	0.036			0.26			use X, Y and e			useX, Y and e	2.40	3.23
		0.056	0.070			0.30			values appli-			values appli-	2.07	2.78
		0.085	0.110			0.34		+	cable to single			cable to	1.87	2.52
Ball Bearing* α = 5°		0.110	0.140	—	—	0.36	1.0	1.2	row, non-fill-	1	0.78	single row,	1.75	2.36
		0.170	0.210			0.40			ing slot assem-			non-filling	1.58	2.13
		0.280	0.350			0.45			bly, radial			slot assemb-	1.39	1.87
		0.420	0.530			0.50			contact,			ly, radial	1.26	1.69
		0.560	0.700			0.52			groove bail			contact,	1.21	1.63
									bearing			groove ball		
												bearing		
		0.014	0.018			0.29						1.88	2.18	3.06
		0.029	0.035			0.32						1.71	1.98	2.78
		0.057	0.070			0.36						1.52	1.76	2.47
		0.086	0.110			0.38		+				1.41	1.63	2.29
Angular contact groove	α = 10°	0.110	0.140	—	—	0.40	1.0	1.2	0.46	1	0.75	1.34	1.55	2.18
		0.170	0.210			0.44						1.23	1.42	2.00
		0.290	0.350			0.49						1.10	1.27	1.79
		0.430	0.530			0.54						1.01	1.17	1.64
		0.570	0.700			0.54						1.00	1.16	1.63
		0.015	0.018			0.38						1.47	1.65	2.39
		0.029	0.035			0.40						1.40	1.57	2.28
		0.058	0.070			0.43		+				1.30	1.46	2.11
	α = 15°	0.087	0.110	—	—	0.46	1.0	1.2	0.44*	1	0.72	1.23	1.38	2.00
		0.120	0.140			0.47						1.19	1.34	1.93
		0.170	0.210			0.50						1.12	1.26	1.82
		0.290	0.350			0.55						1.02	1.14	1.66
		0.440	0.530			0.56						1.00	1.12	1.63
		0.580	0.700			0.56						1.00	1.12	1.63

α = 20°	—	—	—	—	0.57	1	+ 1.2	0.43	1	0.70	1.00	1.09	1.63
= 25°					0.68	1	1.2	0.41	1	0.67	0.87	0.92	1.41
= 30°					0.80	1	1.2	0.39	1	0.63	0.76	0.78	1.24
= 35°					0.95	1	1.2	0.37	1	0.70	0.66	0.66	1.07
= 40°					1.14	1	1.2	0.35	1	0.57	0.57	0.55	0.93
Self-aligning ball bearngs	—	—	—	—	1.5 tan α	1	1	0.40	1	0.65	0.4 cot α	0.42 cot α	0.65 cot α
Single row adial contact separable ball bearings magneto bearings)	—	—	—	—	0.2	1	1	0.5	—	—	2.5	—	—

IS : 38 : 3824 (Part I)—1966

$d_{br} = d_b$ = dia of ball, mm ; $d_{br} = d_r$ = dia of roller, mm ; C_a = static basic load rating.

1. Value of X, Y and *e* for a load or a contact angle other than those shown in the table, are obtained by linear interpolation.
2. For a single row bearing, when $F_a/VF_r \le e$, use X = 1 and Y = 0.
3. Double row bearings are presumed to be symmetrical.
   (*a*) When calculating the equivalent load for a unit consisting of two similar single row angular contact ball bearings in durlex mounting, 'face-to-face' or 'back-to-back', the units shall be considered as one of double row angular contact ball bearing.
   (*b*) When calculating the equivalent load for a unit consisting or two or more single row radial or angular contact ball bearings 'in tandem', X, Y values for single row ball bearings shall be used.

\+ Because experimental data are incomplete, no correct value of factor V for radial and angular contact groove ball bearings with inner ring staionary in rlation to the load can be stated. The values shown in the table are, however well on the safe side.

* Permissible maximum value of F_a/C_0 depends upon the bearing design.

Table; 22.7: Values of X and Y for Thrust Ball and Roller Bearings

Type of Bearing	*X*			*Y*			*e*
	Single direction bearing	*Double direction bearing**		*Single direction bearing*	*Double direction bearing*		
	$\frac{F_a}{F_r} < e$	$\frac{F_a}{F_r} \le e$	$\frac{F_a}{F_r} > e$	$\frac{F_a}{F_r} > e$	$\frac{F_a}{F_r} \le e$	$\frac{F_a}{F_r} > e$	
Thrust ball α = 45°	0.66	1.18	0.66	1.0	0.59	1.0	1.125
Thrust ball α = 60°	0.92	1.90	0.92	1.0	0.54	1.0	2.17
Thrust ball α = 75°	1.66	3.89	1.66	1.0	0.52	1.0	4.67
Thrust ball α = 90°**	—	—	—	—	—	—	—
Self-aligning and tapered thrust roller bearings α = 90°**	tan α	1.5 tan α	tan α	1.0	0.67	1.0	1.5 tan α

*Double direction bearings are presumed to be symmetrical.

**For α = 90°, X = 0, Y = 1.

IS : 3824 (Part III, IV)—1966

Table 22.8: Deep Groove Ball Bearings Series 02 (Fig. 22.1*a*)

Bearing		*Boundary Dimensions mm*				*Basic Load Rating kN*		*Limiting speed rev./min × 10^3*
BIS	*SKF*	*Bore d*	*Outside dia D*	*Width B*	*Corner radius r*	*Static C_0*	*Dynamic C*	
10BC02	6200	10	30	9		2.24	3.9	24
12BC02	6201	12	32	10		3.1	53	22
15B002	6202	15	35	11	1	355	6	19
17BO02	6203	17	40	12		45	735	17
20BC02	6204	20	47	14		62	9.8	15
25BO02	6205	25	52	15	15	6.95	10.8	12
30BC02	6206	30	62	16		10	15	10
35BO02	6007	35	72	17		13.7	19.6	9
40BC02	6208	40	80	18		16.6	23.6	85
45B002	6209	45	85	19	2.0	18.6	255	75
50B002	6210	50	90	20		19.6	27	7
55BO02	6211	55	100	21		25	335	63
60BC02	6212	60	110	22		28	365	6
65BC02	6213	65	120	23	25	34	43	53
70BO02	6214	70	125	24		375	475	5
75BO02	6215	75	130	25		405	51	4.8

80BO02	6216	80	140	26		45	54	45
85BC02	6217	85	150	28	3.0	53	64	43
90BO02	6218	90	160	30		62	735	3.8
95BO02	6219	95	170	32		695	83	3.6
100BO02	6220	100	180	34		78	95	3.4
105FC02	6521	105	190	30	3.5	90	102	3.2
110BC02	6222	110	200	38		100	112	3
120 BC02	6224	120	215	40		100	112	2.8

Table 22.9: Deep Groove Ball Bearing Series 03

Bearing No.		*Boundary Dimensions* mm				*Basic Load Rating* kN		*Limiting speed rev. / min* $\times 10^3$
BIS	*SKF*	*d*	*D*	*B*	*r*	*Static* C_0	*Dynamic*	
10BO03	6300	10	35	11	1	3.75	6.2	20
12BC03	6301	12	37	12		4.65	75	19
15BO03	6302	15	42	13	1.5	5.4	8.8	17
17BCD3	6303	17	47	14	655	10.4	16	
20BO03	6304	20	52	15		7.8	12.2	13
25BC03	6305	25	62	17	2	11.4	173	11
30BC03	6306	30	72	19		14.6	21.6	9
35BC03	6307	35	80	21		18	255	8.5
40BC03	6308	40	90	23	2.5	22.4	315	75
4SB003	6309	45	100	25		30	405	6.7
50BC03	6310	50	110	27		36	47.5	63
55BC03	6211	55	120	29	3	415	55	5.6
60BC03	6312	60	130	31		48	63	5
65BC03	6313	65	140	33		56	71	4.8
70BOJ3	6314	70	150	35	35	63	80	4.5
75B003	6315	75	160	37		72	86.5	43
80BC03	6316	80	170	39		80	95	3.8
85BC03	6317	85	180	41		90	102	3.6
90BOD3	6318	90	190	43		98	110	3.4
95BC03	6319	95	200	45	4	110	118	3.2
100BC03	6320	100	215	47		132	134	3.6
105BC03	6321	105	225	49		143	140	2.8
110BC03	6322	110	240	50		166	156	2.6
120BC03	6324	120	260	55		166	160	2.4

Table 22.10: Deep Groove Ball Bearing; Series 04

Bearing No.		Boundary Dimensions mm				Basic Load Rating kN		Limiting speed rev./min × 10^3
BIS	SKF	d	D	B	r	Static C_0	Dynamic	
17BC04	6403	17	62	17	2	11.8	17.6	12
20BC04	6404	20	72	19		16.6	23.6	10
25BC04	6405	25	80	21		19.6	275	9
30BC04	6406	30	90	23	2.5	24	33.5	8.5
35BC04	6407	35	100	25		31	425	7
40BC04	6408	40	110	27	3	36.5	49	6.7
45BG04	6409	45	120	29		455	585	6
50BC04	6410	50	130	31		52	67	5.3
55BC04	6411	55	140	33		63	765	5
60BC04	6412	60	150	35	3.5	69.5	83	4.8
65BC04	6413	65	160	37		78	91.5	4.5
70BC04	6414	70	180	42		104	110	3.8
75BC04	6415	75	190	45	4	114	118	3.6
80BC04	6416	80	200	48		125	125	3.4
85BC04	6417	85	210	52	5	124	134	3.2
90BC04	6418	90	225	54		146	148	3

Table 22.11: Cylindrical Roller Bearings Series 02 (Fig. 22.2*a*)

Bearing No.		Boundary Dimensions mm				Basic Load Rating kN		Limiting speed rev./min × 10^3
BIS	SKF	d	D	B	r	Static C_0	Dynamic	
20RN02	N204	20	47	14		7.35	13.4	15
25RN02	N205	25	52	15	1.5	8.8	15.8	12
30RN02	N206	30	92	16		12	20.4	10
35RN02	N207	35	72	17		17.6	29	9
40RN02	N208	40	80	18	2	24	38	8.5
45RN02	N209	45	85	19		25.5	40	75
50RN02	N210	50	90	20		27.5	42.5	7
55RN02	N211	55	100	21		34	51	6.3
60RN02	N212	60	110	22	2.5	43	62	5.6
65RN02	N213	65	120	23		51	72	5.3
70RN02	N214	70	125	24		51	72	5
75RN02	N215	75	130	25		63	88	4.8
80RN02	N216	80	140	26	3	68	96.5	4.5
85RN02	N217	85	150	28		78	110	4.3
90RN02	N218	90	160	30		100	134	3.8
95RN02	N219	95	170	32		112	1500	03.6
100RN02	N220	100	180	34	3.5	125	166	3.4
110RN02	N222	110	200	38		166	200	3
120RN02	N224	120	215	40		18.3	236	2.8

Table 22.12: Cylindrical Roller Bearings Series 03

Bearing No.		*Boundary Dimensions mm*			*Basic Load Rating kN*		*Limiting speed rev./min × 10^3*
BIS	*SKF*	*d*	*D*	*B*	*Static C_0*	*Dynamic*	
17RN03	N303	17	47	14	8.65	15.6	14
25RN03	N305	25	62	17	15	26	9.5
30RN03	N306	30	72	19	20	34	8.5
35RN03	N307	35	80	21	27	43	8
40KN03	N308	40	90	23	32.5	51	6.7
45RN03	N309	45	100	25	41.5	65	6.3
50RN03	N310	50	110	27	52	80	5.6
55RN03	N311	55	120	29	67	100	5
60RN03	N312	60	130	31	765	112	4.8
65RN03	N313	-65	140	33	85	125	4.5
70RN02	N314	70	150	35	102	146	4
75RN03	N315	75	160	37	125	176	3.8
80RN03	N316	80	170	39	125	176	3.6
85RN03	N317	85	180	41	146	204	3.4
90RN03	N318	90	190	43	160	220	3.2
95RN03	N318	95	200	45	190	220	3.6
100RN03	N320	100	215	47	220	250	2.8
110RN03	N322	110	240	50	290	375	2.4
120RN03	N324	120	260	55	340	440	2.2

Table 22.13: Single Direction Thrust Ball Bearings Series 11 (Fig. 22.2*e*)

Bearing No.		*Boundary Dimensions mm*			*Basic Load Rating kN*		*Limiting speed rev./min × 10^3*
BIS	*SKF*	*d*	*D*	*B*	*Static C_0*	*Dynamic*	
10TA11	51100	10	24	9	8.8	6.7	7
12TA11	51101	12	26	9	19	6.95	7
15TA11	51102	15	28	9	11.2	7.2	6.3
17TA11	51103	17	30	9	12.2	7.5	6.3
20TA11	51104	20	35	10	16.6	9.8	5.6
25TA11	51105	25	42	11	22.8	12.2	4.8
30TA11	51106	30	47	11	26.5	12.9	4.5
35TA11	51107	35	52	12	30	13.4	4.3
40TA11	51108	40	60	13	40	18	3.8
45TA11	51109	45	65	14	45	18.6	3.4

50TA11	51110	50	70	14	50	19.6	3.4
55TA11	51111	55	78	16	62	23.6	3
60TA11	51112	60	85	17	71	275	2.6
65TA11	51113	65	90	18	78	28.5	2.4
70TA11	51114	70	95	18	88	32.5	2.4
75TA11	51115	75	100	19	98	34	2.2
80TA11	51116	80	105	19	102	345	2
85TA11	51117	85	110	19	106	345	2
90TA11	51118	90	120	22	120	39	1.8
100TA11	51120	100	135	25	173	57	1.7
110TA11	51122	110	145	25	190	585	1.6
120TA11	51124	120	155	25	250	68	1.6
130TA11	51126	130	170	30	285	80	1.4
140TA11	51128	140	180	31	305	815	1.3
150TA11	51130	150	190	31	325	83	1.2
160TA11	51132	160	200	31	345	865	1.2
170TA11	51134	170	215	34	415	104	1.1
180TA11	51136	180	225	34	430	104	1
190TA11	51138	190	240	37	530	132	0.95
200TA11	51140	200	250	37	550	132	0.95

H = width.

Table 22.14: Single Direction Thrust Ball Bearings Series 14

Bearing No.		*Boundary Dimensions mm*			*Basic Load Rating kN*		*Limiting speed rev./min × 10^3*
BIS	*SKF*	*d*	*D*	*B*	*Static C_0*	*Dynamic*	
24TA14	51405	25	60	24	63	40	26.00
30TA14	51406	30	70	28	90	52	2200
35TA14	51407	35	SO	32	112	63	1800
40TA14	51408	40	90	36	146	80	1700
45TA14	51409	45	100	39	173	93	1600
50TA14	51410	50	110	43	204	106	1500
S5TA14	51411	55	120	48	265	137	1300
60TA14	51412	60	130	51	285	143	1200
65TA14	51413	65	140	56	360	166	1000
70TA14	51414	70	150	60	400	180	950
75TA14	51415	75	160	65	500	208	900
80TA14	51416	80	170	68	550	224	850
85TA14	51417	85	180	72	610	236	850
90TA14	51418	90	190	77	670	250	800
100TA14	51420	100	210	85	865	305	700

TaMe 22.15: Taper Roller Bearings Series 322 (Fig. 2JU2

Bearing No.		*Boundary Dimensions mm*					*Basic Load Rating kN*		*Limiting speed rev. / min × 10^3*
BIS	*SKF*	*d*	*D*	*B*	*C*	*T*	*Static C_0*	*Dynamic*	
30RK322	32206	30	62	20	17	2125	34	43	63
35RK322	32207	35	72	23	19	24.25	45	56	5.3
40RK322	32208	40	80	23	19	24.75	50	64	4.8
45RK22	32209	45	85	23	19	24.75	56	66	45
50RK322	32210	50	90	23	19	24.75	57	695	4.3
55RK322	32211	55	100	25	21	26.75	75	90	3.8
60RK322	32212	60	110	28	24	29.75	915	108	3.4
6SRK322	32213	65	120	31	27	32.75	112	129	3
70RK322	32214	70	125	31	27	33.25	118	134	2.8
75RK322	32215	75	130	31	27	33.25	120	137	2.6
80RK322	32216	80	140	33	28	35.25	137	160	2.4
85RK322	32217	85	150	36	30	385	163	183	2.2
90RK322	32218	90	160	40	34	42.5	193	216	2
95RK322	32219	95	170	43	37	455	220	240	1.9
100RK322	32220	100	180	46	39	49	250	270	1.8
105RK322	32221	105	190	50	43	53	290	310	1.8
110RK322	32222	110	200	53	46	56	325	340	1.7
120RK322	32224	120	215	58	50	615	340	355	1.6
130RK322	32226	130	230	64	54	67.75	405	425	1.5
140RK322	32228	140	250	68	58	71.75	480	490	1.4
150RK322	32230	150	270	73	60	77	640	630	1.2
160RK322	32232	160	290	80	67	84	780	750	1.1

C as cup length, T = cone length.

Table 22.16: Guide to Values of Life Requirements for Different Classes of Machines

Type of Machine	*Life in Hours of Operation*
Instruments and apparatus that are only infrequently used **Ex.:** Demonstration apparatus, devices for operation of sliding ; doors	500
Aircrafts engines	500—2,000
Machines for service of short duration or intermittent operation, where service interruptions are of minor importance **Ex.:** Hand tools, lifting ihcklc in machinery shops, hand driven machines in general, farm machinery, assembly cranes charging machines, foundry cranes, household machines	4,000—8,000

Machines for intermittent service where dependable operation is of great importance **Ex.:** Auxiliary machines in power stations, conveying equipment in production lines, elevators, general cargo cranes, machine tools less frequently used Machines for 8-h service that are not always fully utilized **Ex.:** Stationary electric motor, gear drives for general purposes	8,000—12,000 12,000—20,000
Machines for 8-h service that are fully utilized **Ex.:** Machines in general in the mechanical industries, cranes for continuous services, blowers, jack shafts	20,000—30,000
Machines for continuous operation (24-h service) **Ex.:** Separators, compressors, pumps, main lines shafting, roller beds and conveyor rollers, mine hoists, stationary electric motors	40,000—60,000
Machines for 24-h service where dependability is of great importance **Ex.:** Pulp and paper machines, public power stations, mine pumps, public pumping stations machines for continuous service aboard ship	1000,000—2000,000

Table 22.17: Loading ratio C/P for different lives expressed in millions of revolutions: ball bearings

Life in millions of revolutions L_n	$\frac{C}{P}$	*Life in millions of revolutions* L_n	$\frac{C}{P}$	*Life in millions of revolutions* L_n	$\frac{C}{P}$	*Life in millions of revolutions* L_n	$\frac{C}{P}$
0.5	0.793	80	4.31	600	8.43	3200	14.7
0.75	0.909	90	4.48	650	8.66	3400	15.0
1.0	1	100	4.64	700	8.88	3600	15.3
1.5	1.14	120	4.93	750	9.09	3800	15.6
2	1.26	140	5.19	800	9.28	4000	15.9
3	1.44	160	5.43	850	9.47	4500	16.5
4	1.59	180	5.65	900	9.65	5000	17.1
5	1.71	200	5.85	950	9.83	5500	17.7
6	1.82	220	5.04	1000	10	6000	18.2
8	2	240	6.21	1100	10.3	6500	18.7
10	2.15	260	6.38	1200	10.6	7000	19.1
12	2.29	280	6.54	1300	10.9	7500	19.6
14	2.41	300	6.69	1400	11.2	8000	20.0
16	2.52	320	6.84	1500	11.4	8500	20.4
18	2.62	340	6.98	1600	11.7	9000	20.8
20	2.71	360	7.11	1700	11.9	9500	21.2

25	2.92	380	7.24	1800	12.2	1000	21.5
30	3.11	400	7.37	1900	12.4	12000	22.9
35	3.27	420	7.49	2000	12.6	14000	24.1
40	3.42	440	7.61	2200	13	16000	25.2
45	3.56	460	7.72	2400	13.4	18000	26.2
50	3.68	480	7.83	2600	13.8	20000	27.1
60	3.91	500	7.94	2800	14.1	25000	29.2
70	4.12	550	8.19	3000	14.4	30000	31.1

Table 22.18: Loading ratio C/P for different lives expressed in millions of revolutions : Roller bearings

Life in millions of revolutions L_n	$\frac{C}{P}$	*Life in millions of revolutions* L_n	$\frac{C}{P}$	*Life in millions of revolutions* L_n	$\frac{C}{P}$	*Life in millions of revolutions* L_n	$\frac{C}{P}$
0.5	0.812	80	3.72	600	6.81	3200	11.3
0.75	0.917	90	3.85	650	6.98	3400	11.5
1.0	1	100	3.98	700	7.14	3600	11.7
1.5	1.13	120	4.20	750	7.29	3800	11.9
2	1.24	140	4.40	800	7.43	4000	12.0
3	1.39	160	4.58	850	7.56	4500	12.5
4	1.52	180	4.75	900	7.70	5000	12.9
5	1.62	200	4.90	950	7.82	5500	13.2
6	1.71	220	5.04	1000	7.94	6000	13.6
8	1.87	240	5.18	1100	8.17	6500	13.9
10	2.00	260	5.30	1200	8.39	7000	14.2
12	2.11	280	5.42	1300	8.59	7500	14.5
14	2.21	300	5.54	1400	8.79	8000	14.8
16	2.30	320	5.64	1500	8.97	8500	15.1
18	2.38	340	5.75	1600	9.15	9000	15.4
20	2.46	360	5.85	1700	9.31	9500	15.6
25	2.63	380	5.94	1800	9.48	10000	15.8
30	2.77	400	6.03	1900	9.63	12000	16.7
35	2.91	420	6.12	2000	9.78	14000	17.5
40	3.02	440	6.21	2200	10.1	16000	18.2
45	3.13	460	6.29	2400	10.3	18000	18.9
50	3.23	480	6.37	2600	10.6	20000	19.5
60	3.42	500	6.45	2800	10.8	25000	20.9
70	3.58	550	6.64	3000	11.0	30000	22.0

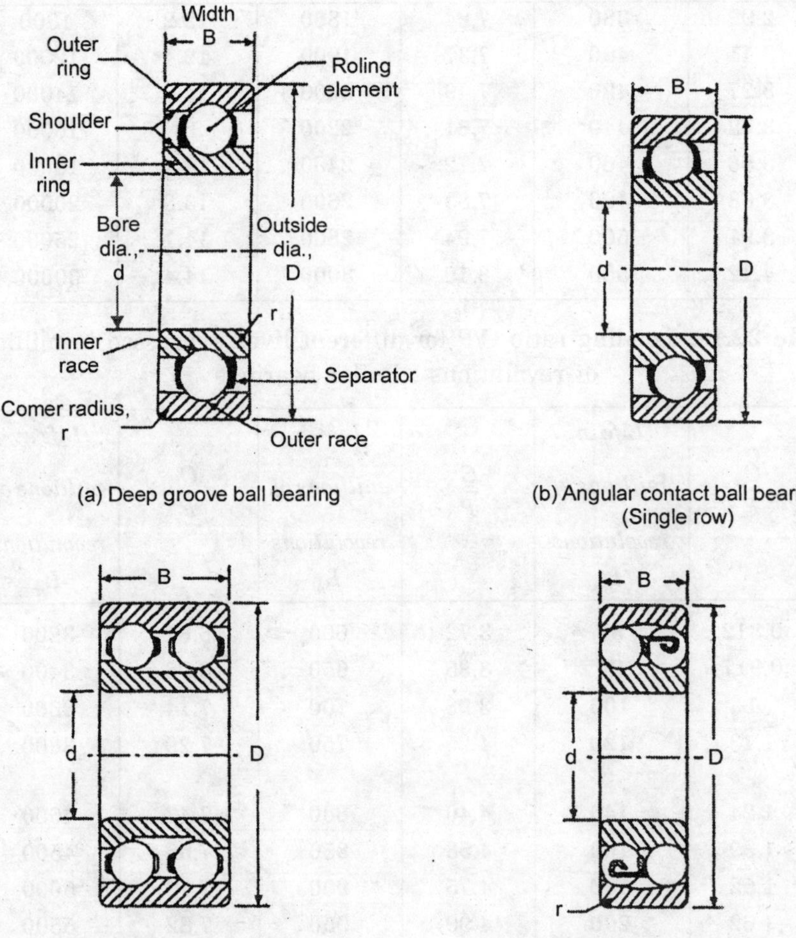

(a) Deep groove ball bearing

(b) Angular contact ball bearing (Single row)

(c) Angular contact ball bearing (Double row)

(d) Self aligning ball bearing

Fig. 22.1 Ball bearings.

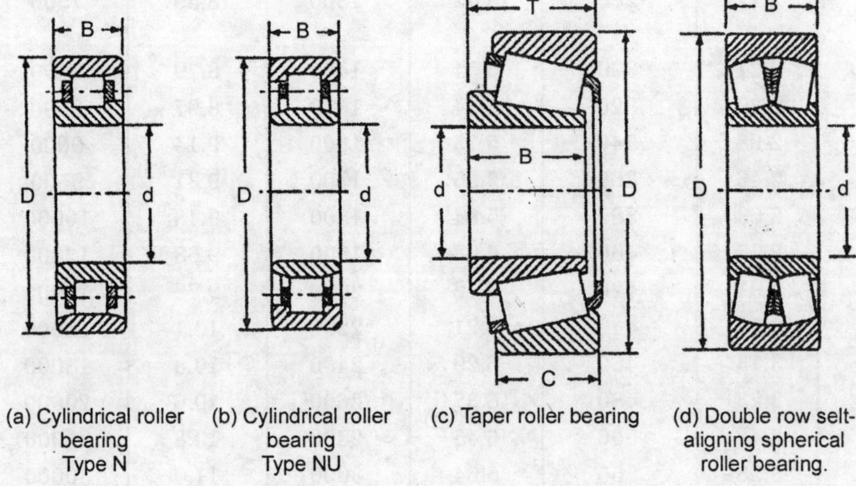

(a) Cylindrical roller bearing Type N

(b) Cylindrical roller bearing Type NU

(c) Taper roller bearing

(d) Double row selt-aligning spherical roller bearing.

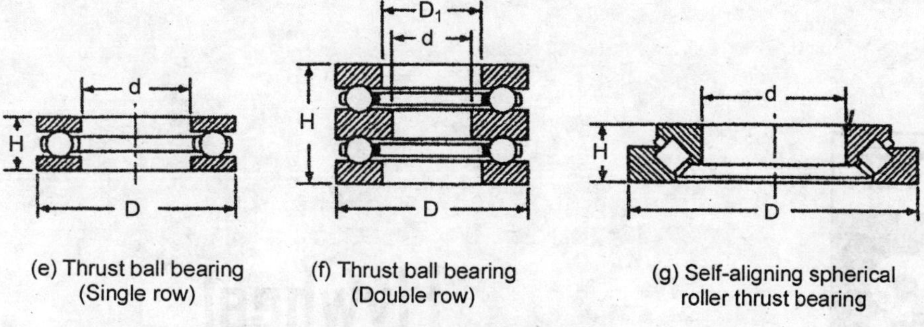

Fig. 22.2 Roller bearings.

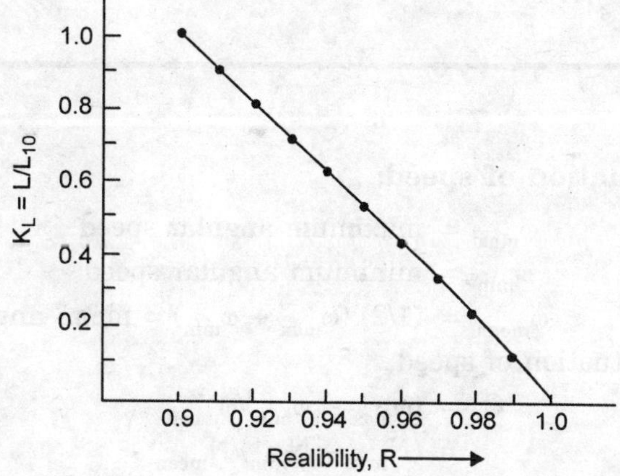

Fig. 22.3 Life adjustment reliability factor.

CHAPTER

23

Flywheel

1. Coefficient of fluctuation of speed;

Let ω_{max} = maximum angular speed

ω_{min} = minimum angular speed

ω_{mean} = (1/2) $(\omega_{max} + \omega_{min})$ = mean angular speed

Coefficient of fluctuation of speed,

$$C_s = (\omega_{max} - \omega_{min})/\omega_{mean} \quad ...(23.1a)$$

$$= (N_{max} - N_{min})/N_{mean} \quad ...(23.1b)$$

Refer to Table 23.1 for values of C_s.

2. Coefficient of fluctuation of energy

$$\Delta E = I\omega^2_{mean} C_s \quad ...(23.2)$$

Fluctuation of kinetic energy,

$$\Delta E = C_e.E \quad ...(23.3)$$

Where C_e = coefficient of fluctuation of energy (see Table 23.2 and 23.3)

3. Solid disc type flywheel.

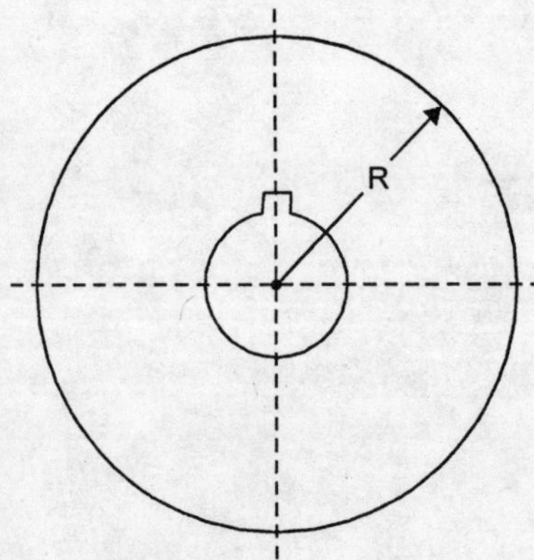

Fig. 23.1 Solid disc type flywheel.

Moment of inertia of a solid disc of radius R and mass m,

$$I = mR^2/2 \quad ...(23.4)$$

Mass of the disc, $m = \pi R^2 t\rho$ ...(23.5)

Where t = thickness of disc

ρ = density of disc material

4. Rimmed flywheel

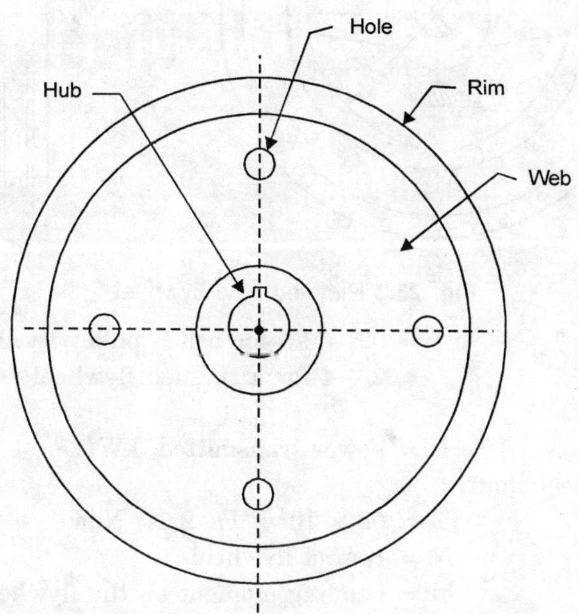

Fig. 23.2 Flywheel with web.

Moment of inertia of the rim,

$$l_r = kl$$

Where k = a factor = 0.9

I = moment of inertia of the entire flywheel

$$l_r = m_r R^2 \quad ...(23.6)$$

Where m_r = mass of rim

R = mean radius of rim

$$R < 30/\omega_{mean} \quad ...(23.7)$$

For limiting rim velocity, refer to Tabte 23.4.

5. Design of Belt - Pulley Rimmed Type Flywheel

(*i*) Rim size

Let b = width of flywheel rim

= belt width + 40 mm

t = thickness of rim

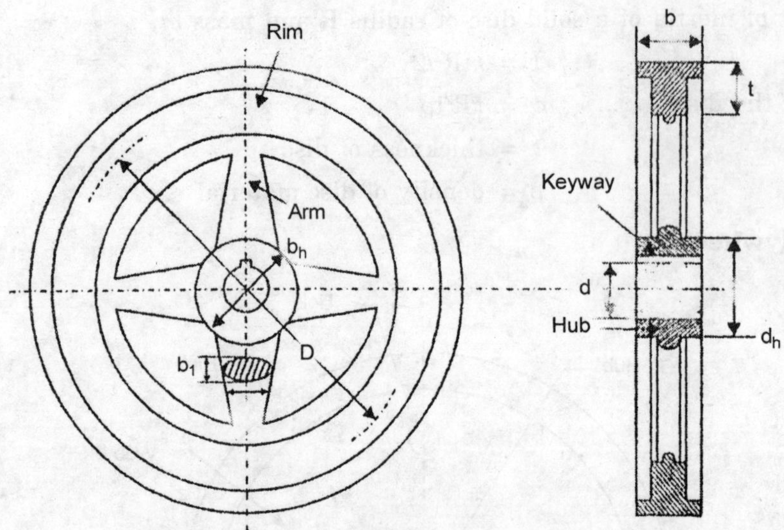

Fig. 23.3 Rimmed type flywheel.

$b/1 = 0.5 - 1.5$ for belt – pulley flywheel

$= 2 - 4$ for a rimmed flywheel

(*ii*) Shaft and hub size

P = power transmitted, kW

Torque on flywheel shaft,

$$T = 60 \times 10^3 \times P/(2\pi N) \text{ N.m}$$

Where N = rpm of flywheel

M = bending moment on the flywheel shaft

= weight of flywheel × overhang

Diameter of flywheel shaft,

$$d = [\{16/(\pi\tau_d)\} \{(K_m M)^2 + (K_t T)^2\}^{1/2}]^{1/3} \quad ...(23.8)$$

Where τ_d = design shear stress for shaft

K_m, K_t = shock and fatigue factors for bending and torsion respectively

Diameter of hub, $d_h = 2d$ ...(23.9*a*)

Length of hub, $l_h = (2 - 2.5)d$ ...(23.9*b*)

6. Stresses in rim

Tensile stress in flywheel rim due to centrifugal force,

$$\sigma_t = \rho v^2 \quad ...(23.10)$$

Bending stress in flywheel rim due to restraint of arms,

$$\sigma_b = 2\pi^2 v^2 \rho R/(n^2 t) \quad ...(23.11)$$

Where R = mean rim radius

n = number of arms on flywheel

Total stress in the rim,

$$\sigma_{total} = (3/4)\,\sigma_t + \sigma_b/4 \quad ...(23.12)$$

$$= \rho v^2 \,[0.75 + 4.935 \times R/(n^2 t)]$$

For a safe design, $\sigma_{total} < 38$ MPa for C.I.

7. Arms size

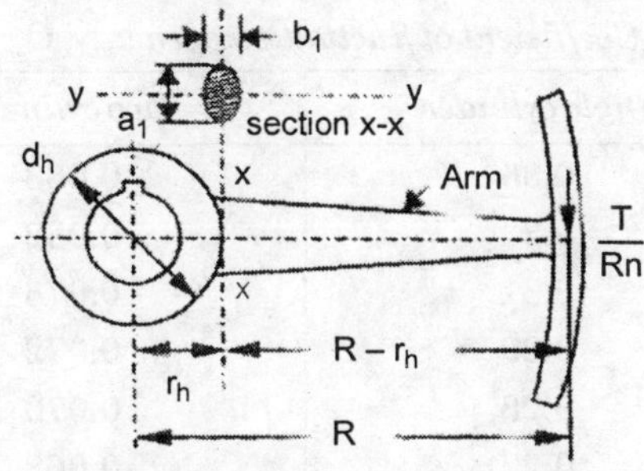

Fig. 23.4 Bending stress in arm.

Stresses in the arm at the hub end,

$$\sigma_{b1} = T(D - d_h)/(nz_y D) \quad ...(23.13)$$

Where z_y = section modulus of arms about y – axis

$$= (\pi/16) b_1 a^2{}_1$$

Major axis of elliptical arm section,

$$\alpha_1 = [64 z_y/\pi]^{1/3} \quad ...(23.14)$$

Minor axis, $b_1 = a_1/2$

For belt – pulley flywheel, stress at the hub,

$$\sigma_{b2} = (F_1 - F_2)(D - d_h)/(2nz_y) \quad ...(23.15)$$

Where F_1, F_2 = belt tensions on tight and slack sides respectively

Direct tensile stress in the arm due to centrifugal force (assuming 75% effect),

$$\sigma_{t1} = 0.75 \rho v^2 \quad ...(23.16)$$

Total stress, $\sigma_{total} = \sigma_{b1} + \sigma_{b2} + \sigma_{t1}$ ...(23.17)

For a safe design, $\sigma_{total} < 20$ MPa for C.I.

< 60 MPa for steel

Table 23.1: Coefficient of fluctuation of speed, C_s

Machine type	*Coefficient of fluctuation of speed C_s*
Punching machine, Pressing machine	0.2 – 0.07
Pumps	0.05 – 0.03
Machine tools	0.03 – 0.025
Textile and paper machinery	0.025
Spinning machinery	0.02
DC generator (lighting)	0.007
AC generator	0.003
DC generator direct drive	0.01
Belt driven DC generator	0.014
AC generator direct drive	0.006
Belt driven AC generator	0.008
Automobile (idling)	0.2
Automobile (normal speed)	0.1
Weaving machine	0.025
Flour mill machinery	0.02
Reciprocating pumps and compressors	0.04

Table 23.2: Coefficient of fluctuation of energy for steam engines, C«

Cutoff percentage	*Coefficient of fluctuation of energy, C_e*		
	Single cylinder	*Two cylinder*	*Three cylinder*
10	0.35	0.088	0.040
20	0.33	0.082	0.037
40	0.31	0.078	0.034
60	0.29	0.072	0.032
80	0.28	0.070	0.031
100	0.27	0.068	0.030

Table 23.3: Coefficient of fluctuation of energy for I.C. engines, C_e

Type of engine	*Number of cylinders*	*Angle between crank(s) Deg*	*Coefficient of fluctuation of energy C_e*	
			Four stroke	*Two stroke*
Single acting	1	–	1.35 – 2.40	0.95 – 1.00
Single acting	2	180	1.50 – 1.60	0.20 – 0.25
Vertical	3	120	0.60 – 0.75	0.15 – 0.18
Vertical	4	180 and 90	0.15 – 0.20	0.075 – 0.10
Vertical	6	180 and 60	0.10 – 0.12	0.016 – 0.02

Table 23.4: Limiting rim velocity of flywheel

Machine type	*Rim velocity, m / min*
Cast iron wheels under 100 kg	1500
Large machinery	2000
Flywheel without blow holes	3500
Cast steel wheel	3500
Large flywheel for steel mills	4500
Cast iron flywheel for automobiles	3100
Cast steel flywheel for automobiles	6200

CHAPTER

24

Clutches and Brakes

1. Disc or Plate Clutch

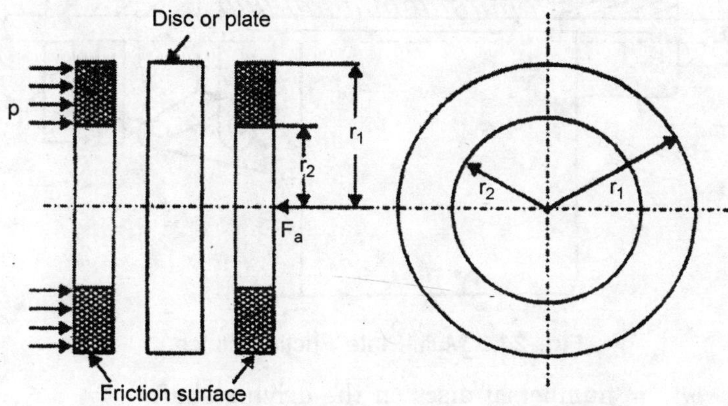

Fig. 24.1 Single plate friction clutch.

Let T = torque transmitted by the clutch

p = intensity of axial pressure

r_1, r_2 = external and internal radii of friction surfaces respectively.

r_m = mean radius of friction surface

μ = coefficient of friction

(*a*) For uniform pressure

$$F_a = \pi(r^2_1 - r^2_2)p \quad \text{...(24.1)}$$

Friction torque,

$$T_f = \mu F_a \, r_m \quad \text{...(24.2)}$$

Where $r_m = (2/3)\,[(r^3_1 - r^3_2)/(r^2_1 - r^2_2)]$

(*b*) For uniform wear

$$T_f = \mu F_a r_m \quad ...(24.3)$$

Where $r_m = (r_1 + r_2)/2$

$$r_2/r_1 = 0.6 - 0.7 \text{ for general machinery} \quad ...(24.4)$$

$$= 0.748 \text{ for tractors}$$

2. Multiple Disc Clutch

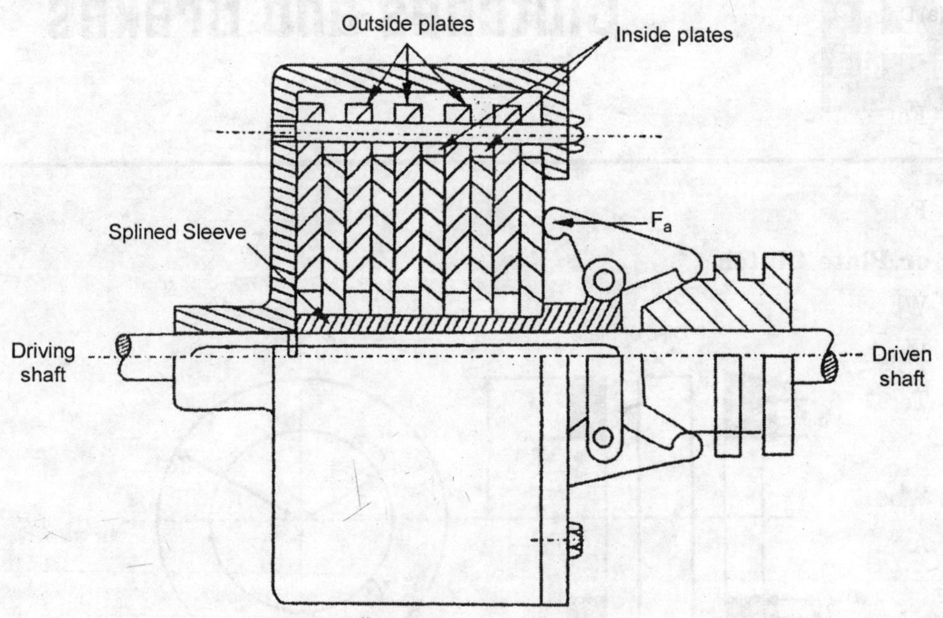

Fig. 24.2 Multi-Plate Friction clutch.

Let n_1 = number of discs on the driving shaft

n_2 = number of discs on the driven shaft

$n = n_1 + n_2 - 1$ = number of pairs of contact surfaces ...(24.5)

Total friction torque acting on the clutch,

$$T_f = n\mu\ F_a\ r_m \quad ...(24.6)$$

Number of plates

= 10 – 20 ...(24.7)

Thickness of core plate

= 0.75 – 3.0 mm (See Table 24.3) ...(24.8)

3. Cone Clutch

r_1, r_2 = external and internal radii of friction surfaces respectively.

r_m = mean radius of friction surface

μ = coefficient of friction

α = semi – cone angle of cone

b = width of the friction surface

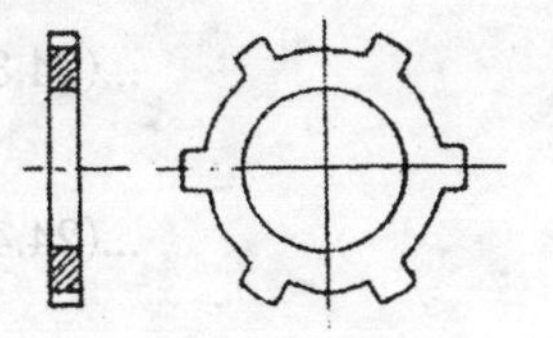

(a) Driving disc

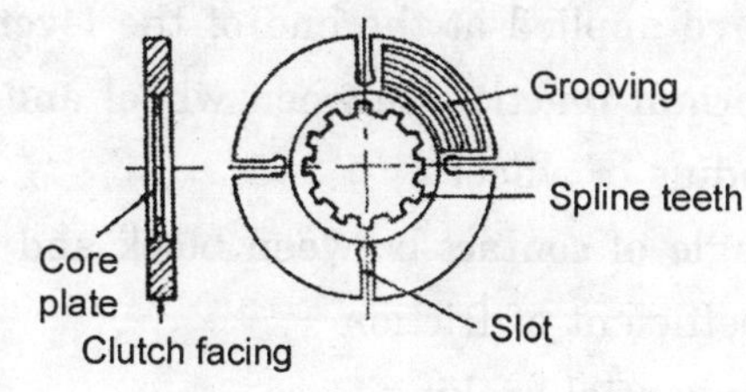

(b) Clutch disc

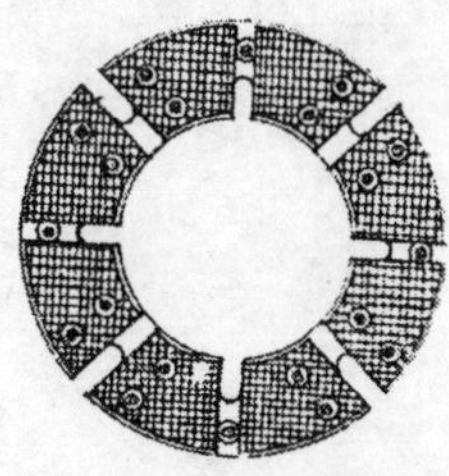

(a) Slots on clutch facing

Fig. 24.3

(*a*) For uniform pressure

$$F_a = \pi(r^2{}_1 - r^2{}_2)\, p_n \quad ...(24.9)$$

Frictional torque

$$T_f = \mu F_a r_m \text{ cosec } \alpha \quad ...(24.10)$$

Where $r_m = (2/3)\, [(r^3{}_1 - r^3{}_2)/(r^2{}_1 - r^2{}_2)]$

(*b*) For uniform wear

Total frictional torque,

$$T_f = \mu F_a r_m \text{ cosec } \alpha$$

Where $r_m = (r_1 + r_2)/2 \quad ...(24.11)$

Axial force required for engaging the clutch,

$$F_e = F_n (\sin \alpha + 0.25\, \mu \cos \alpha) \quad ...(24.12)$$

4. Single Block Brake

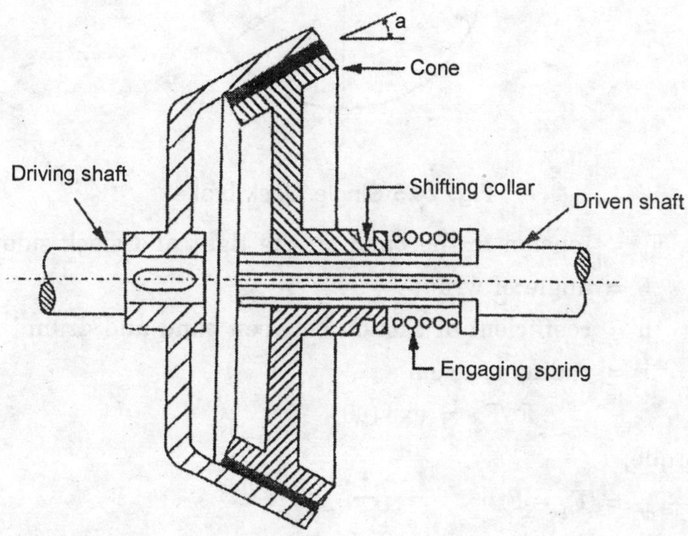

Fig. 24.4 Cone clutch

Let P = force applied at the end of the lever

F_n = normal reaction between wheel and block

r = radius of wheel

2θ = angle of contact between block and wheel

μ = coefficient of friction

F_t = tangential braking force

$F_t = \mu F_n$

Braking torque,

$$T_{br} = F_t \cdot r = \mu F_{n \cdot r} \quad ...(24.13)$$

$$F_n = Pl/(b + a\mu) \quad ...(24.14)$$

Braking torque,

$$T_{br} = \mu Plr/(b + a\mu)$$

5. Simple Band Brake

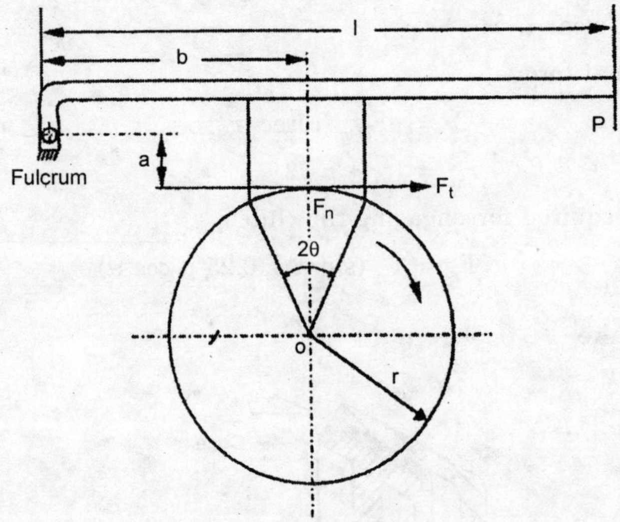

Fig. 24.5 Single block brake

Let F_1, F_2 = tension in the band on the tight and slack side respectively

θ = angle of wrap

μ = coefficient of friction between band and drum

r = radius of drum

Now $F_1/F_2 = \exp(\mu\theta)$

Braking torque,

$$T_{br} = (F_1 - F_2)r \quad ...(24.15)$$

$Pl = F_1 \cdot b$ for cw rotation of drum

$= F_2 \cdot b$ for ccw rotation of drum

Let w = width

t = thickness of band

σ_t = permissible tensile strength of band material

$$F_1 = \sigma_t wt \quad ...(24.16)$$

6. Differential Band Brake

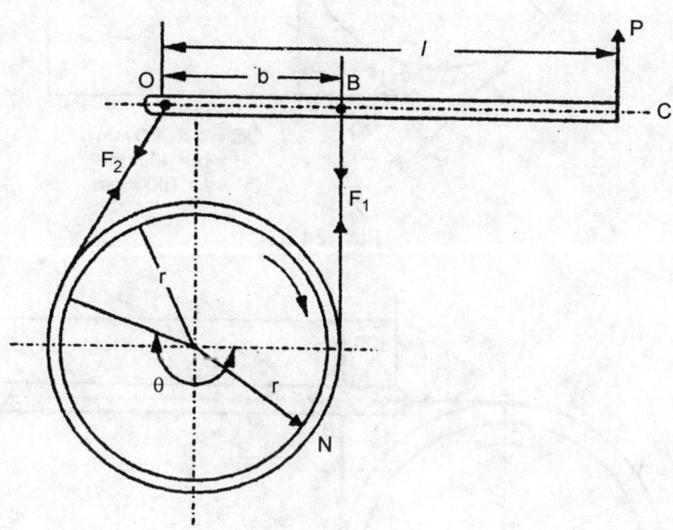

Fig. 24.6 Simple band brake

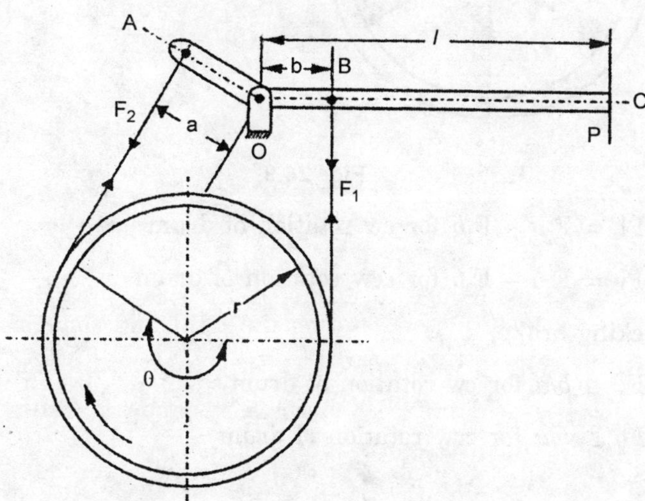

Fig. 24.7 Differential band brake.

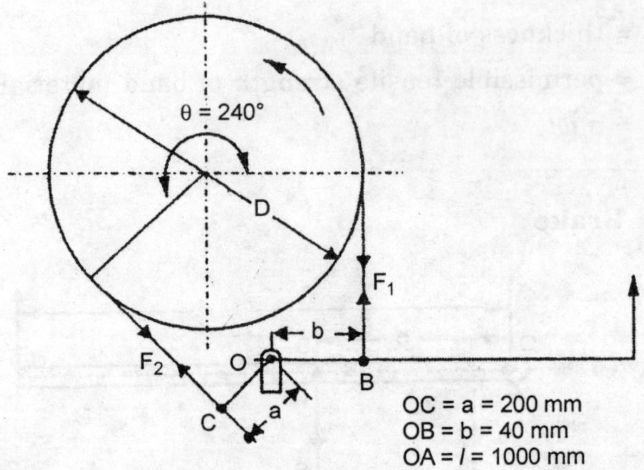

Fig. 24.8

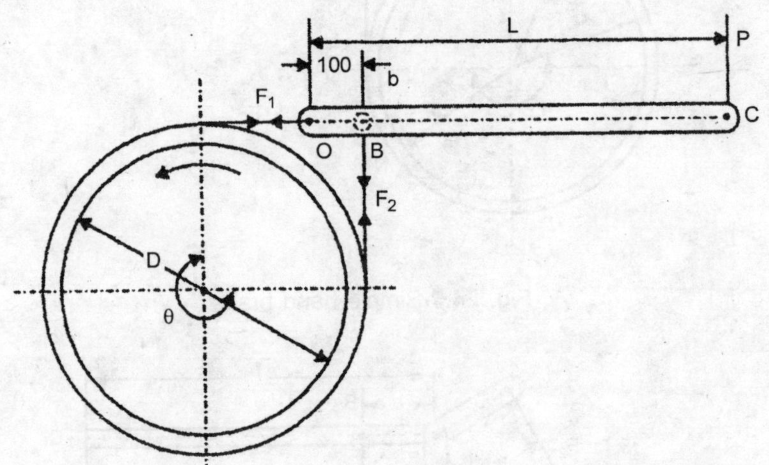

Fig. 24.9

$$Pl = F_2 a - F_1 b \text{ for cw rotation of drum} \quad ...(24.17)$$

$$Pl = F_1 a - F_2 b \text{ for ccw rotation of drum} \quad ...(24.18)$$

For self – locking brake,

$$F_2/F_1 \le b/a \text{ for cw rotation of drum} \quad ...(24.19)$$

$$F_1/F_2 \le b/a \text{ for ccw rotation of drum} \quad ...(24.20)$$

Table 24.1: Service factors for friction clutches

Type of service	*Service factor excluding starting factor*
Driving machine:	
Electric motor, steady load	1.0
Fluctuating load	1.5
Gas engine:	
Single cylinder	1.5
Multiple cylinder	1.0
Diesel engine:	
High speed	1.5
Large, slow speed	2.0
Driven machine:	
Generator, steady load	1.0
Fluctuating load	1.5
Blower	1.0
Compressor, depending on the Number of cylinders	2.0 – 2.5
Pumps, centrifugal:	
Single acting	2.0
Double acting	1.5
Line shaft	1.5
Wood working machinery	1.75
Hoist, elevators, cranes. Shovels	2.0
Hammer mills, ball mills, crushers	2.0
Brick machinery	3.0
Rock crushers	3.0

Table 24.2: Coefficient of friction and unit pressure for clutch facings

Facing material	*Coefficient of friction, μ*	*Unit pressure MPa*
Cast iron on cast iron:		
Dry	0.20	0.28 – 0.42
Oily	0.07	
Cast iron on steel:		
Dry	0.30	0.28 – 0.42
Oily	0.10	
Leather on cast iron:		
Dry	0.50	0.07 – 0.084
Oily	0.15	
Cork on cast iron:		
Dry	0.35	0.00875 – 0.014
Oily	0.30	
Asbestos fabric:		
Dry	0.35 – 0.45	0.21 – 0.42
Oily	0.25	
Moulded asbestos	0.25 – 0.35	0.21 – 0.42
Wood on cast iron	0.30	0.172 – 0.35

Table 24.3: Thickness of core plates

Clutch diameter, mm	Upto 150 mm	150 – 300	300 – 450	450 – 460
Core plate thickness, mm	0.75	1.5	2.25	3.0

Table 24.4: Grooving practice for wet friction clutch plate

Grooves or Slots	*Disc diameter, mm*
No grooves	0 – 230
Spiral grooves	125 – 500
Radial slots	200 – 450
Radial grooves	230 – 450
Alternating radial grooves and slots	350 – 500

Table 24.5: Coefficient of friction and allowable bearing pressure for brakes

Material of contact	*Coefficient of friction*			*Allowable bearing pressure, MPa*
	Dry	*Greasy*	*Lubricated*	
Cast iron on cast iron	0.20 – 0.15	0.10 – 0.06	0.10 – 0.05	1.03 – 1.72
Bronze on cast iron	-	0.10 – 0.06	0.17 – 0.05	0.55 – 0.83
Steel on cast iron	0.35 – 0.25	0.12 – 0.07	0.10 – 0.06	0.83 – 1.38
Wood on cast iron	0.35 – 0.10	0.12 – 0.08	-	0.41 – 0.62
Fibre on metal	-	0.20 – 0.10	-	0.07 – 0.08
Cork on metal	0.35	0.30 – 0.25	0.25 – 0.22	0.06 – 0.10
Leather on metal	0.5 – 0.3	0.20 – 0.15	0.15 – 0.12	0.07 – 0.28
Wired asbestos on metal	0.50 – 0.35	0.30 – 0.25	0.25 – 0.20	0.21 – 0.55
Wired asbestos on metal, short action	-	-	0.25 – 0.20	1.38 – 2.07
Steel on cast iron, or cast iron on cast iron, short action	-	-	0.10 – 0.05	1.38 – 1.08

Table 24.6: Working pressure for shoe material against cast iron

Rubbing velocity m/s	*Pressure, MPa*		
	Wood block	*Asbestos fabric*	*Asbestos blocks*
1.	0.549	0.687	1.103
2.	0.451	0.549	1.035
3.	0.343	0.412	0.896
4.	0.245	0.275	0.687
5.	0.173	0.206	0.483
6.	0.173	0.206	0.275

CHAPTER 25

Internal Combustion Engine Parts

1. Design of Cylinder

Wall thickness based upon the thin cylinder theory,

$$t = pd/(2\sigma_\theta) + c \qquad ...(25.1)$$

Table 25.1: Cylinder bore diameters for internal combustion engines

Preferred bore diameter (mm) – 30, 32, 34, 36, 38, 40, 44, 46, 48, 50, 52, 54, 56, 58, 60, 65, 70, 75, 80, 85, 90, 95, 100, 105, 110, 115, 120, 125, 130, 135, 140, 145, 150, 155, 160, 165, 170, 175, 180, 185, 190, 195, 200, 210, 220, 230, 240, 250, 260, 270, 280, 290, 300, 310, 320, 330, 340, 360

Where d = cylinder bore

p = maximum internal pressure

σ_θ = permissible hoop stress for the cylinder material

= 35 – 70 MPa for cast iron

= 70 – 100 M Pa for alloy steel

c = allowance for reboring

Table 25.2: Reboring allowance for I.C. engine cylinders

Cylinder dia, mm	75	100	150	200	250	300	350	400	450	500
c, mm	1.5	2.4	4.0	6.3	8.0	9.5	11.0	12.5	12.5	12.5

Thickness of cylinder wall based on the thick cylinder theory,

(*i*) Maximum principal theory,

$$t = (d/2)[\ \{(\sigma_\theta + p)/(\sigma_\theta - p)\}^{1/2} - 1] \qquad ...(25.2)$$

This theory is suitable for brittle materials, such as cast iron or cast steel,

(*ii*) Maximum strain theory,

(*a*) Clavarino's Equation

When the cylinder ends are closed.

$$t = (d/2)\ [\{(\sigma_\theta + (1 - 2\nu)p)/(\sigma_\theta - (1 + \nu)p)\}^{1/2} - 1] \qquad \text{...(25.3)}$$

(*b*) Birnie's Equation

For a cylinder with open ends,

$$t = (d/2)\ [\{(\sigma_\theta + (1 - \nu)p)/(\sigma_\theta - (1 + \nu)p)\}^{1/2} - 1] \qquad \text{...(25.4)}$$

This theory is suitable for ductile materials, such as mild steel or alloy steel.

Empirically, the thickness of cylinder wall is given by,

$$t = 0.045\ d + 1.5 \text{ mm} \qquad \text{...(25.5)}$$

2. Design of Liner and Water Jacket Wall

Empirically, the thickness of dry liner is given by:

$$t_1 = (0.030 - 0.035)\ d \qquad \text{...(25.6)}$$

Thickness of water jacket wall,

$$t_2 = 0.032\ d + 1.5 \text{ mm} \qquad \text{...(25.7)}$$

Water space between the outer cylinder wall and inner jacket wall,

$$l_3 = 0.08\ d + 6.5 \text{ mm} \qquad \text{...(25.8)}$$

The hardness of cylinder liners should be as given in Table 25.3.

3. Mean Effective Pressure in the Cylinder

Power developed, $\quad P = p_m\text{LAN}/10^3$ kW ...(25.9)

Where $\quad p_m$ = mean effective pressure, N / mm^2

L = length of stroke, m

A = $(\pi/4)\ d^2$ = area of piston, mm^2

N = number of strokes per second

L/d ratio is generally 1.5 to 2.

4. Cylinder Flange and Head

Thickness of cylinder flange,

$$t_f = (1.2 - 1.4)\ t \qquad \text{...(25.10)}$$

Thickness of cylinder head considering it as a flat circular plate loaded uniformly,

$$t_c = C_1 d(p/\sigma_t)^{1/2} \qquad \text{...(25.11)}$$

Where $\quad C_1$ = a constant = 0.3 – 0.4

σ_t = allowable tensile stress in plate material

= 34.42 MPa for C.I.

= 55 MPa for nickel C.I.

= 42 MPa for Al alloys

= 82.4 MPa for low carbon cast steel

5. Diameter of Studs or Bolts

Core diameter of studs, $\quad d_s = d[\text{p}/(n\sigma_t)]^{1/2}$ ...(25.12)

Where p = combustion pressure, N/mm^2

n = number of studs

$= (d/100 + 40 \text{ to } (d/50 + 4)$ ...(25.13)

d = cylinder bore, mm

σ_t = allowable tensile stress for stud material, MPa

D_s should be more than 16 mm.

Pitch circle diameter of studs or bolts,

$$d_p = d + 3d_s \quad ...(25.14)$$

Outside diameter of flange or cylinder head,

$$d_0 = d_p + 2d_s + 12 \text{ mm} \quad ...(25.15)$$

Pitch of studs, $p_s = (19 - 28)\sqrt{d_s}$ ...(25.16)

6. Piston

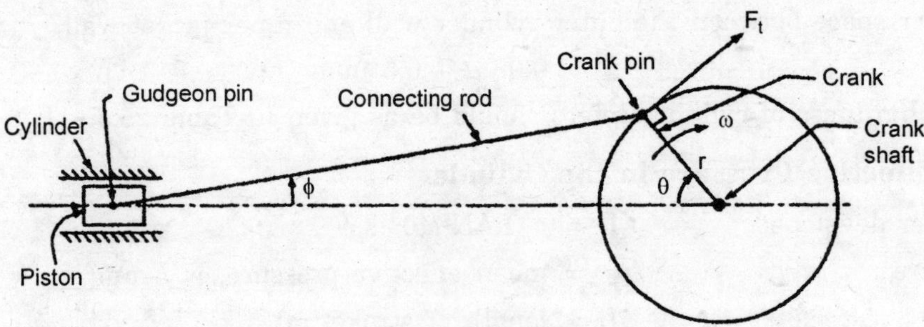

Fig. 25.1. Line diagram of an I.C. Engine.

(*a*) Strength Considerations

The head may be treated as a flat plate with a uniform load and rigidly supported at the outer edge. The head thickness is given by,

$$t_h = 0.433d\,[p/\sigma_t\,)^{1/2} \quad ...(25.17)$$

where σ_t = permissible stress in tension for the piston material (Table 25.4)

Empirically, $t_h = 0.032\,d + 1.5$ mm ...(25.18)

Thickness of wall under the rings,

$$t_f = t_h \quad ...(25.19)$$

Length of piston, $I = (1 - 1.5)\,d$ ...(25.20)

Distance between centre of piston pin and centre of skirt = $(0.02 - 0.04)\,d$ ...(25.21)

(*b*) Heat Transfer Considerations

Head thickness for heat flow,

$$t_h = \text{H}d^2/[16h_c(T_c - T_e)\text{ A}$$

$$= \text{H}/[12.56\ h_c(T_c - T_e) \quad ...(25.22)$$

Where H = heat flow through the head, Watts

= KCW × Brake power per cylinder ...(25.23)

K = fraction of heat absorbed by the piston ≈ 0.05

C = higher heat value of fuel used in kJ/kg = 41840 kJ/kg

W = weight of fuel used in kg/brake power in kW per hour

d = cylinder diameter, m

A = piston head area, m^2

h_c = heat conduction factor, W/(m^2 · °C) per meter length

= 46.5 W/(m^2 · °C) per meter length for grey cast iron

= 162.73 W/(m^2 · °C) per meter length for aluminium.

= 51 W/(m^2 · °C) per meter length for steel

T_c = temperature at centre of head, °C

T_e = temperature at edge of head, °C

$T_c - T_e$ = 220°C for grey cast iron

– 75°C for aluminium alloys

= 250°C for steel

Physical properties of aluminium alloys for I.C. engine pistons are given in Table 25.5.

7. Piston Rings

Radial thickness of piston ring,

$$t_{pr} = d(3p_w/\sigma_{br})^{1/2} \quad ...(25.24)$$

where t_{pr} = radial thickness of piston rings, mm

p_w = radial wall pressure between the piston ring and cylinder wall , MPa

= (0.03 – 0.06) MPa

σ_{br} = allowable bending stress for piston ring

= (90 – 120) MPa for cast iron

Typical materials for piston rings are given in Table 25.6.

Width of rings, t_2 = d/20 for concentric rings ...(25.25a)

= d/225.5 opposite to the joint of eccentric rings ...(25.25b)

= d/55 at the joint of eccentric rings ...(25.25c)

Width of lands on piston between the piston rings,

$$h = (0.75 - 1.0)\ t_2 \quad ...(25.26)$$

Width of top land, $h_0 = (1 - 1.25)\ t_h$ ...(25.27)

For same amount of heat flow,

$$t_r = 0.5[d_r \pm (d_r^2 - 4dt_h)^{1/2}] \quad ...(25.28)$$

Where t_r = thickness under the ring grooves, mm

d_r = diameter at the bottom of the ring grooves, mm

$= d - (2b + 0.006\ d + 0.5\ \text{mm})$ at the compression rings

$= d - (2b + 0.006d + 1.5\ \text{mm})$ at the oil grooves

b = radial depth of the ring section, mm

Gap between free ring ends,

$$g_r = (3 - 4)t_{pr} \quad ...(25.29)$$

8. Piston Skirt

Maximum side thrust, F_s = (0.05 – 0.10) × maximum gas force

$$= (0.05 - 0.10) \times (\pi/4)d^2p_{max} \quad ...(25.30)$$

Length of piston skirt, $h_2 = F_s/(dptc)$ ...(25.31)

where p_{bc} = permissible bearing pressure between piston skirt and cylinder wall

= 0.3 MPa for low speed engines and 0.5 MPa for high speed engines

9. Length of Piston

Length of piston = width of top land + total width of piston rings + total width of lands + length of piston skirt

$$= t_1 + zt_2 + (z - 1)h + h_2 \quad ...(25.32)$$

The empirical proportions of piston are given in Table 25.7.

10. Piston Clearances

Clearance on piston crown,

Δc = (0.006 to 0.008) d, for aluminium piston ...(25.33a)

= (0.004 to 0.006) d, for cast iron piston ...(25.33b)

Clearance on skirt, Δs = (0.001 to 0.002) d, for both aluminium and cast iron pistons ...(25.34)

Where d = piston diameter

The clearances in the hot state can be determined as follows:

$$\Delta c' = d[1 + \sigma_{cyl}\ (T_{cyl} - T_0)] - d_c[1 + \sigma_p(T_c - T_0)]$$
$$> (0.002 \text{ to } 0.0025)\ d \quad ...(25.35)$$

$$\Delta s' = d[1 + \alpha_{cyl}\ (T_{cyl} - T_0\ )] - d_s[1 + \alpha_p\ (\ T_c - T_0)]$$
$$> (0.0005 \text{ to } 0.0015)\ d \quad ...(25.36)$$

Where α_{cyl} = coefficient of expansion for material of cylinder liner

α_p = coefficient of expansion for material of piston

= 22×10^{-6} for aluminium, 1/K

= 11×10^{-6} for cast iron, 1/K

For water cooled engines,

T_{cyl} = 385 K, T_c = 473 to 723 K, T_s = 403 to 473 K, T_0 = 293 K

For air cooled engines,

$$T_{cyl} = 450 \text{ K}, T_c = 573 \text{ to } 873 \text{ K}, T_s = 483 \text{ to } 613 \text{ K}$$

11. Piston (or Gudgeon) Pin

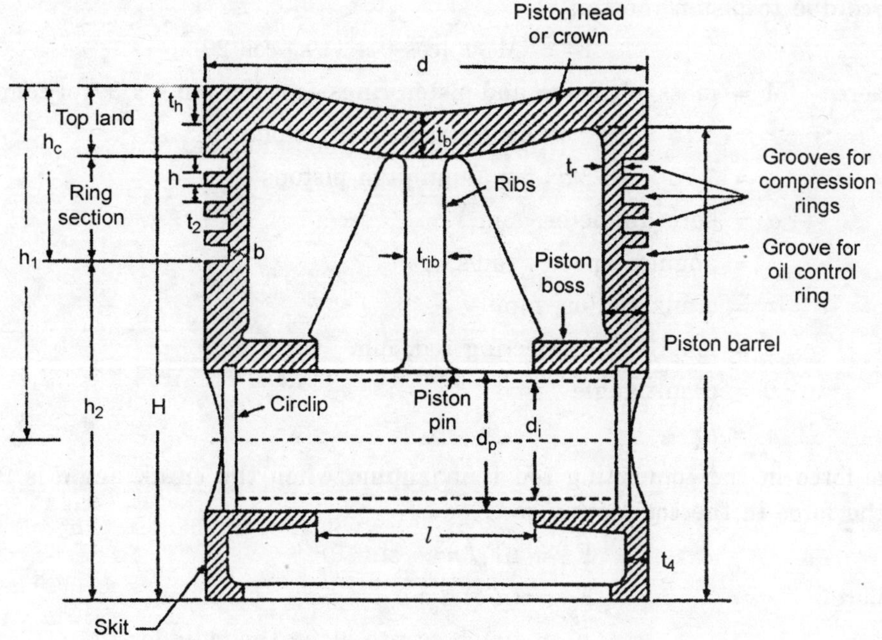

Fig. 25.2 Trunk type piston.

Length between the supports,

$$l' = l + 0.5\ (d - l) = (l + d)/2$$

Maximum bending moment,

$$M_{max} = (F_g/2)\ (l'/2 - 1/4) = F_g d/8 \quad ...(25.37)$$

Where $F_g = \pi_{max} \times$ area of cylinder

Resisting moment, $M_r = (\pi/32)\ [(d_0^{\ 4} - d_i^{\ 4})/d_0]\ \alpha$ ...(25.38)

Where d_i, d_0 = inside and outside diameters of the pin

σ = permissible bending stress for the pin material.

For safe design, $M_{max} = M_r$

The bearing pressure between the piston pin and the connecting rod bearing is given by,

$$F_g = l d_p p_b \quad ...(25.39)$$

Where l = length of the piston pin in connecting rod bearing

d_p = diameter of piston pin

p_b = permissible bearing pressure between the piston pin and connecting rod bearing (Table 25.8) Shear stress in piston pin,

$$\tau = 2F_g/[\pi(d^2_{\ p} - d^2_{\ i})] \quad ...(25.40)$$

The specifications for piston pins are given in Table 25.9

12. Connecting Rod

Maximum force due to gas pressure, $F_g = (\pi/4)\ d^2 pe$ ...(25.41)

Where p_e = explosion pressure

Force due to piston inertia,

$$F_i = M\omega^2 r\ [\cos\theta + (1/n)\cos 2\theta] \quad ...(25.42)$$

Where M = mass of (piston and piston rings + piston pin + 1/3rd of connecting rod)

$\approx 4.15 \times 10^{-6}\ d^3$ kg for cast iron pistons

$\approx 1.78 \times 10^{-6}\ d^3$ for aluminium pistons

d = bore diameter, mm

ω = angular speed , rad/s

r = crank radius, mm

l = length of connecting rod, mm

θ = crank angle

$n = l/r$

The force in the connecting rod is maximum when the crank angle is 90°. At that instant, the force in the connecting rod,

$$F_c = nF_a/(n^2 - \sin^2\theta)^{1/2} \quad ...(25.43)$$

Where $F_g = (n/4)\ d_p^{\ 2}\ p_a$

p_a = net pressure along the cylinder

Force of friction due to piston rings and piston,

$$F_f = \pi hbdip_r\mu \quad ...(25.44)$$

Where h = axial width of rings

i = number of rings

p_r = pressure of rings

μ = coefficient of friction ≈ 0.05 – 0.10

At the crank pin the force per unit length will have a magnitude given by [see Fig. 25.3(a)],

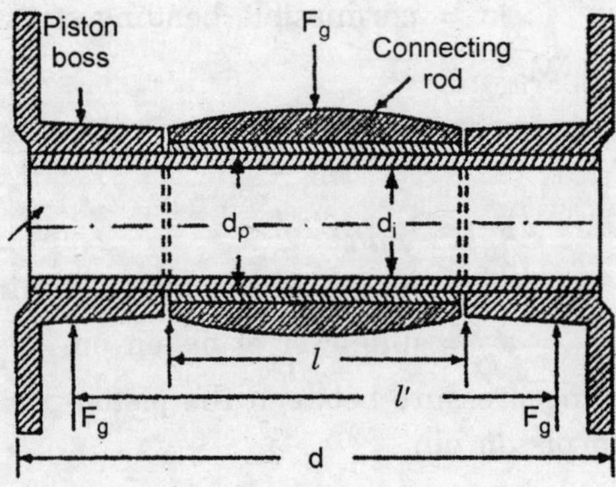

Fig. 25.3 Piston (or gudgeon) pin.

$$F = \rho A\, \omega^2\, r \sin(\theta + \phi)$$

Where ρ = density of connecting rod material

A = cross-sectional area of connecting rod

F will be maximum when the crank and connecting rod are at right angles,

i.e., $\theta + \phi = 90°$.

$$F_{max} = \rho A \omega^2 r$$

The resultant normal force acting on the connecting rod then becomes,

$$F_n = (1/2)\, F_{max} l \qquad ...(25.45)$$

The reactions at the small and big end are [see Fig. 25.3(*b*)],

$$R_1 = F_n/3 \text{ and } R_2 = 2F_n/3$$

The bending moment at a distance from x from the small end is,

$$M_x = F_n \times x/3 - (1/2)F_{max}$$

$$= [F_n \times /(3l^2)]\, (l^2 - x^2)$$

For bending moment to be maximum,

$$dM_x/dx = 0$$

$$[F_n/(3l^2)](l^2 - 3x^2) = 0$$

$$x = 1/\sqrt{3}$$

$$M_{max} = 2Fnl/(9\sqrt{3}) = 0.1283\, F_n l \qquad ...(25.46)$$

Maximum bending stress,

$$\sigma_{max} = M_{max/z} \qquad ...(25.47)$$

Where z = section modulus of the rod.

Due to the gas force, the rod is subjected to buckling. The buckling load is given by the Rankine – Gordon's formula,

$$P_e = \sigma_c A/[1 + a(l_e/K)^2] \qquad ...(25.48)$$

Where σ_c = ultimate compressive strength of rod material

l_e = equivalent length of the rod

a = constant = 1/7500

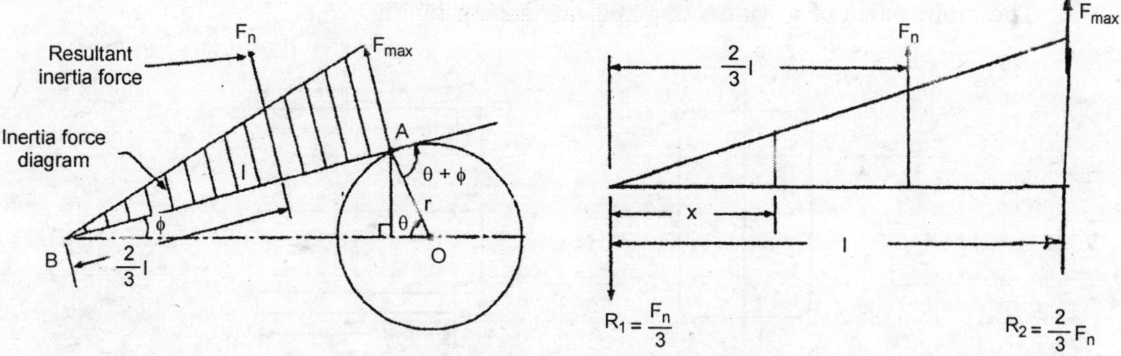

Fig. 25.4 Forces on connecting rod.

K = radius of gyration of the rod about the buckling axis.

$$4I_{yy} = I_{xx}$$

Or

$$4K^2_{yy} = K^2_{xx} \quad ...(25.49)$$

At the piston end the depth may be taken as 0.75 to 0.9 times and at the crank pin end 1.1 to 1.25 times.

13. Dimensions of Small End

Let l_p = length of piston pin

d_p = diameter of piston pin

p_b = permissible bearing pressure ≈ 10 – 14 N/mm²

l_p/d_p = 1.5 to 2

The force on the small end,

$$F_g = l_p d_p p_b \quad ...(25.50)$$

14. Dimensions of Big End

Let l_c = length of crank pin

d_c = diameter of crank pin

p_b = permissible bearing pressure = 5 – 10 N/mm²

l_c/d_c = 1.25 to 1.5

The force on the small end,

$$F_g = l_c d_c p_b \quad ...(25.51)$$

15. Bolts for Big End

$$2 \times (\pi/4)d^2_b\sigma_t = F_i \quad ...(25.52)$$

Where d_b = diameter of bolts

16. Big End Cap

$$M_{max} = F_{ip} l_0/6 \quad ...(25.53)$$

Where l_0 = distance between bolt centres

= d_c + 2 × thickness of liner (≈ 0.05d + 1 mm) + d_b + clearance (≈ 1.5 mm)

If h = cap thickness, and l_c = width of cap = length of bearing, then

$$M_{max} = \sigma_b \times l_c h^2/6 \quad ...(25.54)$$

The main parts of a connecting rod are shown in Fig. 25.5.

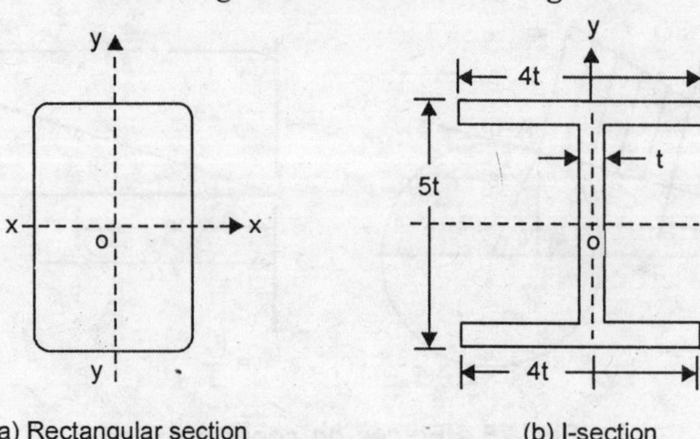

(a) Rectangular section (b) I-section

Fig. 25.5 Proportions of connecting rod.

17. Design of Centre Crankshaft

(a) Crank at maximum bending moment position

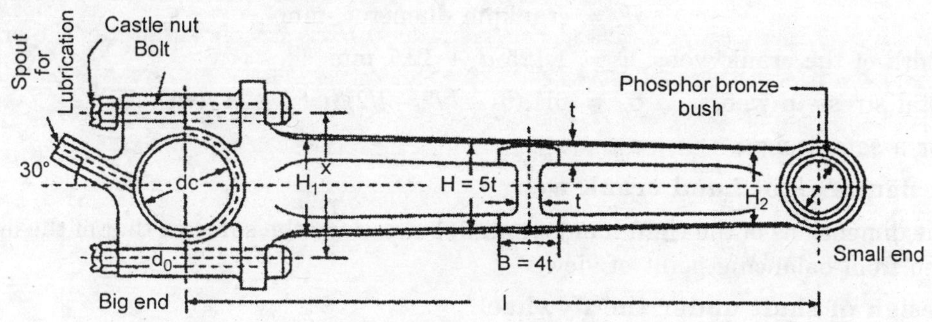

Fig. 25.6 Connecting rod.

d = cylinder bore or piston diameter, mm

p_{max} = maximum pressure on the piston, MPa

W = weight of the flywheel acting downwards, N

$T_1 + T_2$ = resultant belt tension acting horizontally, N

Thrust in the connecting rod = Gas force on the piston acting horizontally

$$F_g = (\pi/4)d^2 p_{max}$$

Horizontal reactions on bearings 1 and 2 are,

$$H_1 = F_g b_1/b, \text{ and } H_2 = F_g b_2/b$$

Vertical reactions due to the weight W of the flywheel on the bearings 2 and 3 are,

$$V_2 = W \cdot c_i/c \text{ and } V_3 = W \times c_2/c$$

Horizontal reactions due to belt tensions at bearings 2 and 3 are,

$$H'_2 = (T_1 + T_2)\, c_1/c, \text{ and } H'_3 = (T_1 + T_2)\, c_2/c$$

Resultant forces at bearings 2 and 3 become,

$$R_2 = [(H_2 + H'_2)^2 + V^2{}_2]^{1/2}$$

$$R_3 = [H^2{}_3 + V^2 3]^{1/2}$$

(i) Design of crankpin

Let d_c = diameter of crankpin, mm

l_c = length of crankpin, mm

σ_b = allowable bending stress for the crankpin, MPa

$$d_c = [32\, H_1 b_2/(\pi\sigma_b)] \quad ...(25.55)$$

$$l_c = F_g/(d_c p_b) \quad ...(25.56)$$

where p_b = permissible bearing pressure

(ii) Design of left hand crank web

Thickness of the crankweb,

$$t = (0.4 - 0.6)\, d_s \quad ...(25.57a)$$

$$= (0.22 - 0.32)\, d \quad ...(25.57b)$$

$$= 0.65\, d_c + 6.35 \text{ mm} \quad ...(25.57c)$$

Where d_s = diameter of shaft, mm

d = bore diameter, mm

d_c = crankpin diameter, mm

Width of the crank web, $w = 1.125\ d_c + 12.7$ mm ...(25.58)

Total stress in web, $\sigma_w = 6H_1(b_2 - l_c/2 - t/2)/(wt^2) + H_1/(wt)$ ...(25.59)

For a safe design, $\sigma_w < \sigma_{allow}$

(iii) Design of right hand crank web

The dimensions of the right hand crank web are made the same as that of the left hand crank web from balancing point of view.

(iv) Design of shaft under the flywheel

Diameter of shaft, $d_s = [\{32/(n \times \sigma_b)\}\ \{(V_3 \cdot c_1)^2 + (H'_3 \cdot c_1)\}^{1/2}]^{1/3}$ ...(25.60)

Where σ_b = allowable bending stress for the shaft, MPa

(b) Crank at maximum twisting moment position

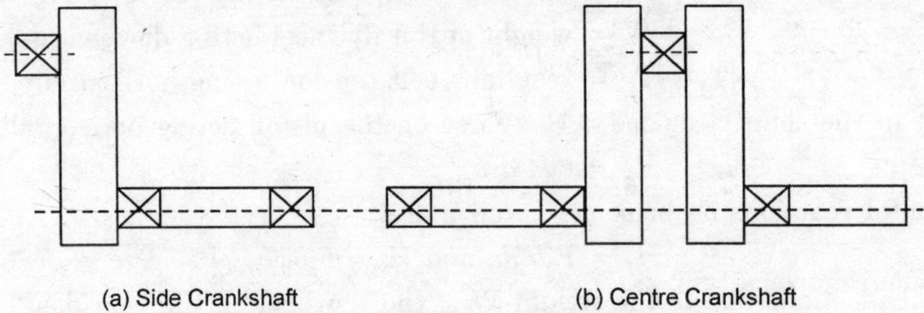

(a) Side Crankshaft (b) Centre Crankshaft

Fig. 25.7 Types of crankshafts.

Gas force on the piston, $F_g = (n/4)\ d^2 p_0$

Thrust in the connecting rod,

$$F_q = F_g/\cos \phi$$

Where ϕ = angle of inclination of the connecting rod with the line of stroke.

Tangential force, $F_t = F_q \sin (\theta + \phi)$

Radial force, $F_r = F_q \cos (\theta + \phi)$

Bearing reactions due to tangential force,

$$H_{t1} = F_t \cdot b_1/b, \text{ and } H_{t2} = F_t \cdot b_2/b$$

Bearing reactions due to radial force,

$$H_{r1} = F_r \cdot b_1/b, \text{ and } H_{r2} = F_r \cdot b_2/b$$

The reactions at the bearings 2 and 3 due to flywheel and belt tensions will be the same as in case (*a*).

(i) Design of crankpin

Crankpin diameter, $d_c = [\{16/(\pi \times \tau)\}\ \{(H_{r1} \cdot b_2)^2 + (H_{t1} \cdot r)^2\}^{1/2}]^{1/3}$ ...(25.61)

Where τ = allowable shear stress in the crankpin

(ii) Design of shaft under the flywheel

Shaft diameter, $d_s = [\{16/(\pi \times \tau)\} \{(R_3 \cdot c_1)^2 + (F_t \cdot r)^2\}^{1/2}]^{1/3}$ ...(25.62)

Where τ = allowable shear stress in the shaft

(iii) Design of shaft at the junction of right hand crank arm

Let d_{s1} = diameter of the shaft at the junction of right hand crank arm.

Bending moment at the junction of right hand crank arm,

$$M_{s1} = R_1(b_2 + l_c/2 + t/2) - F_q(l_c/2 + t/2)$$

Twisting moment at the junction of the right hand crank arm,

$$T_{s1} = F_t \cdot r$$

$$d_{s1} = [\{16/(\pi \times \tau)\} \{M^2_{s1} + T^2_{s1}\}^{1/2}]^{1/3} \quad ...(25.63)$$

Where τ = allowable shear stress in the shaft.

(iv) Design of right hand crank web

Bending moment due to radial component of F_q is,

$$M_r = H_{r2}(b_1 - l_c/2 - t/2) = (wt^2/6) \cdot \sigma_{br}$$

Where σ_{br} = bending stress in the radial direction.

Bending moment due to tangential component of F_q is,

$$M_t = F_t(r - d_{s1}/2) = (tw^2/6) \cdot \sigma_{bt}$$

Where σ_{bt} = bending stress in the tangential direction.

Direct compressive stress,

$$\sigma_d = F_r/(2wt)$$

Maximum compressive stress,

$$\alpha_c = \sigma_{br} + \sigma_{bt} + \sigma_d$$

Twisting moment on the arm,

$$T = H_{t1}(b_2 + l_c/2) - F_t \times l_c/2 = H_{t2}(b_1 - l_c/2)$$

Shear stress in the arm, $\tau = T/z_p = 4.5\ T/(wt^2)$

Maximum combined stress,

$$(\sigma_c)_{max} = \sigma_c/2 + (1/2)[\sigma_c^2 + 4\tau^2]^{1/2} \quad ...(25.64)$$

$$(\sigma_c)_{max} < \sigma_{permissible}$$

(v) Design of left hand crank web

The dimensions for the left hand crank web may be kept the same as for the right hand crank web from balancing point of view.

(vi) Design of crankshaft bearings

The bearing 2 is the most heavily loaded and should be checked for bearing pressure.

$$R_2 = F_g/2 + W/2 + (T_1 + T_2)/2$$

Bearing pressure, $p_{b2} = R_2/(l_2 \cdot d_{s1})$ ...(25.65)

Where l_2 = length of bearing 2.

18. Side Crankshaft

(a) Crank at dead centre position

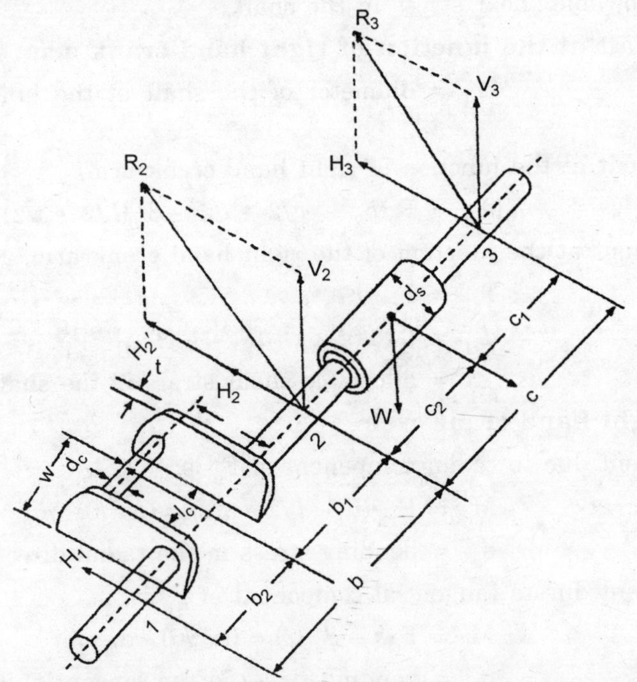

Fig. 25.8 Centre crankshaft at dead centre.

Let d = cylinder bore or piston diameter, mm

p_{max} = maximum pressure on the piston, MPa

W = weight of the flywheel acting downwards, N

$T_1 + T_2$ = resultant belt tension acting horizontally, N

Thrust in the connecting rod = Gas force on the piston = $F_g = (\pi/4)\, d^2 p_{max}$

Horizontal reactions on bearings 1 and 2,

$$H_1 = F_g\,(a + b)/b$$

$$H_2 = F_g\, a/b$$

Vertical reactions due to the weight W of the flywheel on the bearings 1 and 2,

$$V_1 = W \cdot b/b$$

$$V_2 = W \cdot b_2/b$$

Horizontal reactions due to belt tensions at bearings 1 and 2,

(i) Design of crankpin

Let d_c = diameter of crankpin, mm

l_c = length of crankpin, mm

p_b = safe bearing pressure for the crankpin, MPa

Considering bearing of the crankpin, the length of the crankpin is,

$$F_g = l_c d_c p_d \qquad ...(25.66)$$

$$I_c/d_c = 0.6 \text{ to } 1.5$$

Check under bending:

Maximum bending moment (mean value) in the crankpin,

$$M = 3F_9 I_c/4$$

Section modulus, $z_c = (\pi/32)\, d^3{}_c$

Bending stress in crankpin,

$$\sigma_b = M/Z_c < \sigma_{allow} \quad ...(25.67)$$

***(ii)* Design of bearings**

Bending moment at the centre of bearing 1,

$$M_1 = F_9\,(0.75\, l_c + 1 + 0.5\, l_1)$$

Where t = thickness of crank web

$= (0.45 - 0.75)\, d_c$

l_1 = length of bearing 1

$= (1.5 - 2\,)\, d_c$

Now $M = (\pi/32)\, d \quad ...(25.68)$

Bearing 2 may be made of the same size as bearing 1.

***(iii)* Design of crank web**

Bending moment on crank web,

$$M = F_g\,(0.75\, l_c + 0.5\, t)$$

Section modulus, $z_w = wt^{2/6}$

Bending stress, $\sigma_b = M/z_w$

Direct compressive stress,

$$\sigma_d = F_g/(w.t)$$

Total stresses, $\sigma_d = \sigma_b + \sigma_d < \sigma_{allow} \quad ...(25.69)$

***(iv)* Design of shaft under the flywheel**

Let d_s = diameter of shaft, mm

Horizontal bending moment at the flywheel location due to piston gas force,

$$M_1 = F_g(a + b_2) - H_1 \cdot b_2 = H_2 \cdot b_1$$

Horizontal bending moment at the flywheellocation due to belt tensions,

$$M_2 = H'_1 \times b_2 = H'_2 \times b_1 = (T_1 + T_2)\, b_1 \cdot b_2/b$$

Total horizontal bending moment,

$$M_h = M_1 + M_2$$

Vertical bending moment due to flywheel weight,

$$M_v = V_1 \cdot b_2 = V_2 \cdot b_1 = Wb_1 b_2/b$$

Resultant bending moment

$$M_R = (M_h{}^2 + M_y{}^2\,]^{1/2}$$

$$= (\pi/2)\, d^3{}_s\, \sigma_b \quad ...(25.70)$$

Where σ_b = allowable bending stress for the shaft, MPa

(b) Crank at maximum twisting moment position (Fig. 25.9).

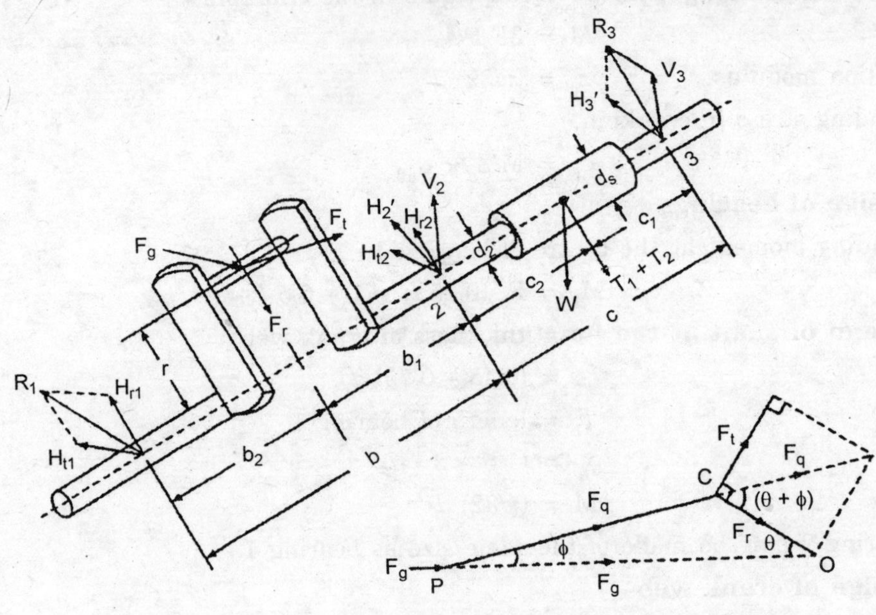

Fig. 25.9. (*a*) Crank at an angle of maximum twisting moment.

Gas force, $F_g = (\pi/4)\, d^2 p_0$

Thrust on the connecting rod,

$$F_q = F_g/\cos\phi$$

Where ϕ = angle of inclination of the connecting rod with the line of stroke.

Tangential force, $F_t = F_q \sin(\theta + \phi)$

Radial force, $F_r = F_q \cos(\theta + \phi)$

Reactions at bearings 1 and 2 due to F_1 are,

$$H_{t1} = F_t \cdot (a + b)/b$$

$$H_{t2} = F_t \cdot a/b$$

The reactions at the bearings 1 and 2 due to flywheel and belt tensions will be the same as in case (*a*).

(*i*) Design of crank web

Bending moment due to tangential component F_t,

$$M_t = F_t(r - d_1/2) = (tw^2/6) \cdot \sigma_{bt}$$

Where σ_{bt} = bending stress in the tangential direction.

d_1 = diameter of bearing 1

Bending moment due to radial component F_r,

$$M_r = F_r\,(0.75\, l_c = 0.5\, t) = (wt^2/6) \cdot \sigma_{br}$$

Where σ_{br} = bending stress in the radial direction.

Direct compressjve stress,

$$\sigma_d = F_r/(wt)$$

Maximum compressive stress,

$$\sigma_c = \sigma_{br} + \sigma_{bt} + \sigma_d$$

Twisting moment on the arm due to

$$F_t,\ T = F_t\,(0.75\,l_c + 0.5\,t)$$

Shear stress in the arm, $\tau = T/z_p = 4.5\ T/(wt^2)$

Maximum combined stress,

$$(\sigma_c)_{max} = \sigma_c/2 + (1/2)\,[\sigma_c^2 + 4\,\tau^2]^{1/2} \qquad ...(25.71)$$

$$(\sigma_c)_{max} < \sigma_{permissible}$$

***(ii)* Design of shaft at the junction of crank**

Let d_{s1} = diameter of the shaft at the junction of right hand crank arm.

Bending moment at the junction of crank,

$$M = F_g\,(0.75\,l_c + t)$$

Twisting moment on the shaft,

$$T = F_t \cdot r$$

Equivalent twisting moment at the junction of right hand crank arm,

$$T_e = [M^2 + T^2]^{1/2}$$

$$(\pi/16)\,d^3_{s1}\,\tau = T_e \qquad ...(25.72)$$

Where τ = allowable shear stress in the shaft.

***(iii)* Design of shaft under the flywheel**

Horizontal bending moment on the shaft due to gas force,

$$M_1 = F_g(a + b_2) - [H^2_{r1} + H^2_{ti}]^{1/2} b_2$$

Horizontal bending moment at the flywheel location due to belt tensions,

$$M_2 = H'_1 \times b_2 = H'_2 \cdot b_1 + (T_1 + T_2)\,b_1 b_2/b$$

Total horizontal bending moment,

$$M_h = M_1 + M_2$$

Vertical bending moment due to flywheel weight,

$$M_v = V_1 \cdot b_2 = V_2 \cdot b_1 = Wb_1b_2/b$$

Resultant bending moment,

$$M_R = [M_h^2 + M_y^2]^{1/2}$$

Twisting moment, $T = F_t \cdot r$

Equivalent twisting moment on the shaft,

$$T_e = [M^2_R + T^2]^{1/2}$$

$$(\pi/16)d^3_s\,\tau = T_e \qquad ...(25.73)$$

Where τ = allowable shear stress in the shaft

Table 25.3: Hardness of cylinder liners for I.C. engines

Wall thickness of liner castings, mm	*Brinell hardness, N/mm^2*
From 8 upto 25	210 – 290
Above 25 upto 50	200 – 280
Above 50	190 – 270

Table 25.4: Allowable tensile stress for piston materials

Material	*Allowable tensile stress, MPa*
Cast iron	35 – 40
Alloy cast iron	50 – 70
Aluminium alloy	50 – 70
Steel	80 – 90

Table 25.5: Physical properties of aluminium alloys for I.C. engine pistons

Alloy designation		*Hardness*	*Tensile strength, MPa*		*Coefficient of thermal expansion*
Casting	*Forging*	*BHN*	*Chill casting*	*Forging*	*(20° – 200° C) per °C × 10^{-6}*
2285	24850	90 – 130	225 – 275	345 – 410	23 – 24
4658	49582	90 – 140	125 – 140	195 – 365	20.5 – 21.5
4928-A	49285	90 – 125	175 – 215	225 – 295	18.5 – 19.5
4928-B	–	90– 125	165 – 205	–	17 – 18

Table 25.6: Typical materials for piston rings

Material symbol	*Material type*	*Modulus of elasticity, GPa*	*Minimum ring tensile strength, MPa*	*Minimum bending strength MPa*	*Minimum hardness*
A_1	Grey cast iron, as cast	100	250	350	200 BHN
A2	Spun grey cast iron, as cast	115	250	350	230 BHN
A3	Alloyed cast iron, heat treated	120	360	500	HV 5 - 400
A4	Alloyed cast iron, heat treated	115	310	1100	260 BHN
A5	Alloyed cast iron, heat treated	145	400	910	270 BHN
A6	S.G. Iron, heat treated	160	600	1300	270 BHN
S_1	Steel	200	-	-	38-44HRC

Table 25.7: Empirical proportions of I.C. engine piston

Dimension	*Petrol engine*	*Diesel engine*
Piston head thickness, t_h	(0.05- 0.10) d	(0.01 – 0.2) d
Piston length, H	(0.8 – 1.3) d	(0.10 – 1.7) d
Height of top part of piston, h_1	(0.45 – 0.75) d	(0.6 – 1.0) d
Skirt length, h_2	(0.6 – 0.8) *d*	(0.7 – 1.1) d
Top land, t_1	(0.06 – 0.12) d	(0.11 – 0.2) d
Radial clearance of ring, Δt	0.7 – 0.95 mm	0.7 – 0.95 mm
Length of pin, l_p	(0.78 – 0.93) d	(0.8 – 0.93) d
Reinforcing ribs	(1/3 – 1/2)t_h	(1/3-1/2)t_h

Table 25.8: Permissible bearing pressure between piston pin and connecting rod bearing

Bearing	p_b, *MPa*
Aluminium	5 – 10
Cast iron	10 – 15
Babbit	15 – 20
Bronze	20 – 25
Needle roller bearing	40 – 60

Tabte 25.9: Specifications for Piston Pins for I.C. Engines

Steel type	*Material*	*Symbol*	*Tensile strength, MPa*	*Core hardness, HRC*
Case hardening	15Cr3	A	325	15 – 35
	18Mn5Cr4	B	432	15 – 35
	15C8	C	233	15 – 35
	15Ni5Cr4Mo1	D	540	25 – 45
	13Ni13Cr3	E	460	25 – 45
Nitriding	40Cr7Al10MO2	F	325	25 – 45

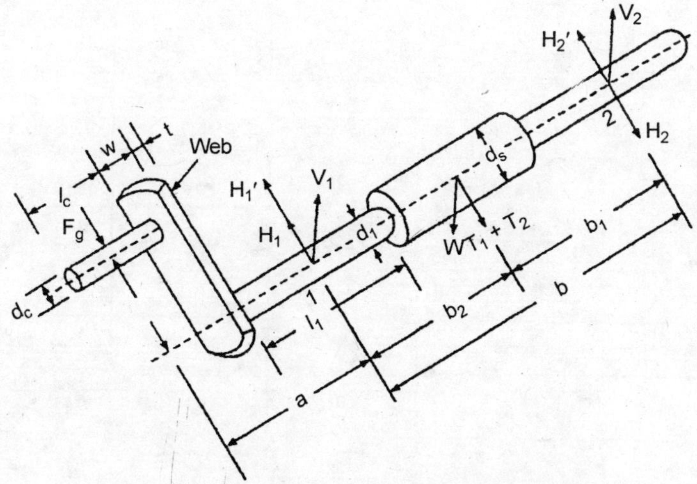

Fig. 25.10 Side crankshaft at dead centre.

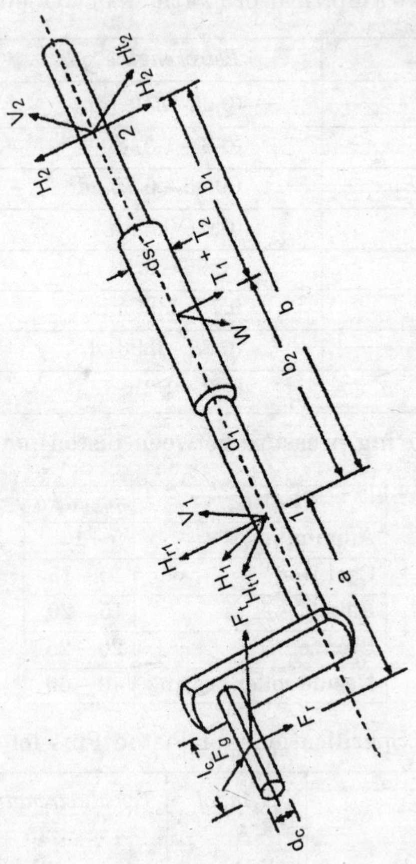

Fig. 25.11 Crank at an angle of maximum twisting moment.

MACHINE DESIGN DATA BOOK

DR. SADHU SINGH

FEEDBACK FORM

1. Which topics from your examination point of view are inadequately or not covered in this book?

____

____

2. Is there any misprints/ mistakes/ factual inaccuracy in this book? If so, please specify.

____

____

3. What is your assessment of this book as regards the presentation of the subject-matter, expression, precision and price in relation to the other recommended books available on this subject?

____

____

4. Which competing books you regard as better than this? Please mention their author and publishers.

____

____

5. Any further suggestions/comments you would like to make for the improvement of the book?

____

____

PS: If needed, please attach a separate sheet.

Name : ______________________

College : ______________________

Address : ______________________

Email : ______________________

Mobile : ______________________

Please mail the feedback form at:

KHANNA BOOK PUBLISHING CO. (P) LTD.

4C/4344, Ansari Road, Daryaganj, New Delhi-110002

You can also email this feedback form at contact@khannabooks.com or Whatsapp at +91 – 99109 09320

BEST FEEDBACK WILL BE REWARDED SUITABLY